GW01044456

Monopoles in Quantum Field Theory

Proceedings of the Monopole Meeting
Trieste, Italy
December 1981

Edited by N.S. Craigie
P. Goddard
W. Nahm

World Scientific

World Scientific Publishing Co Pte Ltd
P.O. Box 128
Farrer Road
Singapore 9128

Editorial Advisory Committee

H. Araki (Kyoto)
S. I. Chan (Caltech)
S. S. Chern (Berkeley)
R. Dalitz (Oxford)
C. C. Hsiung (Lehigh)
K. Huang (MIT)
M. Jacob (CERN)
M. Konuma (Kyoto)
T. D. Lee (Columbia)
B. B. Mandelbrot (IBM)
M. J. Moravcsik (Oregon)
R. Peierls (Oxford)
A. Salam (Trieste)
G. Takeda (Tohoku)
S. Weinberg (Harvard/Texas)
C. W. Woo (San Diego)
Y. Yamaguchi (Tokyo)
C. N. Yang (Stony Brook)

Copyright © 1982 by World Scientific Publishing Co Pte. Ltd. All rights reserved. This book, or parts thereof, may not be reproduced in any form or by any means, electronic or mechanical, including photocopying, recording or any information storage and retrieval system now known or to be invented, without written permission from the publisher.

ISBN 9971-950-29-4 pbk 9971-950-28-6
Printed by Singapore National Printers (Pte) Ltd

T A B L E   O F   C O N T E N T S

F O R E W O R D

The International Centre for Theoretical Physics, Trieste, is an institution devoted to the promotion of research in pure and applied science throughout the world, with special emphasis on developing countries. On 11-15 December 1981, 125 physicists and mathematicians from 37 countries assembled here for a conference. The occasion was the 50th anniversary of the introduction, by Dirac, of the concept of the magnetic monopole.

The meeting, organized by the ICTP in collaboration with the Istituto Nazionale di Fisica Nucleare (INFN), consisted of four days of seminars and discussions on the numerous recent theoretical developments which have emerged out of Dirac's original monopole idea, and on the search for such entities in nature.

We hope these proceedings serve as a useful and comprehensive source of information on the various aspects of a subject which is having a remarkable impact on present-day quantum field theory.

ACKNOWLEDGMENTS

The organizers wish to thank all the speakers and participants for making the meeting an interesting and, we believe, a fruitful one, and the staff of the ICTP for all their help in organizing it. We would like to express our gratitude to L. O'Raifeartaigh for sending us, at our request, a written version of his delightful summary talk, despite his recent illness. Finally we wish to thank Sunil Mukhi for his help in putting the proceedings together.

N.S. Craigie
W. Nahm
P. Goddard

Department of Physics
The Florida State University
Tallahassee, Florida 32306

11 Nov 1981

Dear Abdus,

I am sorry I cannot come to your monopole conference.

It would be too much of a dislocation for me at such short notice

It was very kind of you to invite me.

I am inclined now to believe that monopoles do not exist.

So many years have gone by without any encouragement from the experimental side

It will be interesting to see if your conference can produce any new angles of attack on the problem

With best wishes,

Yours sincerely

Paul Dirac

OPENING ADDRESS BY PROFESSOR ABDUS SALAM

The subject of magnetic monopoles began with Dirac's paper on Quantised singularities in the Electromagnetic field which was received by the Royal Society on 29th March 1931. Dirac's motivation in this paper was to find the reason for the existence of the smallest electric charge. Proposing a generalisation of the formalism of quantum mechanics as used for electromagnetism, Dirac allowed for wave functions with non-integrable phases. He showed that the non-integrability of the phases could be interpreted in terms of the presence of an electromagnetic field, with the possibility of there being singularities of the field. These corresponded to single magnetic poles with their strength restricted by the relation $eg/(4\pi) = (n/2)$. This relation provided Dirac with the basis for the explanation of electric charge quantisation. Already in this paper Dirac commented on the difficulty of creating pole pairs due to their strong binding.

This paper contained much else besides the notion of monopoles and the relation $eg/(4\pi) = (n/2)$. Dirac, in fact, started by commenting on a paper he had written a year before, where he had suggested the idea of filling the negative sea of states and identifying the holes in the negative sea as protons. In the 1931 paper he accepted the results of Weyl about the value of the mass of such a hole. Following a suggestion due to Oppenheimer, he concluded that such a hole should be "a new kind of particle unknown to experimental physics". This he called an anti-electron. Protons, he commented, are unconnected with holes in the electron sea but they should have their own anti-protons. One may recall that this was fully one year before Anderson's discovery of positive electrons, and at a time when unlike to-day theoretical physicists were very reluctant to postulate new particles, at the drop of a hat.

After 1931, and before 1948 when Dirac wrote on the subject again, the theory of monopoles and of bound states of monopoles and electric charges was worked on extensively by Tamm, Fierz, Banderet, Harish-Chandra, and others. In this connection - and I owe this history to a beautiful review by Amaldi and Cabibbo published in Dirac's 70th Birthday Volume - it is good to recall that already around 1895, Poincare and J.J. Thomson had published the result that the static fields generated by a charge $+e$ placed at a distance of r from the pole of charge $+g$ gives rise to an angular momentum $eg/(4\pi)$, with the direction of the angular momentum pointing from $+g$ to $+e$. Saha in 1936 used this result to remark that the Dirac relationship could be obtained by equating this angular momentum $eg/(4\pi)$ to an integral multiple of $(1/2)\hbar$. In modern parlance he was suggesting that a bound state of a spinless charge and a spinless monopole could carry half a unit of spin - an idea which has been advanced recently for motivating supersymmetry dynamically. Saha also suggested that the neutron might consist of bound pole pairs. This anticipated in some ways another idea which has been used by Schwinger, Barut and others to suggest that the basic entities of which all matter may be composed of may be monopoles carrying electric charge (dyons).

The next step in the development of monopole theory came with Dirac's paper of 1948. I can recall, as an undergraduate in Cambridge, a popular evening lecture given by Dirac where he spoke about the Hamiltonian for a system consisting of a fixed number of poles and charges. Introducing electromagnetic potentials, Dirac motivated the well known strings, one attached to each pole. The condition $eg/(4\pi) = (n/2)$ ensured that the coordinates describing the string were ignorable.

I recall that Dirac finished his lecture by passing around the packed hall a set of photographic plates sent to him, if my memory serves me correctly, by Ehrenheft who claimed to have discovered ionising tracks which might, he suggested, be monopoles. Dirac said he did not believe this explanation of the tracks seen.

For our meeting to-day, 50 years after his first paper, Dirac has sent us the following message:

"I am sorry I cannot come to your monopole conference. It would be too much of a dislocation for me at such short notice. It was very kind of you to invite me.

"I am inclined now to believe that monopoles do not exist. So many years have gone by without any encouragement from the experimental side. It will be interesting to see if your conference can produce any new angle of attack on the problem."

Dirac, who concluded his first 1931 paper with the remark "one would be surprised if Nature had not made use" of the monopole concept, is as concerned to-day with the lack of experimental evidence for monopoles as he was in 1948. However, in the meanwhile the theory of monopoles and the experimental prospects of discovering them have undergone a complete transformation since 1974 with the pioneering work of 't Hooft and Polyakov in the context of modern grand unifying gauge theories. There is indeed a "new angle of attack on the problem" which we are assembled to hear about in the next five days, and which posits that monopoles are likely to have been abundant in the early Universe and that if they have survived to the present epoch, we should be looking for objects with masses around 10^{-8} grams. This represents an experimental regime completely different than the one researched on hitherto. A second aspect of this new revolution is the deepening of contact between modern mathematics and modern particle physics, leading to enrichment of both disciplines. This would delight Dirac, since in his 1931 monopole paper his opening remark concerns the abstraction in mathematics to be expected as a consequence of interaction with physics: "... It seems likely that this process of increasing abstraction will continue in the future and that the advance in physics is to be associated with a continual modification and generalisation of the axioms at the base of the mathematics..."

As you know Dirac will be eighty on 8th August 1982. I am sure you will wish me to cable him the following message on behalf of all those here to-day:

"Around one hundred and twenty physicists and mathematicians from 37 countries assembled at the International Centre for Theoretical Physics at Trieste, for a symposium on developments in Monopole Theory and Experiment, wish to express their deepest appreciation to you on the fiftieth anniversary of your seminal paper on the subject and extend to you and to your family their warmest greetings."

Copy of Paul Dirac's original article published in Proceedings of the Royal Society of London, A, Vol.133 (1931).

Quantised Singularities in the Electromagnetic Field.

By P. A. M. Dirac, F.R.S., St. John's College, Cambridge.

(Received May 29, 1931.)

§ 1. *Introduction.*

The steady progress of physics requires for its theoretical formulation a mathematics that gets continually more advanced. This is only natural and to be expected. What, however, was not expected by the scientific workers of the last century was the particular form that the line of advancement of the mathematics would take, namely, it was expected that the mathematics would get more and more complicated, but would rest on a permanent basis of axioms and definitions, while actually the modern physical developments have required a mathematics that continually shifts its foundations and gets more abstract. Non-euclidean geometry and non-commutative algebra, which were at one time considered to be purely fictions of the mind and pastimes for logical thinkers, have now been found to be very necessary for the description of general facts of the physical world. It seems likely that this process of increasing abstraction will continue in the future and that advance in physics is to be associated with a continual modification and generalisation of the axioms at the base of the mathematics rather than with a logical development of any one mathematical scheme on a fixed foundation.

There are at present fundamental problems in theoretical physics awaiting solution, *e.g.*, the relativistic formulation of quantum mechanics and the nature of atomic nuclei (to be followed by more difficult ones such as the problem of life), the solution of which problems will presumably require a more drastic revision of our fundamental concepts than any that have gone before. Quite likely these changes will be so great that it will be beyond the power of human intelligence to get the necessary new ideas by direct attempts to formulate the experimental data in mathematical terms. The theoretical worker in the future will therefore have to proceed in a more indirect way. The most powerful method of advance that can be suggested at present is to employ all the resources of pure mathematics in attempts to perfect and generalise the mathematical formalism that forms the existing basis of theoretical physics, and *after* each success in this direction, to try to interpret the new mathematical features in terms of physical entities (by a process like Eddington's Principle of Identification).

A recent paper by the author* may possibly be regarded as a small step according to this general scheme of advance. The mathematical formalism at that time involved a serious difficulty through its prediction of negative kinetic energy values for an electron. It was proposed to get over this difficulty, making use of Pauli's Exclusion Principle which does not allow more than one electron in any state, by saying that in the physical world almost all the negative-energy states are already occupied, so that our ordinary electrons of positive energy cannot fall into them. The question then arises as to the physical interpretation of the negative-energy states, which on this view really exist. We should expect the uniformly filled distribution of negative-energy states to be completely unobservable to us, but an unoccupied one of these states, being something exceptional, should make its presence felt as a kind of hole. It was shown that one of these holes would appear to us as a particle with a positive energy and a positive charge and it was suggested that this particle should be identified with a proton. Subsequent investigations, however, have shown that this particle necessarily has the same mass as an electron† and also that, if it collides with an electron, the two will have a chance of annihilating one another much too great to be consistent with the known stability of matter.‡

It thus appears that we must abandon the identification of the holes with protons and must find some other interpretation for them. Following Oppenheimer,§ we can assume that in the world as we know it, *all*, and not merely nearly all, of the negative-energy states for electrons are occupied. A hole, if there were one, would be a new kind of particle, unknown to experimental physics, having the same mass and opposite charge to an electron. We may call such a particle an anti-electron. We should not expect to find any of them in nature, on account of their rapid rate of recombination with electrons, but if they could be produced experimentally in high vacuum they would be quite stable and amenable to observation. An encounter between two hard γ-rays (of energy at least half a million volts) could lead to the creation simultaneously of an electron and anti-electron, the probability of occurrence of this process being of the same order of magnitude as that of the collision of the two γ-rays on the assumption that they are spheres of the same size as classical

* 'Proc. Roy. Soc.,' A, vol. 126, p. 360 (1930).

† H. Weyl, 'Gruppentheorie und Quantenmechanik,' 2nd ed. p. 234 (1931).

‡ I. Tamm, 'Z. Physik,' vol. 62, p. 545 (1930); J. R. Oppenheimer, 'Phys. Rev.,' vol. 35, p. 939 (1930); P. Dirac, 'Proc. Camb. Philos. Soc.,' vol. 26, p. 361 (1930).

§ J. R. Oppenheimer, 'Phys. Rev.,' vol. 35, p. 562 (1930).

electrons. This probability is negligible, however, with the intensities of γ-rays at present available.

The protons on the above view are quite unconnected with electrons. Presumably the protons will have their own negative-energy states, all of which normally are occupied, an unoccupied one appearing as an anti-proton. Theory at present is quite unable to suggest a reason why there should be any differences between electrons and protons.

The object of the present paper is to put forward a new idea which is in many respects comparable with this one about negative energies. It will be concerned essentially, not with electrons and protons, but with the reason for the existence of a smallest electric charge. This smallest charge is known to exist experimentally and to have the value e given approximately by*

$$hc/e^2 = 137. \tag{1}$$

The theory of this paper, while it looks at first as though it will give a theoretical value for e, is found when worked out to give a connection between the smallest electric charge and the smallest magnetic pole. It shows, in fact, a symmetry between electricity and magnetism quite foreign to current views. It does not, however, force a complete symmetry, analogous to the fact that the symmetry between electrons and protons is not forced when we adopt Oppenheimer's interpretation. Without this symmetry, the ratio on the left-hand side of (1) remains, from the theoretical standpoint, completely undetermined and if we insert the experimental value 137 in our theory, it introduces quantitative differences between electricity and magnetism so large that one can understand why their qualitative similarities have not been discovered experimentally up to the present.

§ 2. *Non-integrable Phases for Wave Functions.*

We consider a particle whose motion is represented by a wave function ψ, which is a function of x, y, z and t. The precise form of the wave equation and whether it is relativistic or not, are not important for the present theory. We express ψ in the form

$$\psi = Ae^{i\gamma}, \tag{2}$$

where A and γ are real functions of x, y, z and t, denoting the amplitude and phase of the wave function. For a given state of motion of the particle, ψ will be determined except for an arbitrary constant numerical coefficient, which must be of modulus unity if we impose the condition that ψ shall be normalised.

* h means Planck's constant divided by 2π.

The indeterminacy in ψ then consists in the possible addition of an arbitrary constant to the phase γ. Thus the value of γ at a particular point has no physical meaning and only the difference between the values of γ at two different points is of any importance.

This immediately suggests a generalisation of the formalism. We may assume that γ has no definite value at a particular point, but only a definite difference in values for any two points. We may go further and assume that this difference is not definite unless the two points are neighbouring. For two distant points there will then be a definite phase difference only relative to some curve joining them and different curves will in general give different phase differences. The total change in phase when one goes round a closed curve need not vanish.

Let us examine the conditions necessary for this non-integrability of phase not to give rise to ambiguity in the applications of the theory. If we multiply ψ by its conjugate complex ϕ we get the density function, which has a direct physical meaning. This density is independent of the phase of the wave function, so that no trouble will be caused in this connection by any indeterminacy of phase. There are other more general kinds of applications, however, which must also be considered. If we take two different wave functions ψ_m and ψ_n, we may have to make use of the product $\phi_m\psi_n$. The integral

$$\int \phi_m \psi_n \, dx \, dy \, dz$$

is a number, the square of whose modulus has a physical meaning, namely, the probability of agreement of the two states. In order that the integral may have a definite modulus the integrand, although it need not have a definite phase at each point, must have a definite phase difference between any two points, whether neighbouring or not. Thus the change in phase in $\phi_m\psi_n$ round a closed curve must vanish. This requires that the change in phase in ψ_n round a closed curve shall be equal and opposite to that in ϕ_m and hence the same as that in ψ_m. We thus get the general result:— *The change in phase of a wave function round any closed curve must be the same for all the wave functions.*

It can easily be seen that this condition, when extended so as to give the same uncertainty of phase for transformation functions and matrices representing observables (referring to representations in which x, y and z are diagonal) as for wave functions, is sufficient to insure that the non-integrability of phase gives rise to no ambiguity in all applications of the theory. Whenever a ψ_n appears, if it is not multiplied into a ϕ_m, it will at any rate be multiplied into something of a similar nature to a ϕ_m, which will result in the uncertainty of phase cancelling out, except for a constant which does not matter. For example, if ψ_n is to be transformed to another representation in which, say, the observables ξ are diagonal, it must be multiplied by the transformation function $(\xi \mid xyzt)$ and integrated with respect to x, y and z. This transformation function will have the same uncertainty of phase as a ϕ, so that the transformed wave function will have its phase determinate, except for a constant independent of ξ. Again, if we multiply ψ_n by a matrix $(x'y'z't \mid \alpha \mid x''y''z''t)$, representing an observable α, the uncertainty in the phase as concerns the column [specified by x'', y'', z'', t] will cancel the uncertainty in ψ_n and the uncertainty as concerns the row will survive and give the necessary uncertainty in the new wave function $\alpha\psi_n$. The superposition principle for wave functions will be discussed a little later and when this point is settled it will complete the proof that all the general operations of quantum mechanics can be carried through exactly as though there were no uncertainty in the phase at all.

The above result that the change in phase round a closed curve must be the same for all wave functions means that this change in phase must be something determined by the dynamical system itself (and perhaps also partly by the representation) and must be independent of which state of the system is considered. As our dynamical system is merely a simple particle, it appears that the non-integrability of phase must be connected with the field of force in which the particle moves.

For the mathematical treatment of the question we express ψ, more generally than (2), as a product

$$\psi = \psi_1 e^{i\beta}, \tag{3}$$

where ψ_1 is any ordinary wave function (*i.e.*, one with a definite phase at each point) whose modulus is everywhere equal to the modulus of ψ. The uncertainty of phase is thus put in the factor $e^{i\beta}$. This requires that β shall not be a function of x, y, z, t having a definite value at each point, but β must have definite derivatives

$$\kappa_x = \frac{\partial\beta}{\partial x}, \qquad \kappa_y = \frac{\partial\beta}{\partial y}, \qquad \kappa_z = \frac{\partial\beta}{\partial z}, \qquad \kappa_0 = \frac{\partial\beta}{\partial t},$$

at each point, which do not in general satisfy the conditions of integrability $\partial\kappa_x/\partial y = \partial\kappa_y/\partial x$, etc. The change in phase round a closed curve will now be, by Stokes' theorem,

$$\int (\boldsymbol{\kappa}, \mathbf{ds}) = \int (\operatorname{curl} \boldsymbol{\kappa}, \mathbf{dS}), \tag{4}$$

where **ds** (a 4-vector) is an element of arc of the closed curve and **dS** (a 6-vector) is an element of a two-dimensional surface whose boundary is the closed curve. The factor ψ_1 does not enter at all into this change in phase.

It now becomes clear that the non-integrability of phase is quite consistent with the principle of superposition, or, stated more explicitly, that if we take two wave functions ψ_m and ψ_n both having the same change in phase round any closed curve, any linear combination of them $c_m\psi_m + c_n\psi_n$ must also have this same change in phase round every closed curve. This is because ψ_m and ψ_n will both be expressible in the form (3) with the same factor $e^{i\beta}$ (*i.e.*, the same κ's) but different ψ_1's, so that the linear combination will be expressible in this form with the same $e^{i\beta}$ again, and this $e^{i\beta}$ determines the change in phase round any closed curve. We may use the same factor $e^{i\beta}$ in (3) for dealing with all the wave functions of the system, but we are not obliged to do so, since only curl $\boldsymbol{\kappa}$ is fixed and we may use κ's differing from one another by the gradient of a scalar for treating the different wave functions.

From (3) we obtain

$$-ih\frac{\partial}{\partial x}\psi = e^{i\beta}\left(-ih\frac{\partial}{\partial x} + h\kappa_x\right)\psi_1, \tag{5}$$

with similar relations for the y, z and t derivatives. It follows that if ψ satisfies any wave equation, involving the momentum and energy operators **p** and W, ψ_1 will satisfy the corresponding wave equation in which **p** and W have been replaced by $\mathbf{p} + h\boldsymbol{\kappa}$ and $\mathrm{W} - h\kappa_0$ respectively.

Let us assume that ψ satisfies the usual wave equation for a free particle in the absence of any field. Then ψ_1 will satisfy the usual wave equation for a particle with charge $-e$ moving in an electromagnetic field whose potentials are

$$\mathbf{A} = hc/e\,.\,\boldsymbol{\kappa}, \qquad \mathrm{A}_0 = -\,h/e\,.\,\kappa_0. \tag{6}$$

Thus, since ψ_1 is just an ordinary wave function with a definite phase, our theory reverts to the usual one for the motion of an electron in an electromagnetic field. This gives a physical meaning to our non-integrability of phase. We see that we must have the wave function ψ always satisfying the same wave equation, whether there is a field or not, and the whole effect of the field when there is one is in making the phase non-integrable.

The components of the 6-vector curl $\boldsymbol{\kappa}$ appearing in (4) are, apart from numerical coefficients, equal to the components of the electric and magnetic fields **E** and **H**. They are, written in three-dimensional vector-notation,

$$\text{curl}\,\boldsymbol{\kappa} = \frac{e}{hc}\,\mathbf{H}, \qquad \text{grad}\,\kappa_0 - \frac{\partial \boldsymbol{\kappa}}{\partial t} = \frac{e}{h}\,\mathbf{E}. \tag{7}$$

The connection between non-integrability of phase and the electromagnetic field given in this section is not new, being essentially just Weyl's Principle of Gauge Invariance in its modern form.* It is also contained in the work of Iwanenko and Fock,† who consider a more general kind of non-integrability based on a general theory of parallel displacement of half-vectors. The present treatment is given in order to emphasise that non-integrable phases are perfectly compatible with all the general principles of quantum mechanics and do not in any way restrict their physical interpretation.

§ 3. *Nodal Singularities.*

We have seen in the preceding section how the non-integrable derivatives κ of the phase of the wave function receive a natural interpretation in terms of the potentials of the electromagnetic field, as the result of which our theory becomes mathematically equivalent to the usual one for the motion of an electron in an electromagnetic field and gives us nothing new. There is, however, one further fact which must now be taken into account, namely, that a phase is always undetermined to the extent of an arbitrary integral multiple of 2π. This requires a reconsideration of the connection between the κ's and the potentials and leads to a new physical phenomenon.

The condition for an unambiguous physical interpretation of the theory was that the change in phase round a closed curve should be the same for all wave functions. This change was then interpreted, by equations (4) and (7), as equal to (apart from numerical factors) the total flux through the closed curve of the 6-vector **E, H** describing the electromagnetic field. Evidently these conditions must now be relaxed. The change in phase round a closed curve may be different for different wave functions by arbitrary multiples of 2π and is thus not sufficiently definite to be interpreted immediately in terms of the electromagnetic field.

To examine this question, let us consider first a very small closed curve. Now the wave equation requires the wave function to be continuous (except in very special circumstances which can be disregarded here) and hence the change in phase round a small closed curve must be small. Thus this change cannot now be different by multiples of 2π for different wave functions. It must have one definite value and may therefore be interpreted without

* H. Weyl, 'Z. Physik,' vol. 56, p. 330 (1929).

† D. Iwanenko and V. Fock, 'C. R.,' vol. 188, p. 1470 (1929); V. Fock, 'Z. Physik,' vol. 57, p. 261 (1929). The more general kind of non-integrability considered by these authors does not seem to have any physical application.

ambiguity in terms of the flux of the 6-vector **E**, **H** through the small closed curve, which flux must also be small.

There is an exceptional case, however, occurring when the wave function vanishes, since then its phase does not have a meaning. As the wave function is complex, its vanishing will require two conditions, so that in general the points at which it vanishes will lie along a line.* We call such a line a *nodal line*. If we now take a wave function having a nodal line passing through our small closed curve, considerations of continuity will no longer enable us to infer that the change in phase round the small closed curve must be small. All we shall be able to say is that the change in phase will be close to $2\pi n$ where n is some integer, positive or negative. This integer will be a characteristic of the nodal line. Its sign will be associated with a direction encircling the nodal line, which in turn may be associated with a direction along the nodal line.

The difference between the change in phase round the small closed curve and the nearest $2\pi n$ must now be the same as the change in phase round the closed curve for a wave function with no nodal line through it. It is therefore this difference that must be interpreted in terms of the flux of the 6-vector **E**, **H** through the closed curve. For a closed curve in three-dimensional space, only magnetic flux will come into play and hence we obtain for the change in phase round the small closed curve

$$2\pi n + e/hc \, . \int (\mathbf{H}, \mathbf{dS}).$$

We can now treat a large closed curve by dividing it up into a network of small closed curves lying in a surface whose boundary is the large closed curve. The total change in phase round the large closed curve will equal the sum of all the changes round the small closed curves and will therefore be

$$2\pi \Sigma n + e/hc \, . \int (\mathbf{H}, \mathbf{dS}), \qquad (8)$$

the integration being taken over the surface and the summation over all nodal lines that pass through it, the proper sign being given to each term in the sum. This expression consists of two parts, a part $e/hc \, . \int (\mathbf{H}, \mathbf{dS})$ which must be the same for all wave functions and a part $2\pi \Sigma n$ which may be different for different wave functions.

* We are here considering, for simplicity in explanation, that the wave function is in three dimensions. The passage to four dimensions makes no essential change in the theory. The nodal lines then become two-dimensional nodal surfaces, which can be encircled by curves in the same way as lines are in three dimensions.

F 2

Expression (8) applied to any surface is equal to the change in phase round the boundary of the surface. Hence expression (8) applied to a closed surface must vanish. It follows that Σn, summed for all nodal lines crossing a closed surface, must be the same for all wave functions and must equal $-e/2\pi hc$ times the total magnetic flux crossing the surface.

If Σn does not vanish, some nodal lines must have end points inside the closed surface, since a nodal line without such end point must cross the surface twice (at least) and will contribute equal and opposite amounts to Σn at the two points of crossing. The value of Σn for the closed surface will thus equal the sum of the values of n for all nodal lines having end points inside the surface. This sum must be the same for all wave functions. Since this result applies to *any* closed surface, it follows that *the end points of nodal lines must be the same for all wave functions. These end points are then points of singularity in the electromagnetic field.* The total flux of magnetic field crossing a small closed surface surrounding one of these points is

$$4\pi\mu = 2\pi nhc/e,$$

where n is the characteristic of the nodal line that ends there, or the sum of the characteristics of all nodal lines ending there when there is more than one. Thus at the end point there will be a magnetic pole of strength

$$\mu = \tfrac{1}{2}nhc/e.$$

Our theory thus allows isolated magnetic poles, but the strength of such poles must be quantised, the quantum μ_0 being connected with the electronic charge e by

$$hc/e\mu_0 = 2. \tag{9}$$

This equation is to be compared with (1). The theory also requires a quantisation of electric charge, since any charged particle moving in the field of a pole of strength μ_0 must have for its charge some integral multiple (positive or negative) of e, in order that wave functions describing the motion may exist.

§ 4. *Electron in Field of One-Quantum Pole.*

The wave functions discussed in the preceding section, having nodal lines ending on magnetic poles, are quite proper and amenable to analytic treatment by methods parallel to the usual ones of quantum mechanics. It will perhaps help the reader to realise this if a simple example is discussed more explicitly.

Let us consider the motion of an electron in the magnetic field of a one-quantum pole when there is no electric field present. We take polar co-ordinates r, θ, ϕ with the magnetic pole as origin. Every wave function must now have a nodal line radiating out from the origin.

We express our wave function ψ in the form (3), where β is some non-integrable phase having derivatives κ that are connected with the known electromagnetic field by equations (6). It will not, however, be possible to obtain κ's satisfying these equations all round the magnetic pole. There must be some singular line radiating out from the pole along which these equations are not satisfied, but this line may be chosen arbitrarily. We may choose it to be the same as the nodal line for the wave function under consideration, which would result in ψ_1 being continuous. This choice, however, would mean different κ's for different wave functions (the difference between any two being, of course, the four-dimensional gradient of a scalar, except on the singular lines). This would perhaps be inconvenient and is not really necessary. We may express all our wave functions in the form (3) with the same $e^{i\beta}$, and then those wave functions whose nodal lines do not coincide with the singular line for the κ's will correspond to ψ_1's having a certain kind of discontinuity on this singular line, namely, a discontinuity just cancelling with the discontinuity in $e^{i\beta}$ here to give a continuous product.

The magnetic field **H**, lies along the radial direction and is of magnitude μ_0/r^2, which by (9) equals $\frac{1}{2}hc/er^2$. Hence, from equations (7), curl $\boldsymbol{\kappa}$ is radial and of magnitude $1/2r^2$. It may now easily be verified that a solution of the whole of equations (7) is

$$\kappa_0 = 0, \qquad \kappa_r = \kappa_\theta = 0, \qquad \kappa_\phi = 1/2r \,.\, \tan \tfrac{1}{2}\theta, \tag{10}$$

where κ_r, κ_θ, κ_ϕ are the components of $\boldsymbol{\kappa}$ referred to the polar co-ordinates. This solution is valid at all points except along the line $\theta = \pi$, where κ_ϕ becomes infinite in such a way that $\int (\boldsymbol{\kappa}, \mathbf{ds})$ round a small curve encircling this line is 2π. We may refer all our wave functions to this set of κ's.

Let us consider a stationary state of the electron with energy W. Written non-relativistically, the wave equation is

$$- h^2/2m \,.\, \nabla^2\psi = W\psi.$$

If we apply the rule expressed by equation (5), we get as the wave equation for ψ_1

$$- h^2/2m \,.\, \{\nabla^2 + i\,(\boldsymbol{\kappa}, \nabla) + i\,(\nabla, \boldsymbol{\kappa}) - \boldsymbol{\kappa}^2\}\, \psi_1 = W\psi_1. \tag{11}$$

The values (10) for the κ's give

$$(\kappa, \nabla) = (\nabla, \kappa) = \kappa_\phi \frac{1}{r \sin\theta}\frac{\partial}{\partial\phi} = \frac{1}{4r^2}\sec^2\tfrac{1}{2}\theta\frac{\partial}{\partial\phi}$$

$$\kappa^2 = \kappa_\phi{}^2 = \frac{1}{4r^2}\tan^2\tfrac{1}{2}\theta,$$

so that equation (11) becomes

$$-\frac{h^2}{2m}\left\{\nabla^2 + \frac{i}{2r^2}\sec^2\tfrac{1}{2}\theta\frac{\partial}{\partial\phi} - \frac{1}{4r^2}\tan^2\tfrac{1}{2}\theta\right\}\psi_1 = W\psi_1.$$

We now suppose ψ_1 to be of the form of a function f of r only multiplied by a function S of θ and ϕ only, *i.e.*,

$$\psi_1 = f(r)\,S\,(\theta\phi).$$

This requires

$$\left\{\frac{d^2}{dr^2} + \frac{2}{r}\frac{d}{dr} - \frac{\lambda}{r^2}\right\}f = -\frac{2mW}{h^2}f, \tag{12}$$

$$\left\{\frac{1}{\sin\theta}\frac{\partial}{\partial\theta}\sin\theta\frac{\partial}{\partial\theta} + \frac{1}{\sin^2\theta}\frac{\partial^2}{\partial\phi^2} + \tfrac{1}{2}i\sec^2\tfrac{1}{2}\theta\frac{\partial}{\partial\phi} - \tfrac{1}{4}\tan^2\tfrac{1}{2}\theta\right\}S = -\lambda S, \tag{13}$$

where λ is a number.

From equation (12) it is evident that there can be no stable states for which the electron is bound to the magnetic pole, because the operator on the left-hand side contains no constant with the dimensions of a length. This result is what one would expect from analogy with the classical theory. Equation (13) determines the dependence of the wave function on angle. It may be considered as a generalisation of the ordinary equation for spherical harmonics.

The lowest eigenvalue of (13) is $\lambda = \frac{1}{2}$, corresponding to which there are two independent wave functions

$$S_a = \cos\tfrac{1}{2}\theta, \quad S_b = \sin\tfrac{1}{2}\theta\, e^{i\phi},$$

as may easily be verified by direct substitution. The nodal line for S_a is $\theta = \pi$, that for S_b is $\theta = 0$. It should be observed that S_a is continuous everywhere, while S_b is discontinuous for $\theta = \pi$, its phase changing by 2π when one goes round a small curve encircling the line $\theta = \pi$. This is just what is necessary in order that both S_a and S_b, when multiplied by the $e^{i\beta}$ factor, may give continuous wave functions ψ. The two ψ's that we get in this way are both on the same footing and the difference in behaviour of S_a and S_b is due to our having chosen κ's with a singularity at $\theta = \pi$.

The general eigenvalue of (13) is $\lambda = n^2 + 2n + \frac{1}{2}$. The general solution of this wave equation has been worked out by I. Tamm.*

* Appearing probably in ' Z. Physik.'

§ 5. *Conclusion.*

Elementary classical theory allows us to formulate equations of motion for an electron in the field produced by an arbitrary distribution of electric charges and magnetic poles. If we wish to put the equations of motion in the Hamiltonian form, however, we have to introduce the electromagnetic potentials, and this is possible only when there are no isolated magnetic poles. Quantum mechanics, as it is usually established, is derived from the Hamiltonian form of the classical theory and therefore is applicable only when there are no isolated magnetic poles.

The object of the present paper is to show that quantum mechanics does not really preclude the existence of isolated magnetic poles. On the contrary, the present formalism of quantum mechanics, when developed naturally without the imposition of arbitrary restrictions, leads inevitably to wave equations whose only physical interpretation is the motion of an electron in the field of a single pole. This new development requires *no change whatever* in the formalism when expressed in terms of abstract symbols denoting states and observables, but is merely a generalisation of the possibilities of representation of these abstract symbols by wave functions and matrices. Under these circumstances one would be surprised if Nature had made no use of it.

The theory leads to a connection, namely, equation (9), between the quantum of magnetic pole and the electronic charge. It is rather disappointing to find this reciprocity between electricity and magnetism, instead of a purely electronic quantum condition, such as (1). However, there appears to be no possibility of modifying the theory, as it contains no arbitrary features, so presumably the explanation of (1) will require some entirely new idea.

The theoretical reciprocity between electricity and magnetism is perfect. Instead of discussing the motion of an electron in the field of a fixed magnetic pole, as we did in § 4, we could equally well consider the motion of a pole in the field of fixed charge. This would require the introduction of the electromagnetic potentials B satisfying

$$\mathbf{E} = \operatorname{curl} \mathbf{B}, \quad \mathrm{H} = \frac{1}{c}\frac{\partial \mathrm{B}}{\partial t} + \operatorname{grad} \mathrm{B}_0,$$

to be used instead of the A's in equations (6). The theory would now run quite parallel and would lead to the same condition (9) connecting the smallest pole with the smallest charge.

There remains to be discussed the question of why isolated magnetic poles are not observed. The experimental result (1) shows that there must be some cause of dissimilarity between electricity and magnetism (possibly connected with the cause of dissimilarity between electrons and protons) as the result of which we have, not $\mu_0 = e$, but $\mu_0 = 137/2 \,.\, e$. This means that the attractive force between two one-quantum poles of opposite sign is $(137/2)^2 = 4692\frac{1}{4}$ times that between electron and proton. This very large force may perhaps account for why poles of opposite sign have never yet been separated.

I: RECENT PROGRESS ON MULTIMONOPOLE SOLUTIONS

GEOMETRY OF MONOPOLES

M.F. Atiyah

Mathematical Institute, University of Oxford,
24-29 St. Giles, Oxford, UK.

§1. Introduction

In recent years remarkable explicit solutions have been found for some of the non-linear partial differential equations arising in gauge theories. Although these equations are fully four-(or three)-dimensional they have many similarities with the large class of two-dimensional equations which exhibit "soliton" solutions, such as the KdV and Sine-Gordon equations. The main feature which enables such non-linear equations to be successfully solved is that they all have interpretations as the integrability (or compatibility) conditions for an associated linear system. Using the full power of the theory for linear operators, notably spectral theory, one then has a strong hold on the non-linear equation.

Another remarkable feature of these soliton type solutions is that they frequently involve the classical theory of abelian functions, depending on some algebraic curve associated to the spectrum of the linear system.

In this lecture I shall report on the recent construction of explicit multi-monopole solutions by R.S. Ward and others and on the geometrical interpretation of these solutions by N.J. Hitchin [4]. I shall also comment briefly on the relation of Hitchin's approach to the very recent results of W. Nahm which will be reported on in this conference.

The techniques for solving the multi-monopole equation are essentially the same as those previously used for the construction of multi-instantons, although the precise details are of course different. I shall begin therefore with a general review of the self-duality equations and in particular I shall try to elucidate the analogy with the integrable two-dimensional equations referred to earlier.

§2. The self-duality equations

Let G be a compact Lie group which for simplicity we may take to be $SU(n)$. If A_α are the four components of a Yang-Mills potential on Euclidean 4-space R^4 the Yang-Mills field has components

$$F_{\alpha\beta} = \partial_\alpha A_\beta - \partial_\beta A_\alpha + [A_\alpha, A_\beta].$$

The field is said to be self-dual if $*F = F$, i.e. $F_{12} = F_{34}, F_{13} = F_{42}, F_{14} = F_{23}$. It is anti-self-dual if $*F = -F$: the difference between the two notions is just one of orientation.

These equations, which are non-linear partial differential equations of first order in the A_α, imply the second-order Yang-Mills equations. The construction of all solutions of the equations $*F = \pm F$ is beautifully solved by use of the Penrose twistor space as first point out by R.S. Ward. Essentially this gives the interpretation in terms of the integrability of an associated linear system with the added advantage that this linear system is essentially the Cauchy-Riemann equation so that we are reduced to holomorphic function theory.

The Penrose construction can be introduced and motivated from many points of view. I shall present it here in a manner which brings out its relation to the "inverse scattering method" used for two-dimensional problems. The first step is to introduce a complex "spectral parameter" λ, thus enlarging R^4 to the 6-dimensional space $R^4 \times P_1$ (where P_1 is the complex projective line or Riemann sphere). Next we observe that the complex structures on R^4, i.e. identifications $R^4 = C^2$ compatible with metric and orientation, are naturally parametrized by $\lambda \in P_1$. This is because a complex structure is determined by an orthogonal 4×4 matrix J with $J^2 = -1$ and $\det J = 1$, and the set of such J is just the homogeneous space $SO(4)/U(2) = S^2 = P_1$. For each λ we can therefore introduce the covariant Cauchy-Riemann operator $\bar{\nabla}^\lambda$ defined by the given potential A. The equations $*F = -F$ can then be rewritten as the commutation relation

$$[\bar{\nabla}^\lambda, \bar{\nabla}^\lambda] = 0.$$

When this equation is satisfied the associated linear equation

$$\bar{\nabla}^\lambda f = 0 \qquad (1)$$

has solutions. This means that, after a complex gauge transformation, $\bar{\nabla}^\lambda$ becomes the usual Cauchy-Riemann operator $\bar{\partial}^\lambda$ (for the complex structure determined by λ). In geometric terms we say that we have a holomorphic bundle on C^2, the local sections of which are the solutions of the equation (1). This holds for each λ and the dependence on λ is itself holomorphic. This is clarified by noting that we can naturally identify $R^4 \times P_1$ with the 3-dimensional complex manifold $P_3 - P_1$.

This identification can be set up as follows. Fix complex homogeneous coordinates (x,y,u,v) for the complex projective 3-space P_3 and distinguish the two lines $P^o (x = y = o)$ and $P^\infty (u = v = o)$. In the open set $P_3 - P^\infty$ consider the family of planes $u = \lambda v$ and the family of lines

$$\begin{aligned} x &= au + bv \\ y &= -\bar{b}u + \bar{a}v \end{aligned} \qquad (2)$$

It is easy to verify that through each point of $P_3 - P^\infty$ there passes a unique plane and a unique line of these families. This identifies $P_3 - P^\infty$ with $R^4 \times P_1$. Moreover it shows how $R^4 \times \lambda$ acquires its complex structure, namely from the (complex) plane $u = \lambda v$.

Thus a solution of $*F = -F$ on R^4 leads to a holomorphic bundle on $P_3 - P_1$. Moreover this holomorphic bundle alone completely determines the original Yang-Mills field (up to gauge equivalence). This is essentially a consequence of Liouville's theorem that a holomorphic function of λ on the whole of P_1 must be constant.

Thus to construct all solutions of the equations $*F = -F$ we need only construct holomorphic bundles on $P_3 - P_1$. These bundles have to satisfy the condition that they are trivial when restricted to the lines (2) : this condition may fail for certain lines and these correspond to singularities of the Yang-Mills field. All this holds for $SL(n,C)$ potentials but the unitary condition is easily dealt with as an additional constraint.

Comparing this with the usual two-dimensional theory we see that here the extra spectral parameter λ is linked to the original 4-space and that the resulting 3 complex parameters of $P_3 - P_1$ are all on an equal footing for the anti-self-duality euqations. We may view this statement as expressing a "hidden symmetry" of these equations of a very far reaching kind.

A holomorphic bundle on $P_3 - P_1$ can always be specified by giving a holomorphic function, with values in $SL(n,C)$, defined on an appropriate open set of $P_3 - P_1$ intersecting all lines (2) in an annular strip around a suitable equator.

For the instanton problem one is interested in finite-action solutions over the whole of R^4 and these correspond to holomorphic bundles which extend to the whole of P_3. Two explicit constructions have been employed in this case. The first, valid for $G = SU(2)$, was used in [1] and consists in constructing bundles whose group is the group of triangular 2 x 2-matrices (of the form $\begin{pmatrix} a & b \\ o & a^{-1} \end{pmatrix}$). Since this group is soluble we are reduced in two steps to abelian groups and the corresponding bundle problems are <u>linear</u>. The second construction, used in [2], produced bundles as sub-quotients of trivial bundles.

The monopole problem which I shall discuss in the next section corresponds mathematically to a finite-energy static solution in R^4. Both the constructions employed for the instanton problem have been successfully used for the monopole problem: the first by Ward and others, the second by Nahm. I shall comment on the respective merits of these two approaches at the end.

§3. Magnetic Monopoles

The magnetic monopoles I shall be discussing, and which will be treated in more detail by subsequent lecturers, are the non-abelian static monopoles which arise in Yang-Mills-Higgs theories in the Prasad-Summerfield limit. They are given by an SU(2) Yang-Mills potential A and a Higgs field ϕ in the adjoint representation, both defined on R^3 which satisfy the Bogomolny differential equation

(3.1) $\nabla\phi = *F$

and the boundary condition

(3.2) $|\phi| \to 1$ as $|x| \to \infty$.

in (3.1) F is the Yang-Mills field of A and ∇ the covariant derivative, while $*$ is the duality operator of R^3. In (3.2) we shall actually assume a uniform asymptotic expansion of the form:

(3.3) $|\phi| \sim 1 - \frac{m}{r} + O(r^{-2})$

The constant m (up to a factor 4π) coincides with the topological degree k of the map $S^2 \to S^2$ given by restricting ϕ to a large sphere in R^3. The integer k is called the magnetic or topological charge.

An explicit spherically symmetric solution for $k = 1$ has been known for some time (the Bogomolny-Prasad-Summerfield monopole). At large distances it looks like a classical Dirac monopole, but at finite distances it differs from the latter in being non-singular - the price for this "regularization" being the non-linear (and non-abelian) nature of the equations.

A natural question was whether solutions existed for $k > 1$. If we imagine k BPS monopoles at large distances from one another repulsive magnetic forces will tend to make the configuration unstable. On the other hand the non-linear forces might produce a compensating attraction. In [8] R.S. Ward exhibited explicit solutions with $k = 2$. Also C. Taubes [7] proved, by methods of functional analysis, that solutions exist for all k. Moreover E. Weinberg [9] had already computed the number of real parameters on which such solutions should depend and found it to be $4k - 1$.

After this the problem was to extend Ward's approach to all k, so as to get explicit solutions and finally to prove a <u>completeness theorem</u> asserting that this procedure generates <u>all</u> solutions. Much progress has now been made in this direction and will be reported on by other lecturers. It seems likely that, putting together the joint efforts of different workers a complete solution will shortly emerge. I shall concentrate now on describing Hitchin's contribution [4], which essentially establishes the completeness theorem.

As I already indicated the Bogomolny equation in R^3 can be formally reinterpreted as the self-duality equation in R^4, where everything is independent of the fourth coordinate and ϕ is replaced by A_4.

§4. Hitchin's results

Because the Bogomolny equations are time-independent it is possible to present the twistor approach in a more elementary manner than for the full self-duality equations. I shall explain this, following Hitchin.

Consider the space T of all oriented lines ℓ in R^3. Now any ℓ is uniquely specified by a unit vector u (parallel to ℓ) and a perpendicular vector v representing its displacement from the origin.

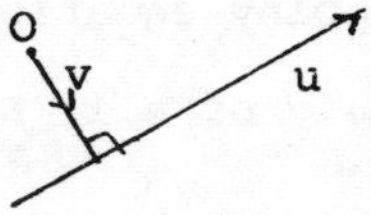

Thus T is described by all pairs $u, v \in R^3$ with

$$|u| = 1, \langle u,v \rangle = 1$$

and so can be identified with the tangent bundle of $S^2 = P_1$. In particular T is naturally a complex surface. In fact it can be realized as the projective quadric cone

$$z_1^2 + z_2^2 + z_3^2 = 0$$

in P_3, with the vertex (0,0,0,1) removed.

The point $\xi = (\xi_1, \xi_2, \xi_3)$ of R^3 corresponds to the section of this cone by the plane $\Sigma \xi_i z_i = 1$. The points of this conic section (intrinsically an S^2) represent the oriented lines through ξ. Note also that the points of a generating line of the cone correspond to a family of oriented parallel lines of R^3.

The twistor transformation in this context consists of translating problems from R^3 to T. For the Bogomolny equation Hitchin proceeds as follows. Given A, ϕ we consider on each oriented line ℓ the ordinary differential equation

$$(4.1) \quad (\nabla_\ell - i\phi)f = 0$$

where ∇_ℓ denotes the covariant derivative in the direction of ℓ. Here f is a 2-component function on ℓ and so we have a 2-dimensional space E_ℓ of solutions of (4.1). Hitchin proves that, if (A, ϕ) satisfy the Bogomolny equation (3.1), the bundle E over T given by the family of all E_ℓ has a natural holomorphic structure.

Hitchin's most significant observation is that if (A, ϕ) satisfies (3.1) and in addition the asymptotic condition (3.3) then one can associate to it an algebraic curve $\Gamma \subset T$. The points of Γ represent those lines ℓ for which (4.1) has a square-integrable solution. For this reason Hitchin calls Γ the spectral curve. Note that Γ is compact, i.e. it does not go through the vertex of the cone T. This corresponds to the fact that, for lines ℓ which lie entirely in regions where ϕ is close to its asymptotic value, the equation (4.1) has no square-integrable solution.

The curve Γ meets the generators of the cone T in k points so that Γ is a k-fold branched cover of P_1 (the "base" of Γ). Back in R^3 this means that any parallel family of oriented lines ℓ contains k "spectral" lines. Also through every point of R^3 there are k spectral lines (giving 2k oriented lines).

Hitchin proves the important result that <u>the spectral curve</u> Γ <u>determines uniquely the solution</u> (A,ϕ) (up to gauge equivalence).

As simple examples consider first the case $k = 1$. Then Γ is a plane conic section of T (corresponding to the location of the BPS monopole). For axially-symmetric solutions the curve Γ has to admit rotational symmetry and so can consist only of rational curves: in fact Γ breaks up into k plane conic sections parametrized by imaginary points on the axis of symmetry.

General solutions involve curves Γ of high genus and hence the theory of abelian functions. When $k = 2$ Γ is an elliptic curve and the explicit solutions involve elliptic functions.

As I mentioned earlier Hitchin also proves that Ward's construction (using triangular matrices) gives all monopoles. This arises as follows. For each line ℓ the equation (4.1) has one solution (unique up to a constant multiple) which decays exponentially at $+\infty$. This defines a one-dimensional subspace V_ℓ of the space E_ℓ of all solutions. As we vary ℓ it turns out that the V_ℓ vary holomorphically and so give a holomorphic sub-bundle of E. Thus E has a canonical reduction to triangular form and this explains why Ward's construction gives all solutions.

There is also a subspace V'_ℓ of E_ℓ given by the solutions decaying at $-\infty$. Clearly ℓ is a spectral line precisely when $V'_\ell = V_\ell$, and this is a holomorphic condition which explains why Γ in a holomorphic curve.

The curve Γ is algebraic but Ward's construction also involves an exponential function. From Hitchin's point of view this arises from a basic line-bundle L on T (given by a simple exponential transition function). Moreover when restricted to Γ we must have L^2 trivial. This imposes constraints on the coefficients of the polynomial defining Γ and reduces the number of free parameters to 4k - 1, consistently with Weinberg's result.

§5. Relation with Nahm's work

Nahm's approach to the monopole problem, which is based on the method used for instantons in [2], involves solving a system of ordinary non-linear differential equations. There are the analogues of the algebraic constraints for instantons in [2]. The equations take the form

$$(5.1)\qquad \begin{aligned} T_1' &= [T_2, T_3] \\ T_2' &= [T_3, T_1] \\ T_3' &= [T_1, T_2] \end{aligned}$$

where T_1, T_2, T_3 are $k \times k$ matrix functions of a variable z and T' denotes $\frac{dT}{dz}$.

If we put $A = T_1 + iT_2$ then (5.1) implies the Lax type equation:

$$A' = [P, A]$$

where $P = -iT_3$. In particular this implies that <u>the eigenvalues of</u> A <u>are constant, independent of</u> z. Now the equations (5.1) are orthogonally invariant so we get a matrix like A for each oriented plane through O in R^3, and all these have constant eigenvalues. As we vary the plane these eigenvalues trace out a compact Riemann surface which is a k-fold cover of the 2-sphere (parametrizing the planes). This Riemann surface is then a constant of the flow given by (5.1).

By comparing Nahm's procedure with that of Hitchin one can verify that this Riemann surface coincides with Hitchin's spectral curve Γ. Moreover the flow (5.1) can be translated into Hitchin's framework, from which one sees that it is "integrable".

To explain this in more detail let $\zeta \in S^2 = P_1$ be a complex parameter and let A_ζ be the associated matrix constructed from the T_i and the plane defined by ζ. Thus Γ is the union of the eigenvalues of all A_ζ. Now for each point γ of Γ we have the corresponding eigenspace F_γ and the family of all F_γ gives a holomorphic line-bundle over Γ. The classical Abel-Jacobi theory in its modern form tells us that F is determined (up to holomorphic isomorphism) by an integer topological invariant and a point of the Jacobian J of Γ. We recall that J is a complex torus of complex dimension equal to the genus of Γ. So far we have simply associated to any set of 3 matrices T_1, T_2, T_3 the curve and the line-bundle F(and so a point $f \in J(\Gamma)$). One can show that the pair Γ, f are essentially equivalent data to the triple T_1, T_2, T_3. The flow (5.1) can therefore be reinterpreted as a flow on $J(\Gamma)$: recall that Γ is fixed under the flow, only f moves. The important fact about (5.1) is that <u>this flow on</u> $J(\Gamma)$ <u>is actually linear</u>. This means that (5.1) can be explicitly solved using abelian functions, so that in this sense it is "integrable". Similar though slightly different situations of this type have been discussed in [5]. In any case this is the basic way in which integrable systems of soliton type have been solved: they are interpreted as linear Jacobian flows for an algebraic curve determined by the spectrum of a suitable linear system.

In Hitchin's framework this Jacobian flow is simply given by raising the basic line-bundle L to the power z: if L is given by the exponential transition function $\exp(g)$ then L^z is given by $\exp(zg)$.

Nahm has also shown that every monopole given by his construction has a distinguished "centre". In Hitchin's set-up this point of R^3 corresponds to a plane section of the cone T which is the "centroid" of the spectral curve Γ. That is, on each generator of T (which is a complex affine line) we take the centroid of the points of intersection with Γ. The locus of these centroids then turns out to be a plane section. Returning to R^3 this means that, fixing a direction, the k spectral lines in that direction have a "central line" and all such central lines (as we vary the direction) pass through a fixed point which is then the "centre" of the monopole.

§6. Conclusion

I shall now indicate the respective merits of the different approaches to the monopole problem and discuss what still remains to be done.

As far as getting useful explicit formulae the method initiated by Ward [8] and developed further by Corrigan-Goddard [3] and Prasad-Rossi [6] appears to be the most effective. Hitchin's approach proves that all solutions are obtained this way and it also gives an attractive geometrical interpretation of the construction, and in particular of the spectral curve. However, these methods are not very convenient in deciding whether a given solution is everywhere regular.

Nahm's method on the other hand, as for the corresponding instanton method, is well-suited to decided regularity. It may also be useful for constructing Green's functions. The non-linear equation which enters is, as explained in §5, integrable - effectively by reverting to Hitchin's viewpoint.

At present there is still some work to be done in describing in detail the parameter space of monopoles. The difficulty arises from the constraints on the curve Γ. As mentioned in §4 Hitchin formulates this by saying that L^2 is trivial when restricted to Γ. This amounts to specifying an equation on $J(\Gamma)$ and there integer choices are involved (since a complex torus is C^g modulo an integer lattice). It is not yet known how to make these choices so as to guarantee regularity of the solution. Nahm's approach appears to give some guidance on this question and may help to solve the problem.

So far I have restricted myself to SU(2). In fact all the methods used work also in principle for other groups, though there are extra complications and many details to be worked out. Hitchin's method leads now to several spectral curves and Nahm's differential equation (5.1) now has k as an integer function of z with jumps at integer points which can be poles of the T_i.

Finally let me just say that I believe we have much to learn yet from the mathematics of monopoles. The remarkable connection between the Bogomolny equation and Nahm's "integrable" equation (5.1) is only one of many fascinating aspects.

REFERENCES

1. M.F. Atiyah and R.S. Ward, Comm. Math.Phys. 55 (1977), 117.
2. M.F. Atiyah, N.J. Hitchin, V.G. Drinfeld and Yu. I. Manin, Phys. Lett 65A (L978), 185.
3. E. Corrigan and P. Goddard, Comm. Math. Phys. 80 (1981), 575.
4. N.J. Hitchin, Monopoles and Geodesics, Comm. Math. Phys.
5. P. van Moerbeke and D. Mumford, Acta Math. 143 (1979), 93.
6. M.K. Prasad and P. Rossi, MIT preprint CTP 903 (1980).
7. C. Taubes, Harvard preprint HUTMP 79/B94.
8. R.S. Ward, Comm. Math. Phys. 79 (1981), 317.
9. E. Weinberg, CERN preprint T.H. 2779. CERN (1979).

PHYSICIST'S TECHNIQUES FOR MULTIMONOPOLE SOLUTIONS*

Peter Forgács
Central Research Institute for Physics
H-1525 Budapest 114, P.O.B. 49., Hungary

and

Zalán Horváth and László Palla
Institute for Theoretical Physics
Roland Eötvös University
H-1088 Budapest, Puskin u. 5-7, Hungary

ABSTRACT

The generation of SU/2/ axially symmetric multimonopoles is reviewed. A Riemann-Hilbert problem is set up for the SU/N/ Bogomolny equations yielding all local solutions. The 4n-1 parameter family of SU/2/ monopoles is derived.

1. INTRODUCTION

As it was shown for the first time by 'tHooft and Polyakov [1] spontaneously broken gauge theories with a simple gauge group possess classical solutions which may be identified as magnetic monopoles. In the limit of vanishing Higgs potential, when the Higgs field becomes massless, a considerable simplification arises and the analytic form of the static singly charged and spherically symmetric 'tHooft-Polyakov monopole was found by Prasad and Sommerfield [2] and by Bogomolny [3]. In this case the theory is considerably simplified and the static minimal energy configurations are solutions of a first order system of equations /Bogomolny equations/. After a long search multimonopoles were constructed and we are now near finding the general n-monopole solution. Three different lines of attack turned out to be successful; the Atiyah-Ward /AW/ ansätze[4] made explicit by Corrigan et al. [5], the Atiyah-Drinfeld-Hitchin-Manin /ADHM/ [6] construction as applied by Nahm [7] and soliton theoretic methods applied by the present authors [8].

Following the construction of the axially symmetric 2 monopole /2MP/ [9] and n monopole [10] in which case the monopoles were situated at the same point, Ward [11] outlined how to construct a separated 2 monopole solution using the AW ansätze. The proposed solution is necessarily not axially symmetric [12] and was argued to be nonsingular when the monopoles are sufficiently close to each other. Ward's solution was immediately generalized by Corrigan and Goddard [13] to an n monopole solution depending on 4n-1 parameters. As it was shown [14] any n monopole solution must belong to a 4n-1 parameter class so the proposed solutions in Ref. [13] have the maximal number of degrees of freedom, although it was argued to be regular for the case when it is "close" to the axially symmetric n monopole solution.

Independently of the above groups we also generalized the axially symmetric monopoles /depending on 5 parameters/ to a 4n-1 parameter family using an "inverse scattering" method developed in [15] for the SU(N) self duality and for the Bogomolny equations.

Our soliton theoretic methods are based on the fact that the Bogomolny equations can be linearized. This is a well known property of all two dimensional completely integrable systems. For these systems various solution generating techniques have been found e.g. Bäcklund transformations BT's and an "inverse scattering" method /Riemann-Hilbert problems/. It was shown in [8] that the axially and mirror symmetric Bogomolny equations are equivalent to the Ernst equations. These elliptic equations share a lot of properties with completely integrable systems. What is more important for us, the solution generating techniques , especially an appropriate BT could be succesfully used for generating the axially symmetric multimonopoles in an SU(2) theory. This approach was generalized for any gauge group in the axially symmetric case [16].

We deduced a Riemann-Hilbert problem (RHP) for the SU(N) SDE which provided the key for us to understand the structure of these equations and paved the way to get the multimonopoles without any symmetry.

An advantage of our method that the solutions we obtain are completely explicit. The main drawback is that the solutions are local in the sense that regularity is not automatically guaranteed. This applies to AW ansätze as well. The ADHM construction as applied by Nahm [7] (ADHMN) was a real tour de force. He solved this problem completely since regularity is guaranteed by the ADHMN method. However, to get the explicit form of the solutions is perhaps simpler using the other methods.

It seems that there still remains some inherent nonlinearity despite the fact that the differential equations could be linearized. These nonlinearities are algebraic in nature and can be succesfully tackled by the methods of algebraic geometry. The most recent result is due to Hitchin [17] who has shown completeness i.e. that <u>every</u> static monopole of charge n can be constructed in the AW Ansatz A_n. We refer to prof. Atiyah's lectures about this beautiful result.

It is our opinion that although these different methods surely have a common background and they lead to the same solutions in different ways they complement each other rather nicely and all of them bring up something interesting and unexpected.

What we found is perhaps not so geometrical and elegant but it may show the way how to apply the methods of inverse scattering /via the RHP/ in four dimensions which led to striking results in 2 dimensions.

Our paper is organized as follows in sect. 2 we formulate the problem and introduce the notations. The axially symmetric SU(2) monopoles are generated in sect. 3 by iterating a Bäcklund transformation. In sect. 4 we set up a Riemann-Hilbert problem for the Bogomolny equations which is used in sect. 5 to construct the separated multimonopoles.

2. Formulation of the problem

We consider an SU(2) gauge theory with an isotriplet Higgs field in the limit of vanishing Higgs potential. The Lagrangian density is

$$L = -\frac{1}{4} F^a_{\mu\nu} F^{a\mu\nu} - \frac{1}{2}\left(D_\mu \Phi\right)^a \left(D^\mu \Phi\right)^a \tag{1}$$

where $F^a_{\mu\nu} = \partial_\mu A^a_\nu - \partial_\nu A^a_\mu - \varepsilon^{abc} A^b_\mu A^c_\nu$

$$\left(D_\mu \Phi\right)^a = \partial_\mu \Phi^a - \varepsilon^{abc} A^b_\mu \Phi^c$$

/ We chose the coupling constant e = 1 /.

The Hamiltonian density for static configurations with no electric fields / $A^a_o = 0$ / is

$$H = \frac{1}{4} F^a_{ij} F^{aij} + \frac{1}{2}\left(D_i \Phi\right)^a \left(D^i \Phi\right)^a \tag{2}$$

$$a,i,j = 1,2,3$$

The field equations of this theory are solved by configurations satisfying the Bogomolny equations [3]:

$$F^a_{ij} = -\varepsilon_{ijk}\left(D_k \Phi\right)^a \tag{3}$$

The energy, E , can be written using the Bogomolny equations as

$$E = \int H d^3x = \int \left(D_k \Phi\right)^a \left(D^k \Phi\right)^a d^3x \tag{4}$$

Now

$$\left(D_k\Phi\right)^a \left(D^k\Phi\right)^a = \frac{1}{2}\,\partial^k\partial_k\Phi^a\Phi^a = \frac{1}{2}\Delta|\Phi|^2 \tag{5}$$

using the equations of motion for the Higgs field, $D^kD_k\Phi = 0$, so the energy can be calculated from the Higgs field alone

$$E = \frac{1}{2}\int \Delta|\Phi|^2\, d^3x \tag{6}$$

The topological charge, n, is given by

$$n = \lim_{r\to\infty} \frac{1}{8\pi V} \int_{r=\text{const.}} dS^i\partial_i|\Phi|^2 \tag{7}$$

Since the asymptotic boundary condition $\lim_{r\to\infty}|\Phi| = V$ is imposed, for a solution with topological charge n

$$|\Phi| \to V - \frac{n}{r} + O\left(\frac{1}{r^2}\right) \quad \text{as } r\to\infty \tag{8}$$

However, condition (8) alone does not guarantee that the topological charge is indeed n due to the presence of possible singularities. From now on, we take the vacuum expectation value of the Higgs field V = 1.

3. Bäcklund transformations in the case of axially symmetry

When one is looking for exact solutions of complicated equations, there is a need for a simplifying ansatz consistent with the configuration in quest. In the present case the simplest ansatz would be a spherically symmetric one. However, the assumption of spherical symmetry is too strong, it excludes all but the singly charged 'tHooft-Polyakov monopole [18] . The next simplest thing one can do is to assume axial symmetry. In gauge theories by a symmetry we mean that the change of the gauge field, A_μ^a , under this symmetry transformation / e.g. a rotation around any axis / can be compensated by a gauge transformation [19] .

Manton [20] constructed an axially and mirror symmetric ansatz which can be written in polar coordinates as

$$A_0^a = 0\ , \quad \Phi^a = (0,\phi_1,\phi_2)\ , \quad A_\phi^a = -(0,\eta_1,\eta_2)$$
$$A_z^a = -(W_1,0,0)\ , \quad A_\rho^a = -(W_2,0,0) \tag{9}$$

where $x_1 = \rho\cos\phi$, $x_2 = \rho\sin\phi$, and ϕ_i,η_i,W_i are functions of ρ,z only. Plugging (9) into the Bogomolny equations (3) they simplify to

$$\partial_\rho\phi_1 - W_2\phi_2 = -\rho^{-1}(\partial_z\eta_1 - W_1\eta_2) \tag{10a}$$

$$\partial_\rho\phi_2 + W_2\phi_1 = -\rho^{-1}(\partial_z\eta_2 + W_1\eta_1) \tag{10b}$$

$$\partial_\rho W_1 - \partial_z W_2 = \rho^{-1}(\phi_1\eta_2 - \phi_2\eta_1) \tag{10c}$$

$$\partial_z\phi_1 - W_1\phi_2 = \rho^{-1}(\partial_\rho\eta_1 - W_2\eta_2) \tag{10d}$$

$$\partial_z\phi_2 + W_1\phi_1 = \rho^{-1}(\partial_\rho\eta_2 + W_2\eta_1) \tag{10e}$$

We have five equations for the six unknown variables, however, equations (10a-e) still possess a residual U(1) gauge invariance:

$$W_i' = W_i + \partial_i \Lambda$$

$$\begin{pmatrix} \phi_i' \\ \eta_i' \end{pmatrix} = \begin{pmatrix} \phi_i \\ \eta_i \end{pmatrix} \cos\Lambda + \varepsilon_{ij} \begin{pmatrix} \phi_j \\ \eta_j \end{pmatrix} \sin\Lambda \ , \quad \varepsilon_{12} = 1 = -\varepsilon_{21} \tag{11}$$

This residual gauge freedom enables us to reduce the number of unknown functions from six to five. In fact, we can do even more than that, namely, we can satisfy one equation in (10) by the following trick: it is possible to find such a Λ that

$$W_1' = -\phi_1' \tag{12a}$$

$$W_2' = \frac{1}{\rho} \eta_1' \tag{12b}$$

are simultaneously true.

Now, we observe that (10b) implies the existence of a function f , such that

$$\phi_2 = -\frac{f_{,z}}{f} \ , \quad \frac{1}{\rho} \eta_2 = \frac{f_{,\rho}}{f} \tag{13}$$

Then it is easy to see that (10c) can be also satisfied by putting

$$W_1 = -\frac{\Psi_{,z}}{f} \ , \quad W_2 = \frac{\Psi_{,\rho}}{f} \tag{14}$$

and the remaining two equations (10d-e) reduce to

$$f\Delta f - (\nabla f)^2 + (\nabla\Psi)^2 = 0 \tag{15a}$$

$$f\Delta\Psi - 2\nabla f\nabla\Psi = 0 . \tag{15b}$$

where $\Delta = \partial_\rho^2 + \rho^{-1}\partial_\rho + \partial_z^2$, $\nabla = (\partial_z, \partial_\rho)$

Introducing $\varepsilon = f + i\Psi$ eqs. (15a-b) may be written as

$$\mathrm{Re}\,\varepsilon\Delta\varepsilon - (\nabla\varepsilon)^2 = 0 \tag{16}$$

which is the celebrated form of the Ernst equation of general relativity. A geometrical derivation of eq. (16) from (10a-e) was given in Ref. [8] .

The equivalence of eqs. (10a-e) and (16) is of interest because it reveals a surprising /although completely formal/ connection between general relativity and SU(2) gauge theories, and what is more important for us, it gave the clue to find new exact solutions of the Bogomolny equations describing multiply charged monopoles, using the various solution generating techniques worked out for the Ernst equation.

There are group-theoretic or soliton-theoretic techniques: Bäcklund transformations found by Harrison /HB/ [21] and Neugebauer /NB/ [22] the "inverse scattering" method of Belinsky and Zakharov /BZ/ [23] and the integral equation approach /Riemann-Hilbert problem/ devised by Hauser and Ernst [24] .

The Bäcklund transformations we are going to apply [8] enables us to generate /by iteration/ infinitely many new solutions of (10) from an initial one using algebraic steps only in all but the first step. To implement this transformation one defines from the initial solution η_i°, ϕ_i° the quantities

$$M_2^{\circ} - N_1^{\circ} = \frac{1}{2}\left(\phi_1^{\circ} + i\phi_2^{\circ}\right) \qquad N_2^{\circ} - M_1^{\circ} = \frac{1}{2}\left(\phi_1^{\circ} - i\phi_2^{\circ}\right)$$
$$M_1^{\circ} + N_1^{\circ} = \frac{1}{2\rho}\left(\eta_2^{\circ} + i\eta_1^{\circ}\right) \qquad M_2^{\circ} + N_1^{\circ}. = \frac{1}{2\rho}\left(\eta_2^{\circ} - i\eta_1^{\circ}\right) \tag{17}$$

and solves the total Riccati equation for the pseudopotential $q(\zeta_1, \zeta_2)$:

$$\zeta_1 = \rho + iz \quad , \quad \zeta_2 = \rho - iz$$

$$dq = \left[\left(M_2^{(o)} - M_1^{(o)}\right) q + p(w)\left(M_2^{(o)} - M_1^{(o)} q^2\right)\right] d\,\zeta_1 +$$
$$+ \left[\left(N_1^{(o)} - N_2^{(o)}\right) q + p^{-1}(w)\left(N_1^{(o)} - N_2^{(o)} q^2\right)\right] d\,\zeta_2 \quad , \tag{18}$$

where $p(w) = \sqrt{(w-i\zeta_2)(w+i\zeta_1)^{-1}}$, w being a constant of integration. We remark here that the other integration constant for w will be denoted by β .

The new /transformed/ M_i's and N_i's are given in terms of $M_i^{(o)}$, $N_i^{(o)}$, q and $p(w)$ as follows

$$M_1^{(1)} = BM_1^{(o)} = \frac{p+q}{1+qp}\left(q^{-1}M_2^{(o)} + \frac{p}{4\rho}\right)$$
$$M_2^{(1)} = BM_2^{(o)} = \frac{1+pq}{p+q}\left(q\, M_1^{(o)} + \frac{p}{4\rho}\right)$$
$$N_1^{(1)} = BN_1^{(o)} = \frac{1+pq}{p+p}\left(q^{-1}\, N_1^{(o)} + \frac{1}{4p\rho}\right) \tag{19}$$
$$N_2^{(1)} = BN_2^{(o)} = \frac{p+q}{1+pq}\left(q\, N_2^{(o)} + \frac{1}{4p\rho}\right)$$

A restriction on β and w comes from requiring the resulting ϕ_i, η_i and W_i to be real, which in turn implies that the final M_i's and N_i's have to satisfy

$$M_1^* = N_1 \quad , \quad M_2^* = N_2 \tag{20}$$

We also have to ensure that $\phi^2 = \phi_1^2 + \phi_2^2$ given as

$$\phi^2 = 4\left(M_1 - N_2\right)\left(N_1 - M_2\right) \tag{21}$$

be free of singularities. This completely fixes w and β .

The simplest vacuum state is $M_1^{(o)} = M_2^{(o)} = -\frac{i}{4}$, $N_1^{(o)} = N_2^{(o)} = \frac{i}{4}$. In this case it is easy to integrate (18)

$$q = -\tanh\left(\frac{R(w)}{2} - \beta\right) \tag{22}$$

where $R(w) = \sqrt{(w-z)^2 + \rho^2}$.

We see from (19) that (20) can be satisfied by demanding $q^* = q$ and $p^* = p^{-1}$ which in turn are satisfied if w and β are real numbers. Finally, one obtains from (21) that ϕ^2 has no singularity if $\beta = 0$, when it yields the well known BPS one monopole /1MP/ located at z = w.

At this stage it seems to be very difficult to apply a second BT, because we have to solve eq. (18) again, replacing the M_i's, N_i's by $M_i^{(1)}$'s and $N_i^{(1)}$'s to find q' for the next step. Usually this is a very hard task even for the simplest seed solutions, since after a single BT the new configurations can be rather complicated. The great advantage of the BT is the existence of a composition theorem stating that there is no need to solve analytically this complicated equation as its solution

can be constructed in an appropriate /algebraic/ way from solutions of (18) with the original M_i's and N_i's.
Let (q_1, p_1), (q_2, p_2) denote solutións with different constants W_i, β_i of the Riccati equation (18) with $M_i^{(o)}$, $N_i^{(o)}$. Then $(p_2, \tilde{q}')$ will solve (18) with $M_i^{(1)}$, $N_i^{(1)}$ if

$$\tilde{q}' = \frac{q_1 p_2 - q_2 p_1}{\tilde{q}_1 (q_1 p_1 - q_2 p_2)}$$

where $\tilde{q} = -\frac{p+q}{1+pq}$.
After $n = 2k+\varepsilon$ iterations we get for $|\phi|$:

$$|\phi| = 2 \left| \frac{D_2^{(2k+\varepsilon)}(q_i)}{D_1^{(2k+\varepsilon)}(q_i)} M_{1+\varepsilon}^{(o)} - \frac{D_1^{(2k+\varepsilon)}(q_i)}{D_2^{(2k+\varepsilon)}(q_i)} N_2^{(o)} + \right.$$
$$\left. + \frac{1}{4\rho} \left(\frac{D_3^{(2k+\varepsilon)}(q_i)}{D_1^{(2k+\varepsilon)}(q_i)} - \frac{D_4^{(2k+\varepsilon)}(q_i)}{D_2^{(2k+\varepsilon)}(q_i)} \right) \right| , \qquad (23)$$

where $\varepsilon = 0$ or 1, k is any integer and we used the reality conditions.
D^n 's are determinants of nxn matrices whose i-th row is given as

$$D_1^{(2k+1)}(q_i) = |q_i,\ p_i,\ p_i^2 q_i,\ p_i^3,\ p_i^4 q_i, \ldots,\ p_i^{2k} q_i|$$

$$D_2^{(2k+1)}(q_i) = |1,\ p_i q_i,\ p_i^2,\ p_i^3 q_i,\ p_i^4, \ldots,\ p_i^{2k}|$$

$$D_3^{(2k+1)}(q_i) = |1,\ p_i q_i,\ p_i^2,\ p_i^3 q_i,\ p_i^4 \ldots,\ p_i^{2k-1} q_i,\ p_i^{2k+1} q_i| \qquad (24a)$$

$$D_4^{(2k+1)}(q_i) = |p_i^{-1},\ p_i,\ p_i^2 q_i,\ p_i^3,\ p_i^4 q_i, \ldots,\ p_i^{2k} q_i|$$

where $i = 1, \ldots, 2k+1$, and

$$\begin{aligned}
D_1^{(2k)}(q_i) &= |q_i, p_i, p_i^2 q_i, p_i^3, p_i^4 q_i, \ldots, p_i^{2k-1}| \\
D_2^{(2k)}(q_i) &= |1, p_i q_i, p_i^2, p_i^3 q_i, p_i^4, \ldots, p_i^{2k-1} q_i| \\
D_3^{(2k)}(q_i) &= |1, p_i q_i, p_i^2, p_i^3 q_i, p_i^4, \ldots, p_i^{2k-2}, p_i^{2k}| \\
D_4^{(2k)}(q_i) &= |p_i^{-1}, p_i, p_i^2 q_i, p_i^3, p_i^4 q_i, \ldots, p_i^{2k-1}|
\end{aligned} \tag{24b}$$

where $i = 1, \ldots, 2k$.

Demanding regularity of $|\phi|$ singles out the only possible values of w_i, β_i :

for $n=2k$:

$$\begin{aligned}
w_{2r-1} &= w_{2r}^{*} = i(2r-1)\frac{\pi}{2} \\
q_{4r-3} &= -\tanh\left(\frac{R\left[i\frac{4r-3}{2}\pi\right]}{2}\right) = q_{4r-2}^{*-1} \\
q_{4r-1} &= -\coth\left(\frac{R\left[i\frac{4r-1}{2}\pi\right]}{2}\right) = q_{4r}^{*-1}
\end{aligned} \tag{25}$$

In the case of $n=2k+1$ we obtained

$$\begin{aligned}
&w_1 = 0, \quad w_{2r}^{*} = w_{2r+1} = ir\pi \quad ; \quad q_1 = -\tanh\frac{R(0)}{2} \\
&q_{4r-2} = -\coth\left(\frac{R[i(2r-1)\pi]}{2}\right) = q_{4r-1}^{*} \\
&q_{4r} = -\tanh\left(\frac{R[i2r\pi]}{2}\right) = q_{4r+1}^{*}
\end{aligned} \tag{26}$$

/We must take the first 2k or 2k+1 elements of these sequences of q_i's respectively./ We claim that eqs. (25,26) together with (23,24) yield the solutions of the axially and mirror symmetric Bogomolny equations corresponding to n monopoles superimposed at the origin. It is now easy to get a physical picture of our n-monopole configurations: since

$$|\phi| \underset{r\to\infty}{\to} 1 - \sum_{i=1}^{n} R^{-1}(w_i) + O(e^{-r}) \qquad (27)$$

we see that the polynomial part of $|\phi|$ in the asymptotic region consists of the sum of several rings.

$|\phi|$ drastically simplifies on the z axis:

for n = 2k+1

$$|\phi(\rho=0)| = \left| \coth z - \frac{1}{2} - \sum_{\ell=1}^{k} \frac{2z}{z^2+(\ell\pi)^2} \right| \qquad (28a)$$

for n = 2k

$$|\phi(\rho=0)| = \left| \tanh z - \sum_{\ell=1}^{k} \frac{2z}{z^2+\left[\frac{(2\ell-1)\pi}{2}\right]^2} \right| \qquad (28b)$$

$|\phi|$. of the 2 and 3 monopole was computed explicity in ref [10]. We calculated numerically the norm of the Higgs field for n=4,5 as well.

All these functions show the characteristic behaviour of a monopole with magnetic charge n, situated at the origin. We observe that $|\phi|$ has an n-th order zero on the z=0 plane and a first order zero when approaching the origin from any other direction. This peculiar behaviour is consistent with combined topological and symmetry arguments [12,25,26] . The shape of $|\phi|$ for solutions of different topological charge, n, / $n \neq 1$ / is qualitatively very similar and $|\phi|$ tends to 1 monotonously in every directions. As we obtained the axially symmetric multimonopoles of charge n by applying n BT's to the vacuum, they can be conceived as nonlinear superpositions.

It would be interesting to know whether this nonlinear superposition manifests itself in the "physical" quantities such as the energy density $E = \frac{1}{2}\Delta\phi^2$. For the 1MP E_{1MP} has a maximum at the origin and starting from here it decreases monotonously to zero. Note that E_{1MP} takes on its maximum at the same point where $|\phi|$ has its zero, i.e. the energy is concentrated at the same point where the topological charge resides. Clearly for monopoles with higher charge an analytical investigation of $\Delta\phi^2$ is rather complicated, however, - using the explicit formulae (23,24) with (25,26) - its numerical determination is feasible. We carried out this program for monopoles with charges 2,3,4,5; [10]. From this study one can deduce the following gross features of the energy density of the multimonopoles: In all cases the energy density is fairly well localized, although the region where E differs significantly from zero increases with increasing n. It is interesting that E for the 3,4 and 5MP shows no structure indicating the nonlinear superposition. A very distinct feature of all multi-monopoles generated by our method that E takes on its maximum value on the z = 0 plane <u>away from the origin</u>, i.e. in these solutions the energy is no longer concentrated at the location of the topological charge. The location of E_{max} roughly coincides with the position of the outermost ring.

4. A Rieman-Hilbert problem for the Bogomolny equations

The succes of Bäcklund transformations in obtaining axially symmetric SU(2) monopoles naturally makes one to attempt to generalize them for the non-axially symmetric case and for groups other than SU(2) . The first steps in this direction have been taken [16] by deriving BT's for any group G in the axially symmetric case, but nothing is known yet about BT-s for the non-symmetric equations.

For the Ernst eq. there exists a method devised by Belinsky and Zakharov [23] which is in a certain sense equivalent to the BT and can be generalized for the Bogomolny equations without any symmetry assumption for any gauge group. This generalization - that we are going to describe - works not only for the Bogomolny but for the full selfduality /SDE/ equations. It also can be considered as a special case of a Riemann-Hilbert problem /RHP/ associated to the SDE. The RHP is a unifying principle as any selfdual solution can be obtained from a judiciously set up RHP . As it is well known the self-duality /SDE/ equations for an SU(N) gauge theory can be written as [27]

$$(g,_y g^{-1})_{\bar{y}} + (g,_z g^{-1})_{\bar{z}} = 0 \tag{29}$$

where $y=x_1+ix_2$, $z=x_3+ix_4$, $g= D^+D$, $D\varepsilon SL(N,C)$.
The gauge potentials, B_y, etc are expressed as:

$$B_y = - D,_y D^{-1} \quad , \quad B_{\bar{y}} = D^{+-1}D^+_{,\bar{y}} \quad , \quad \text{etc. [28] .}$$

Note that g is gauge invariant. An intrigueing property of eq. (29) is the existence of an invariance group:

$$g' = \Omega(y,z)\; g\Omega^{+}(y,z) \text{ is again a solution of } (29)$$

where $\Omega \varepsilon SL(N,C)$. Furthermore, there exists a discrete map, I, [15,27]

$$Ig = \tilde{g} \tag{30}$$

giving a new solution, $\tilde{g}$, of (29) with det $\tilde{g} = -1$. Now as the equation for $\tilde{g}$ is invariant under

$$\tilde{g}' = A(y,z)\; \tilde{g}\; A^{+}(y,z)\;, \quad A \varepsilon SL(N,C) \tag{31}$$

The action of $A(y,z)$ on g is not an $\Omega(y,z)$ covariant expression and therefore the product of an A and Ω transformation is not contained in any of these two groups. Thus the repeated applications of these two transformations generate an infinite parameter invariance group first pointed out in [15]. The infinite dimensional Lie algebra of this invariance group was presented in [29].

The key for solving (29) is the following linear eigenvalue problem [15, 30]

$$(\lambda \partial_{\bar{z}} + \partial_y)\,\Psi = A_y \Psi\,, \quad (-\lambda \partial_{\bar{y}} + \partial_z)\Psi = A_z \tag{32}$$

where $A_y = g_y g^{-1}$, $A_z = g_{,z} g^{-1}$

and λ is an arbitrary complex parameter.

We developed a method for solving this system, [15] which we now apply for the Bogomolny equations. The linearized system for the Bogomolny eqs. takes the following form:

$$(\lambda\partial_z + \partial_y)\Psi = A_y\Psi , \quad (-\lambda\partial_{\bar{y}} + \partial_z)\Psi = A_z\Psi , \tag{33}$$

where $y = \frac{1}{2}(x_1+ix_2)$, $z = x_3$.
We demand that $\Psi(\lambda = 0) = g$.

Our strategy for solving (33) is the following: We suppose that there is a known solution, $\Psi_o(\lambda)$, of (33) . We then set up a /matrix/ Riemann-Hilbert problem (RHP), the solution of which enables us to construct infinitely many new solutions from Ψ_o . For two dimensional completely integrable systems the use of the RHP was pioneered by Zakharov et al. [31] and was shown to be equivalent with the usual inverse scattering method. For certain elliptic equations, there is an analogue of the inverse scattering method, namely there exists an associated linear eigenvalue problem which can be reduced to a RHP. This procedure can be generalized to higher dimensions [15] .

The regular matrix RHP is the following:
Consider a closed curve, Γ , in the complex λ plane. Define on Γ a matrix, $G(\lambda)$ and look for $\chi_1(\lambda)$ and $\chi_2(\lambda)$ analytic inside /resp. outside/ Γ and satisfying:

$$\chi_1^{-1}(\lambda)\,\chi_2(\lambda) = G(\lambda). \tag{34}$$

To ensure the uniqueness of the solution, we require $\chi_2(\lambda=\infty) = I$, and demand that $\det \chi_i \neq 0$ in their domain of analiticity.

In our case we take $G(\lambda)$ as:

$$G(\lambda) = \Psi_o(\lambda)G_o\Psi_o^{-1} \tag{35}$$

where $G_o = G_o\left(\lambda, \gamma(\lambda)\right)$ but otherwise arbitrary ,
$\gamma(\lambda) = y\lambda - \frac{\bar{y}}{\lambda} - z$, and

$$\det G_o = 1 \, , \quad G_o^+\left(-\frac{1}{\bar{\lambda}}\right) = G_o(\lambda) \tag{36}$$

Then one can verify that $\Psi(\lambda) = \chi_1(\lambda)\,\Psi_o(\lambda)$ solves again our linear system (33) , with g given by: $g = \chi_1(0)\, g_o$. This g is a new solution of Bogomolny eqs.

To guarantee the hermiticity of g we impose an additional constraint on χ_1 and χ_2 :

$$g = \chi_1\left(-\frac{1}{\lambda}\right) g_o \, \chi_2^+(\bar{\lambda}) \; . \tag{37}$$

The regular RHP can be reduced to a <u>linear</u> singular integral equation in the following way:

We represent $\chi_i(\lambda)$'s as

$$\chi_1(\lambda) = I + \int_\Gamma \frac{\sigma(\zeta)}{\zeta - \lambda}\, d\zeta \qquad \text{inside } \Gamma \tag{38a}$$

$$\chi_2(\lambda) = I + \int_\Gamma \frac{\sigma(\zeta)}{\zeta - \lambda}\, d\zeta \qquad \text{outside } \Gamma \tag{38b}$$

Then from (34) we get the following integral equation for $\sigma(\zeta)$

$$\frac{1}{\pi i}\left[P\int_\Gamma \frac{\sigma(\zeta)}{\zeta - \lambda}\, d\zeta + I\right] T(\lambda) + \sigma(\lambda) = 0, \quad \lambda \varepsilon \Gamma \tag{39}$$

where $T(\lambda) = \left[G(\lambda) - I\right]\left[G(\lambda) + I\right]^{-1}$.

By analogy to the usual inverse scattering method, solutions of (39) would correspond to the so called non solitonic part of solutions of (29) . To get all solutions of (29) we must solve the RHP with zeroes, that is allowing det $\chi_i = 0$ at some points within their domain of analiticity. The corresponding linear equations contain the full information about the original nonlinear equations, that is they would enable us to solve the boundary value problem.

To obtain the monopole solutions it is sufficient to consider the purely solitonic sector, i.e. the RHP with zeroes and $G_o(\lambda) = I$. In this case $\chi(\lambda)$ is a meromorphic function. As it is known a meromorphic function can be constructed from its zeroes and poles if we normalize it , $\chi(\infty) = I$. In this case however we have to gnarantee that det g=1, which can be done without any difficulty.

5. Construction of the multimonopoles

Based on the above we look for new solutions in the form $\Psi(\lambda) = \chi(\lambda)\,\Psi_o$ where in the "purely solitonic" case χ has simple poles in λ :

$$\chi(\lambda) = I + \sum_{k=1}^{n} \frac{R_k}{\lambda - \mu_k} \tag{40}$$

and R_k, μ_k are independent of λ . It is of importance to note that the poles, μ_k, are in fact functions satisfying

$$\mu_k\,\mu_{k,z} + \mu_{k,y} = 0\ , \qquad -\mu_k\mu_{k,\bar{y}} + \mu_{k,z} = 0. \tag{41}$$

Finally the new solution g is given in terms of g_o, Ψ_o and μ_k /k=1,...,n/ for n poles as

$$g_{ab} = \prod_{k=1}^{n} |\mu_k|^{\frac{2S_k}{N}} \left\{ (g_o)_{ab} - \sum_{i,j,\alpha_i,\alpha_j} (\mu_i\bar{\mu}_j)^{-1} (\Gamma^{-1})_{i\alpha_i, j\alpha_j} (g_o)_{ac}\, \bar{m}_c^{(j,\alpha_j)} m_d^{(i,\alpha_i)} (g_o)_{db} \right\}$$

where $$\Gamma_{i\alpha_i, j\alpha_j} = \frac{m^{(i,\alpha_i)} g_o \bar{m}^{(j,\alpha_j)}}{1 + \mu_i \bar{\mu}_j} \qquad \begin{array}{l} a,b = 1,\ldots N \\ i,j = 1 \ldots n \\ \alpha_i = 1,\ldots s_i \end{array}$$

and $$m^{(i,\alpha_i)} = M^{(i,\alpha_i)} (\gamma(\mu_i), \mu_i)\ \Psi_o^{-1}(\mu_i, y, \bar{y}, z) \tag{42}$$

As we are considering a SU(N) gauge theory here for each pole μ_k one associates an $s_k \le N-1$ dimensional subspace. The residua R_k project on these subspaces:

$$(R_k)_{ab} = \sum_{\alpha_k=1}^{s_k} n_a^{(k,\alpha_k)} \cdot m_b^{(k,\alpha_k)} \tag{43}$$

Furthermore $M^{(k,\alpha_k)}$ form s_k linearly independent N dimensional vectors for fixed k. In constructing (42) we also guaranteed that |det g| = 1.

Choosing in (42) the arbitrary vectors $M^{(k,\alpha_k)}$ and the μ_k poles in an appropriate way one can construct monopole solutions in SU(N) gauge theory. Here we show how the general expression (42) can be used in two special cases: in the axially symmetric and in the general SU(2) case, when all subspaces are one dimensional.

Axial symmetry implies that $g = g(\rho, z)$. However, there is a subtlety, namely Ψ depends on ϕ also, in the following way:

$$\Psi(\lambda, \rho, z, y) = \Psi(\lambda e^{i\phi}, \rho, x_3) \qquad (44)$$

Now introducing $\hat{\lambda} = \lambda e^{i\phi}$ we obtain a linear eigenvalue problem for $\Psi(\hat{\lambda}, \rho, z)$ [32] . It is important to note that in this case the pole equations hawe a unique solution

$$\mu_k = -e^{-i\phi} \frac{w_k - z + \sqrt{(w_k - z)^2 + \rho^2}}{\rho} \qquad (45)$$

and both $M^{(k,\alpha_k)}$ and w_k are arbitrary constants.

To obtain the axially symmetric $SU(2)$ multimonopoles we start from the same ground state that we used in the Bäcklund transformations: $g_o = \mathrm{diag}(e^z, e^{-z})$. For this g_o it is easy to solve (33) yielding

$$\Psi_o(\hat{\lambda}) = \mathrm{diag}\left(\exp\left[z - \frac{\rho}{2}\hat{\lambda}\right], \quad \exp\left[-z + \frac{\rho\hat{\lambda}}{2}\right]\right) \qquad (46)$$

One can show by direct calculations that choosing the w_k's in (45) according to equ's (25,26) and taking appropriate $M^{(k)}$ vectors /e.g. $M^{(1)} = (1,1)$ for the 1MP, $M^{(1)} = M^{(2)*} = i(1,-1)$ for the 2MP/ in (42) one reelly obtains the axially symmetric multimonopoles.

It has been shown [12] that axial symmetry is so strong that it excludes the possibility of having separated monopoles. Therefore to obtain separated monopoles one has to abandon axial symmetry. Practically, it means that we keep g_o and Ψ_o of (46) but we take $M^{(k)}$ and μ_k different from the axially symmetric case [33] .

First we clarify the structure of the arbitrary vectors $M_a^{(k)}(\gamma(\mu),\mu)$ /a=1,2/ appearing in (42) . It is not difficult to prove that g depends only on the ratios $M_1^{(k)} M_2^{(k)-1}$, therefore, without loss of generality we may take

$$M_a^{(k)} = \left(e^{-f_k(\gamma,\lambda)} \; , \; e^{f_k(\gamma,\lambda)}\right) . \; .$$

Furthermore, we assume that $f_k(\gamma,\lambda)$ are analytic at $\lambda = 0$, i.e. $f_k(\gamma,\lambda) = \sum_{i=0}^{\infty} f_k^{(i)}(\lambda)(\gamma\lambda)^i$. We also suppose that all $M^{(k)}$-s are described <u>by the same function f</u>; i.e. $f_k(\gamma,\lambda) = c_k + f(\gamma,\lambda)$, where c_k's are constants.

It is an intrigueing possibility to work with several different functions, f_k's. Nevertheless, as it will be shown we get a 4n-1 parameter family of solutions /with appears to be the most general case for SU(2) . So it seems that these more general choices /for the f_k's/ would not give anything new.

The general solution of the pole equation (41) is the following

$$h(\gamma(\mu) \, , \, \mu) = 0 \tag{47}$$

where h is an arbitrary /nice/ function. An essential step to generate the multimonopoles is to choose suitable poles. In the case of axially symmetric n monopoles we found that the μ_k's must satisfy $\prod_{i=1}^{n} (\gamma - w_i) = 0$, where the w_i's are constants completely fixed by imposing regularity. To introduce free parameters it is natural to take also for h a polinomial of degree n in γ

$$h = \sum_{i=o}^{n} a_i(\mu)\gamma^i \qquad (48)$$

where $a_i(\mu)$'s are restricted by the demand that h be a polinomial of degree n in both μ and μ^{-1} as in the axially symmetric case. That is $a_i(\mu) = \sum_{j=0}^{n-i}\left(b_j\mu^j + c_j\mu^{-j}\right)$, so h contains $2(n+1)^2 - 2$ free parameters.

As in (42) only $\Psi_o(\mu_k)$ and $M^{(k)}(\mu_k, \gamma(\mu_k))$'s are present, we effectively need the values of the arbitrary function, f, at the pole sites, μ_k's. Since the algebraic eq. (47) defining the poles is of degree n in γ, we can express all γ^m's when $m > n-1$ in terms of γ^i /i=0,...,n-1/ at the pole sites, μ_k. Consequently one can uniquely associate with any power series in γ a polinomial of degree n-1 /in γ/ yielding the same values at all pole sites, μ_k.

Therefore, without any loss of generality we can take f as a polinomial of degree n-1 in γ.

A necessary condition for the regularity of the solution is the nonsingularity of Γ_{ij} when $\mu_i = -\bar{\mu}_j^{-1}$. Computing Γ_{ij} for our seed solution (46) we get

$$\Gamma_{ij} = \left(1 + \mu_i\bar{\mu}_j\right)^{-1}\left(e^{\alpha_{ij}} + \varepsilon_{ij}^{(n)}\, e^{-\alpha_{ij}}\right),$$

where

$$\alpha_{ij} = \gamma(\mu_i) + \bar{y}\left(\mu_i^{-1} + \bar{\mu}_j\right) - f\left(\mu_i, \gamma(\mu_i)\right) - \bar{f}\left(\mu_j, \gamma(\mu_j)\right) \qquad (49)$$

$$\varepsilon_{ij}^{(n)} = \pm 1,$$

and the bar denotes the complex conjugation.

Obtaining (49) we fixed the values of the constants, c_k's, to be the same as in the axially symmetric case. The condition of regularity implies that

$$\alpha(\mu_i, \gamma(\mu_i)) = i\pi\tilde{w}_i^{(n)} \quad , \tag{50}$$

where $\alpha(\lambda,\gamma(\lambda)) = \gamma - f(\lambda,\gamma(\lambda)) - f^*$, $f^* = \bar{f}(-\bar{\lambda}^{-1},\gamma(-\bar{\lambda}^{-1}))$. α is a polinomial of degree n-1 in γ /with coefficients depending only on λ/: $\alpha = \sum_{i=o}^{n-1} \alpha_i(\lambda)\, \gamma^i$. Assuming that $\alpha_i(\lambda)$'s are analytic in an annular domain, using Cauchy's formula we get for $f(\lambda,\gamma)$

$$f(\lambda,\gamma) = \frac{1}{2\pi i} \oint_{|\lambda|<|\zeta|} \frac{\alpha(\zeta,\gamma)}{\lambda-\zeta}\, d\zeta + \gamma \, . \tag{51}$$

The next step is to compute α explicitly which is not difficult. Since α is a polinomial of degree n-1 with its values given at n points /in eq. 50 /, it is uniquely determined by Lagrange's interpolation formula

$$\alpha(\lambda,\gamma(\lambda)) = i\pi \sum_{j=1}^{n} \tilde{w}_j^{n} \prod_{i\neq j} \frac{\gamma(\lambda) - \gamma_i(\lambda)}{\gamma_j(\lambda) - \gamma_i(\lambda)} \tag{52}$$

where we wrote h as $\prod_{i=1}^{n} (\gamma - \gamma_i(\lambda))$.

Since $\alpha(\mu,\gamma(\mu)) = \bar{\alpha}\left(-\bar{\mu}^{-1}, \gamma(-\bar{\mu}^{-1})\right)$

by construction this implies extra constraints for h, namely $h(\gamma,\mu) = \bar{h}(\gamma(-\bar{\mu}^{-1}), -\bar{\mu}^{-1})$. Furthermore, in eq. (52) the $\tilde{w}_i^n$'s should satisfy

$$\tilde{w}_i^{(n)} = -\tilde{w}_j^{(n)} \quad \text{if} \quad \gamma_i(\lambda) = \overset{*}{\gamma}_j(\lambda) = \bar{\gamma}_j(-\bar{\lambda}^{-1}) \tag{53}$$

$\tilde{w}_i^n$ 'a are determined from the axially symmetric case yielding $\tilde{w}_i^{(n)} = \frac{n+1-2i}{2}$, $1 \leq i \leq n$. This means that the number of free parameters would be halved, so h would contain only $(n+1)^2 -1$ free parameters. From eq. (51) we obtain the following constraints for $\alpha_i(\lambda)$'s

$$\oint \frac{\alpha_i(\zeta)}{\zeta^{j+1}} d\zeta = 0, \quad j = -i+1, \ldots, i-1; \quad 2 \leq i \leq n-1 \; ; \tag{54}$$

$$\frac{1}{2\pi i} \oint \frac{\alpha_1(\zeta)}{\zeta} d\zeta = 1. \tag{55}$$

Eqs. (54) , (55) express the fact that f /f^*/ is a polinomial of $\lambda\gamma(\gamma\lambda^{-1})$, and they represent $\sum_{i=2}^{n-1} (2i-1)+1$

additional relations among the $(n+1)^2 -1$ free parameters in h. Consequently our solution depends on 4n-1 parameters. It reduces to the axially symmetric case when $a_i(\mu)$'s are constants independent of μ /uniquely determined by w_i's/. At this point we remark that our construction of f from α was the same as the one applied by Corrigan and Goddard to obtain their χ in Ref. [13].

This strongly suggests that we ended up with the same solution, however, since our method appears to be radically different from theirs we have no analytic proof of the equivalence.

Here we argue that our solution really describes n monopoles. One can verify using (42) and the fact that asymptotically $\mu_{i',z} = -r^{-1}\mu_i$ that

$$|\phi| \underset{r\to\infty}{\sim} 1 - \frac{n}{r} . \tag{56}$$

For the n=2 case we have numerically computed $|\phi|$ and explicitly proved that it is nowhere singular.

In what follows we describe the separated two monopole solution in some detail.

We take h as

$$h = \gamma^2 + A\left(\mu^{-2} + \mu^2\right) + B \quad , \tag{57}$$

where A,B are real constants. We note that using similar arguments as in Ref. [13] h can always be reduced to this form. Eq. (55) implies the following relation between A and B

$$\sqrt{B} = \frac{1}{\sqrt{1+\beta}} K\left(\frac{2\beta}{1+\beta}\right) , \quad A = \frac{1}{2}\beta B \ , \ -1<\beta\leq 0, \tag{58}$$

where K(m) is a complete elliptic integral of the first kind with parameter m [34] .

For the choice (57) of h f is given by

$$f(\lambda,\gamma) = \frac{\gamma}{\sqrt{-A\tilde{\beta}}}\left[K(\tilde{\beta}^{-4}) - \Pi(\lambda^2\,\tilde{\beta}^{-2}\backslash\delta)\right] ,$$

$$\sin\delta = \tilde{\beta}^{-2} , \quad \tilde{\beta} = \sqrt{-\beta^{-1} + \sqrt{\beta^{-2} - 1}} , \tag{59}$$

and $\Pi(n\backslash\delta)$ is the complete elliptic integral of the third kind [34].

In the case when β is negative the integrand in (51) has four branch points on the real axis. We deformed the contour of integration in (51) to the cuts chosen to run on the real axis from $-\infty$ to the smallest branch point and from the largest to ∞ .

We still have to show that our solution given by eqs. (59), (58), (57), (42) is nowhere singular. In the present case the most straightforward way is to determine $|\phi|$ on a computer using

$$\phi^2 = \frac{1}{2}\,\mathrm{Tr}\left(g_z g^{-1}\right)^2 \tag{60}$$

This way one should also see the two zeroes of $|\phi|$.

As a result of our numerical investigations we indeed established the regularity of our 2MP for a wide range of the free parameter β corresponding to both small and large separations. The axonometric view of $|\phi|$ for our solution on the x_1, z plane is presented on Fig. 3

The only remaining free parameter, β , in our solution determines the location of the zeroes of $|\phi|$ on the x_1 axis as

$$x_1 = \pm\sqrt{-m}\,K(m) , \quad \text{where} \quad m = \frac{2\beta}{1+\beta} \tag{61}$$

One easily sees that the distance between the MP's goes to zero as $\beta \to 0$, and our solution gives back the axially symmetric 2MP [9] . On the other hand the separation of the monopoles, d, tends to infinity as $\beta \to -1$: $d \sim -\ln(1+\beta)$.

Conclusion

In this paper we applied soliton theoretic methods to construct multimonopole solutions of the Bogomolny equations. We reviewed the use of Bäcklund transformations to get the axially symmetric SU(2) monopoles depending on 5 parameters. So far this lead to the most transparent form of these monopoles [10] . We computed the energy density of these configurations as well and found that it is concentrated in a doughunt-like structure surrounding the symmetry axis.

To obtain monopoles depending on more free parameters we developed a RHP which in principle enables us to find all solutions of the Bogomolny /or SDE/ equations in SU(N)gauge theory . We solved in a special case this RHP : /with zeroes only/ in a closed form.

For SU(2) this solution gave the most general multimonopole configurations depending on 4n-1 parameters. In two dimensional completely integrable models this special RHP yields the purely solitonic solutions. As in the axially symmetric SU(2) case the RHP gives the same monopoles we obtained by Bäcklund **transformations it is tempting to consider them as solitons.**

However to establish their true solitonic nature one should study monopole-monopole scattering that is outside the realm of Bogomolny equations.For small velocities an interesting line of attack on this question was proposed by Manton using the concept of geodesics on the manifold of parameters [35] . As the most general n monopole configuration depends on 4n-1 parameters of which 3n corresponds to positional degrees of freedom the interpretation of the remaining n-1 parameters poses a problem. The usual interpretation of them as relative U(1) phases doesn't help in understanding the peculiar features of the only gauge invariant local quantity $|\phi|^2$.

As it turned out these parameters have a probably deep geometrical significance. The whole remaining nonlinearity of the problem is reduced to solving an algebraic equation(48) . These parameters determine those algebraic curves in R^3 where the roots of eq. (48) coincide, which are very characteristic for the n monopole configuration as $|\phi|^2$ depends on these curves crucially [17] .

The multimonopole solutions are not only interesting in their own right but they may shed some light on the intriqueing conjecture of Montonen and Olive [36] .

Finally we would like to mention that we were able to linearize the full Yang-Mills equations [37] .

The success of our construction raises the hope that these powerful methods would find an application in exploring the structure of four dimensional gauge theories.

REFERENCES

[1] G.'tHooft, Nucl.Phys. B79 /1974/ 276 ;
A.M. Polyakov, JETP Lett. 20 /1974/ 194.

[2] M.K.Prasad and C.M. Sommerfield, Phys.Rev.Lett. 35 /1975/ 760.

[3] E.B. Bogomolny, Sov.J. Nucl. Phys. 24 /1976/ 861.

[4] M.F. Atiyah and R.S. Ward, Commun. Math.Phys. 55 /1977/ 117

[5] E. Corrigan, D.B. Fairlie, P.Goddard and R.G.Yates, Comm.Math.Phys. 58 /1978/ 223.

[6] M.F. Atiyah, V.G. Drinfeld, N.J. Hitching and Yu. I. Manin, Phys.Lett. 65A /1978/ 185.

[7] W. Nahm, talk presented at this conference CERN preprint, TH. 3172 /1981/.

[8] P.Forgács, Z.Horváth and L.Palla, Phys.Rev.Lett. 45, /1980/ 505;
P.Forgács, Z.Horváth and L.Palla, Ann. Phys. /N.Y./ 136 /1981/ 371.

[9] R.S.Ward, Comm.Math.Phys. 80 /1981/ 137;
P.Forgács, Z.Horváth and L.Palla, Phys.Lett. 99B /1981/ 232.

[10] M.K. Prasad and P.Rossi, Phys.Rev.Lett. 46 /1981/ 806;
P.Forgács, Z.Horváth and L.Palla, Phys. Lett. 102B /1981/;
P.Forgács, Z. Horváth and L.Palla, Nucl.Phys. 192B /1981/ 141.

[11] R.S.Ward, Phys.Lett. 102B /1981/ 136 .

[12] P.Houston and L. O'Raifeartaigh, Phys.Lett. 93B /1980/ 151.

[13] E. Corrigan and P.Goddard, Comm.Math.Phys. 80 /1981/ 575.

[14] E.Weinberg, Phys.Rev. D20 /1979/ 936;
W.Nahm, Phys.Lett. 85B /1979/ 373.

[15] P.Forgács, Z.Horváth and L.Palla, Phys.Rev. D23 /1981/ 1876.

[16] F.A.Bais and R.Sasaki, Utrecht preprints /1981/;
F.A.Bais, lecture given at this conference
Z.Horváth and T.Kiss-Tóth, Wuppertal preprint, Wu.B.81-23 /1981/.

[17] N.I.Hitchin, Oxford preprint /1981/.

[18] E.J.Weinberg and A.H.Guth, Phys.Rev. D14 /1976/ 1660.

[19] P.Forgács and N.S.Manton, Comm.Math.Phys. 72 /1980/ 15.

[20] N.S.Manton, Nucl.Phys. B135 /1978/ 319.

[21] B.K.Harrison, Phys.Rev.Lett. 41 /1978/ 1197.

[22] G.Neugebauer, J.Phys. A 12 /1979/ L67.

[23] V.A.Belinski and V.E.Zakharov, Sov.Phys. JETP 50 /1979/ 1.

[24] I.Hauser and F.J.Ernst, Phys.Lev. D 20 /1979/ 362 ;
Phys.Rev. D 20 /1979/ 1783; J.Math.Phys. 21 /1980/ 1126.

[25] C.Rebbi and P.Rossi, Phys. D22 /1980/ 2010.

[26] J.Arafune, P.G.O. Freund and C.J. Goebel, J.Math.Phys. 16 /1975/ 433.

[27] Y.Brihaye, D.B.Fairlie, I.Nuyts and R.G.Yates, I.Math.Phys. 19 /1978/ 2528.

[28] C.N.Yang, Phys.Rev.Lett. 38 /1977/ 1377.

[29] K.Ueno and Y.Nakamura, Kyoto University preprint, RIMS-376 /1981/.

[30] The same equations were obtained using different methods by G.Sartori, Nuovo Cimento 56A, 73 /1980/; L.L.Wang, BNL-27617 preprint, Conference on Theoretical Particle Physics, January 5-14. 1980, Guangzhou /Canton/; K.Pohlmeyer, Commun.Math.Phys. 72, 37 /1980/.

[31] V.E.Zakharov, S.V.Manakov, S.P.Novikov and S.P.Pitaevski, Theory of solitons /Nauka, Moscow, 1980/ /in Russian/.

[32] L.Palla, lectures given at ICTP, Trieste and at the Visegrad conference /1981/; D.Maison, lecture given at this conference

[33] P.Forgács, Z.Horváth and L.Palla, KFKI report, KFKI-1981-92, to appear in Phys.Lett.B.

[34] M.Abramowitz and I.A.Stegun, Handbook of Mathematical Functions, Dover Publ. /N.Y./ /1972/.

[35] N.S.Manton, talk presented at this conference.

[36] C.Montonen and D.Olive, Phys.Letters 72B, 117 /1977/.

[37] P.Forgács, Z.Horváth and L.Palla, in preparation.

FIGURE CAPTIONS

Fig. 1. Axonometric view of $|\phi|$ of the 5 monopole configuration. The plane of projection inclines at 45° with both the ρ and z axis. The coordinates of O are $\rho=0.22$, $z=0$ respectively.

Fig.2. The energy density of the 5 monopole configuration. The coordinates of O are $\rho=0.62$, $z=0$ respectively.

Fig.3. The axonometric view of $|\phi|$ on the x_1x_3 plane with $\beta=-0.5/d=3.2/$. The plane of projection inclines at 75° degrees with that plane. The coordinates of O are $x_1=0$, $x_3=0.01$.

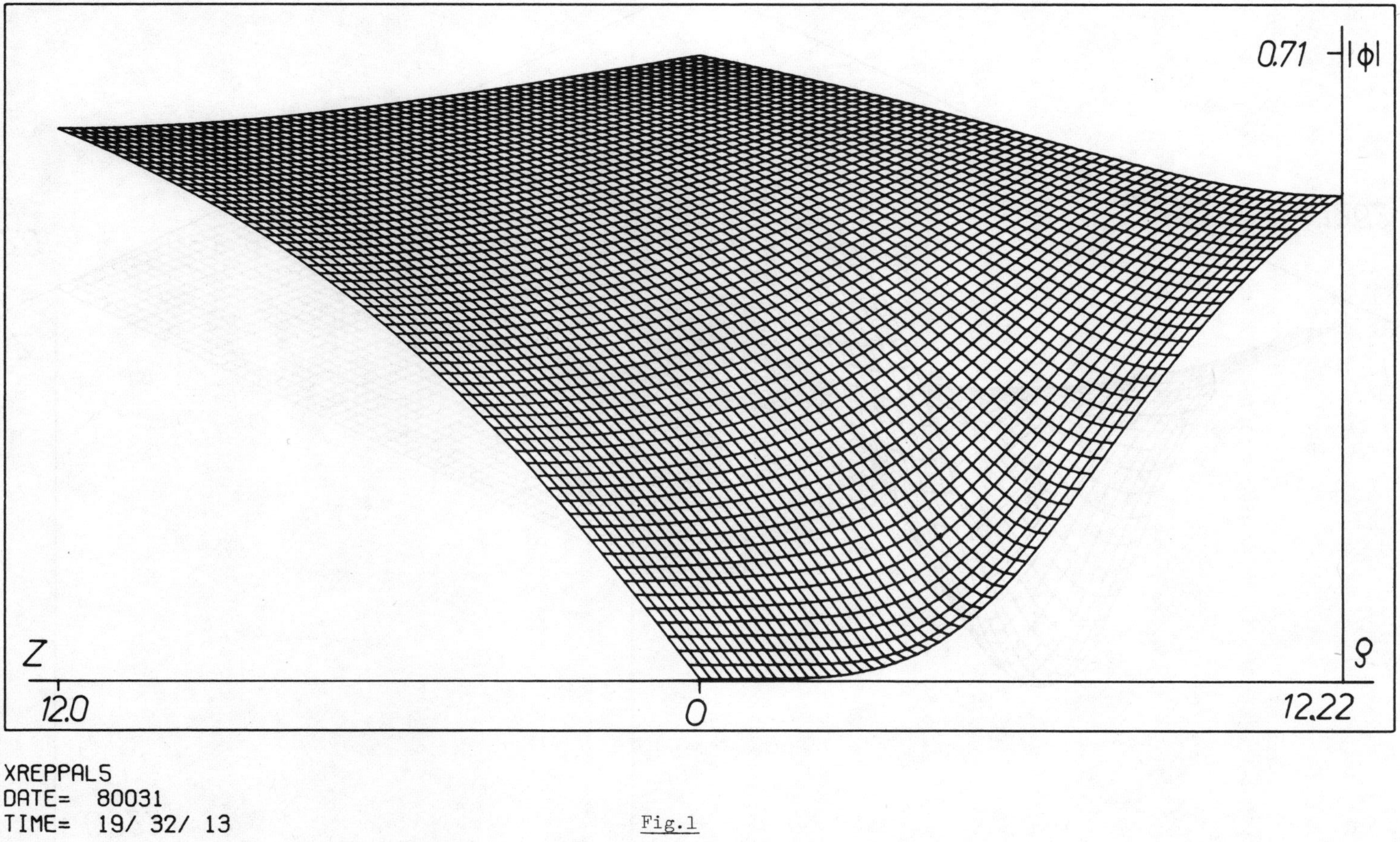

Fig.1

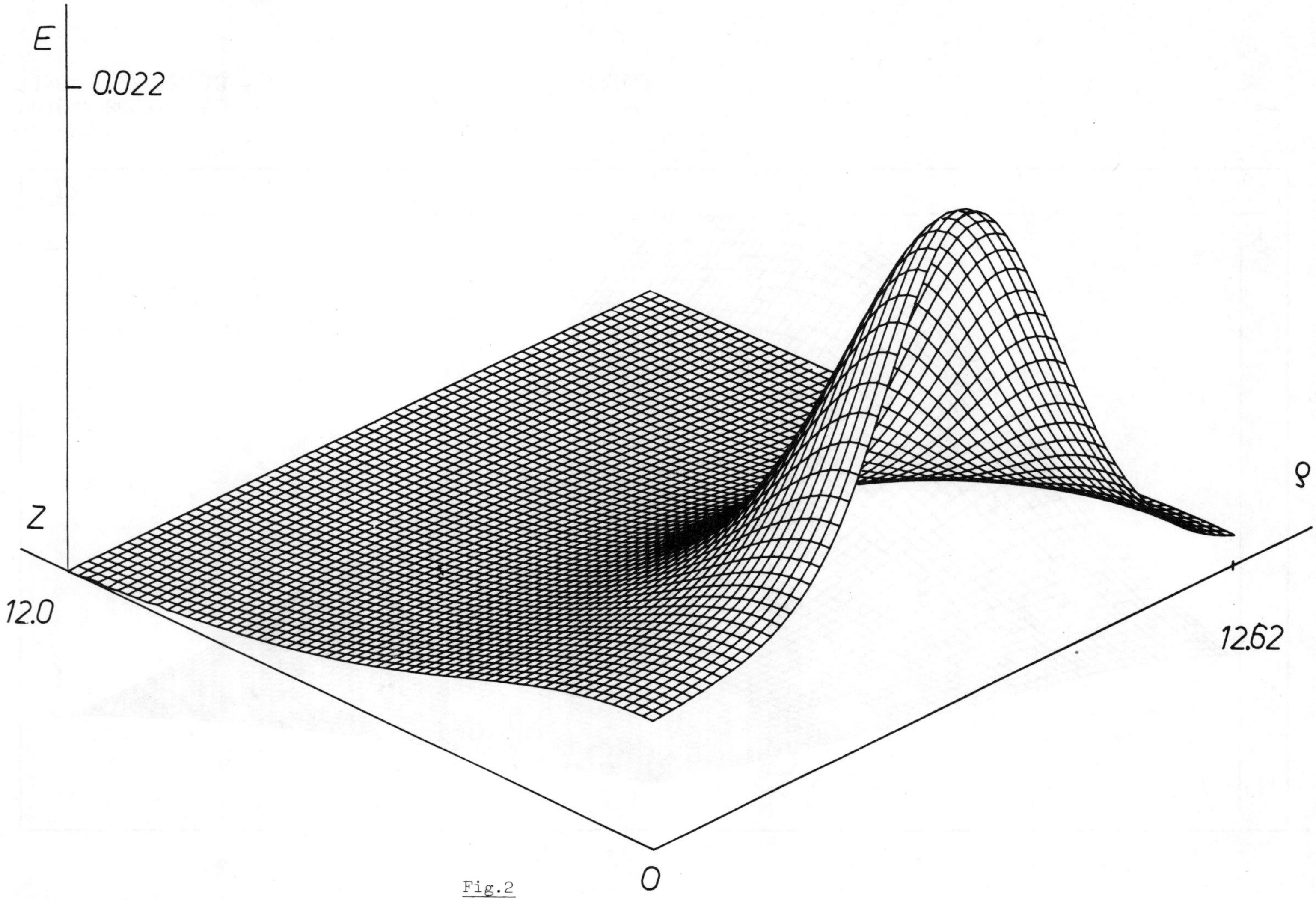

Fig.2

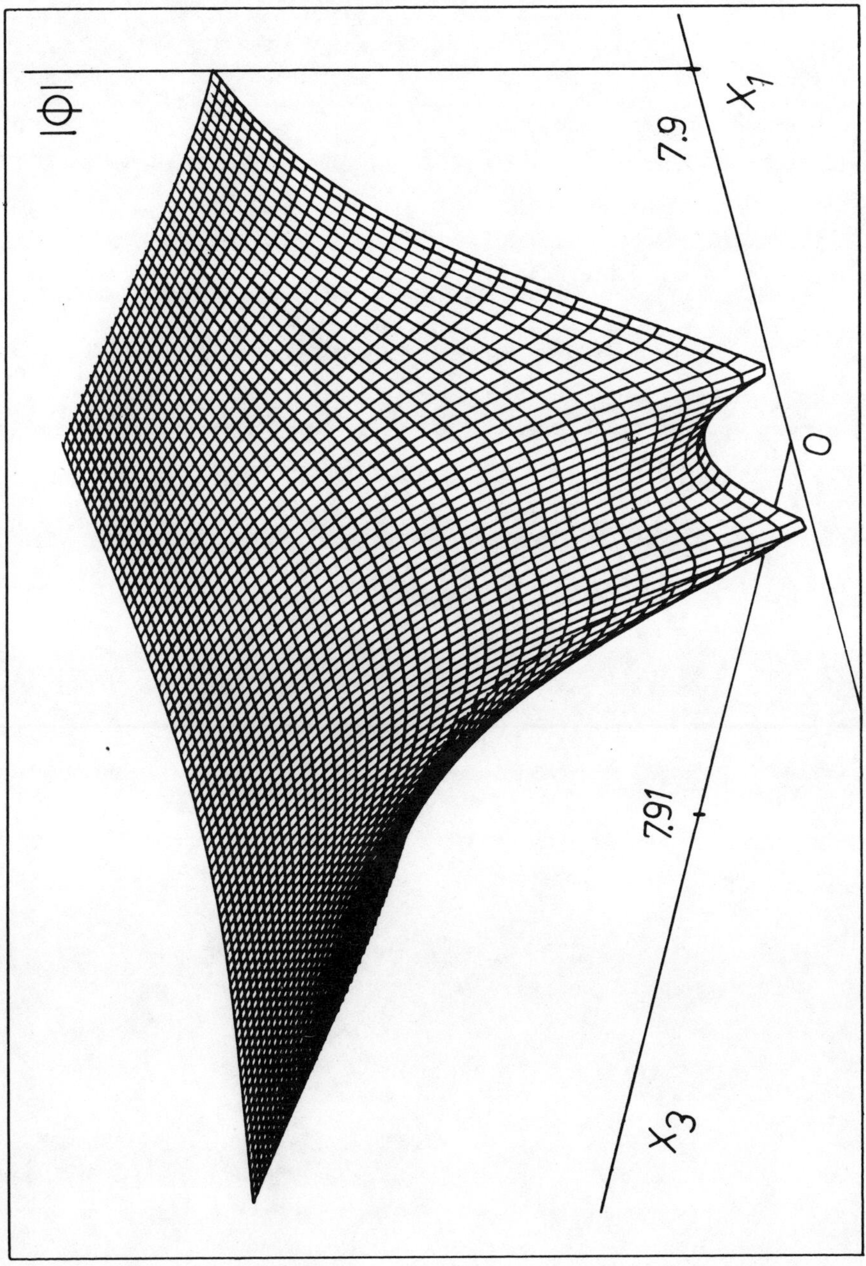

Fig.3

CONSTRUCTION OF SU(3) MONOPOLES

R.S. Ward

Dept. of Mathematics,
Trinity College,
Dublin,
IRELAND.

1. Introduction

In the past few years, there has been a spate of activity in the field of non-abelian magnetic monopoles, and especially in those that possess the following properties:

(i) The gauge potential A_j is static ($\partial_0 A_j = 0$) and purely magnetic ($A_0 = 0$).

(ii) The Higgs field Φ is in the adjoint representation, is static, and does not interact with itself (i.e. the Prasad-Sommerfield limit is imposed).

(iii) The Bogomolny equations

$$F_{jk} = -\varepsilon_{jk\ell} D_\ell \Phi \qquad (1)$$

hold, together with the boundary condition $\mathrm{tr}\,\Phi^2 \to \mathrm{const} + O(r^{-1})$ as $r \to \infty$ in $\mathbb{R}^3$, which guarantees finite energy.

Such "special" monopole solutions appear to be considerably easier to construct than more general monopole configurations (for example, those involving a non-vanishing Higgs potential). For several years, spherically symmetric solutions have been known [1,2,3]. More recently, the "twistor" method was applied to the problem. This involves transforming from $\mathbb{R}^3$ to a two-dimensional complex manifold, the twistor space, whereupon the non-linear partial differential equation (1) miraculously "disappears". The details of all this are described in Atiyah's lecture in these proceedings.

The first case to be tackled was, naturally, that of finding SU(2) monopoles. The first new solution to be constructed using the twistor method was an axially symmetric solution of charge 2 [4]. This led to axially symmetric solutions of arbitrary charge [5], then to the general solution of charge 2 [6], and finally to the general solution of arbitrary charge n, containing the expected 4n-1 parameters [7]. It still has to be proved that the $n \geqslant 3$ solutions are non-singular, but it is known that the twistor construction does give all solutions [8]. In other words, some of the solutions constructed in the above-listed papers may have singularities (their explicit form is so complicated that it is difficult to see whether they do or not), but we can be sure that all solutions have been included.

The twistor method can also be applied in a somewhat different way, and this variant is described in the lecture by Nahm. It seems that here the problem of singularities is easier to deal with, at the expense of difficulties elsewhere. This holds out the possibility that an amalgamation of the two variants may overcome the difficulties inherent in each. This possibility is currently being investigated.

The twistor method applies not only when the gauge group is SU(2); indeed, it can be used with any gauge group at all. The subject of this lecture is to investigate some of the interesting new features that appear when one goes from SU(2) to SU(3). Section 2 below is a review of the topological classification of SU(3) monopoles, section 3 a description of the twistor method as it applies to this case, and section 4 a summary of some new results.

2. Topological Charges and SU(2) Imbeddings

The topological classification of SU(3) monopoles may be summarized as follows; see [9] for more details. Choose a gauge so that, on the positive z-axis, the Higgs field Φ has the form

$$\Phi = \mathrm{diag}(\lambda_1, \lambda_2, \lambda_3) - \tfrac{1}{2}z^{-1}\mathrm{diag}(n_1, n_2-n_1, -n_2) + O(z^{-2}) \qquad (2)$$

as $z \to +\infty$. The λ_i are constants such that $\lambda_1 + \lambda_2 + \lambda_3 = 0$ and $\lambda_1 \geqslant \lambda_2 > \lambda_3$. If $\lambda_1 > \lambda_2$, then the isotropy group of $\Phi(x = y = 0, z = +\infty)$ is $U(1) \times U(1)$; in other words, the Higgs field breaks the symmetry from SU(3) down to $U(1) \times U(1)$. If, on the other hand, $\lambda_1 = \lambda_2$, then the isotropy group is $U(1) \times SU(2)$, and so this is the residual symmetry group.

The numbers n_1 and n_2 appearing in the z^{-1} term in (2) are necessarily integers [9]. (This depends on the conventions in use, which are as follows: Φ and A_j are 3×3 Hermitian trace-free matrices, $D_j \Phi = \partial_j \Phi + i[A_j, \Phi]$, and equation (1) holds.) In the case where the little group is $U(1) \times U(1)$, both n_1 and n_2 are topological invariants of the configuration, and we may refer to them as <u>topological charges</u>. In the other case, namely that of $U(1) \times SU(2)$, the fact that λ_1 and λ_2 are equal means that we cannot distinguish in a gauge-invariant way between n_1 and $n_2 - n_1$. So there is only one topological charge, namely n_2. The number n_1, modulo the ambiguity $n_1 \mapsto n_2 - n_1$, appears to be an analytic rather than a topological invariant. We may refer to it as a <u>magnetic weight</u> [10].

This, then, is the topological classification of SU(3) monopoles. It is worth remarking that the energy $E = \frac{1}{2}\int \mathrm{tr}(D\Phi)^2 d^3x$ of a monopole solution depends only on the topological charges and the asymptotic eigenvalues of Φ: it is not difficult to prove that E is given by $E = (\lambda_1 n_1 + \lambda_2 n_2 - \lambda_2 n_1 - \lambda_3 n_2)\pi$.

In the case of maximal symmetry breaking, i.e. where the little group is $U(1) \times U(1)$, the number of zero-modes of a monopole solution with fixed $(\lambda_1, \lambda_2, \lambda_3)$ and (n_1, n_2) has been computed: it is $N = 4(n_1 + n_2) - 2 + k$, where k is a non-negative integer less than or equal to the number of n_a which vanish [9]. The conjecture is that equality holds; in other words, $k = 1$ if either $n_1 = 0$ or $n_2 = 0$, and $k = 0$ otherwise [9].

One can construct some examples of SU(3) monopoles by imbedding SU(2) monopoles, using the trivial (reducible) imbedding of SU(2) in SU(3). (Yet more examples arise if one uses the maximal, irreducible imbedding; cf. [2]. We will not pursue this possibility further here.) The subject of imbedding is discussed in some detail in [9], where use is made of the machinery of root systems. What one finds is the following.

Given $\lambda_1 > \lambda_2 > \lambda_3$, there are three distinct ways of imbedding an SU(2) n-monopole into SU(3), in such a way that the SU(3) Higgs field has the desired form (2). These three possibilities are:

$$\begin{bmatrix} * & * & 0 \\ * & * & 0 \\ 0 & 0 & * \end{bmatrix} \qquad \begin{bmatrix} * & 0 & * \\ 0 & * & 0 \\ * & 0 & * \end{bmatrix} \qquad \begin{bmatrix} * & 0 & 0 \\ 0 & * & * \\ 0 & * & * \end{bmatrix}$$

$n_1 = n$	$n_1 = n$	$n_1 = 0$
$n_2 = 0$	$n_2 = n$	$n_2 = n$
$N = 4n-1$	$N = 8n-2$	$N = 4n-1$

In each case, Φ and A_j have the indicated form, where * denotes a non-zero entry. The topological charges and the number of zero-modes are listed below each case. It is a rather curious feature that the number of zero-modes is greater in one case than in the other two. This will be investigated further in section 4.

The above discussion of zero-modes and imbeddings applies to the case of maximal symmetry breaking. The other case, of non-abelian symmetry breaking, is a bit more complicated [9], and will not be dealt with here.

3. The Twistor Construction

The basic theorem which underlies the construction says that solutions (Φ, A_j) of equation (1) correspond to certain holomorphic vector bundles over twistor space (see [8] and Atiyah's lecture in these proceedings). This section describes one way in which the theorem can be applied to construct solutions.

The starting-point is a 3×3 matrix $g(\gamma,\zeta)$ of functions of the two complex variables γ and ζ satisfying

(i) g is analytic for all γ, and for ζ near $|\zeta| = 1$;
(ii) $\det(g) = 1$;
(iii) $g(\bar{\gamma}, -\bar{\zeta}^{-1}) = g(\gamma,\zeta)^{\dagger}$. (3)

(In fact, g is the "patching matrix" which determines the vector bundle mentioned above.) First one "splits" the matrix g:

$$g(\xi\zeta - 2z - \bar{\xi}\zeta^{-1},\zeta) = hk^{-1}, \tag{4}$$

where $\xi = x + iy$ (x,y,z are the usual coordinates on $\mathbb{R}^3$), $h = h(\xi,\bar{\xi},z,\zeta)$ is a 3×3 matrix analytic for $|\zeta| \geqslant 1$ (including $\zeta = \infty$), and $k = k(\xi,\bar{\xi},z,\zeta)$ is a 3×3 matrix analytic for $|\zeta| \leqslant 1$. Now put $H(\xi,\bar{\xi},z) = h(\xi,\bar{\xi},z,\infty)$ and $K(\xi,\bar{\xi},z) = k(\xi,\bar{\xi},z,0)$. Then Φ and A_j are given by

$$\Phi = \tfrac{1}{2}K^{-1}K_z - \tfrac{1}{2}H^{-1}H_z ,$$

$$A_z = -\tfrac{1}{2}iK^{-1}K_z - \tfrac{1}{2}iH^{-1}H_z ,$$

$$A_\xi = -iK^{-1}K_\xi$$

$$A_{\bar{\xi}} = -iH^{-1}H_{\bar{\xi}} ,$$

where the subscripts on the right-hand sides denote partial differentiation. The (Φ,A_j) defined in this way is automatically a smooth SU(3) monopole solution, provided only that the boundary condition tr $\Phi^2 \to$ const is satisfied (this is not guaranteed a priori, although all the other required conditions are). Another point that should be mentioned (and which usually involves hard work) is that the splitting (4) may not be possible; one has to choose g in such a way that it is. And even if the splitting <u>is</u> possible (that is, possible in principle), it may be very difficult to carry out explicitly.

4. An Example

There are, however, a class of matrices g for which the splitting can be effected in a fairly explicit way, namely the upper triangular matrices:

$$\begin{bmatrix} * & * & * \\ 0 & * & * \\ 0 & 0 & * \end{bmatrix} \tag{5}$$

Actually, the form (5) is inconsistent with the reality condition (3), unless the matrix is diagonal, which is too severe a restriction to give anything interesting. Instead, we impose the condition that the matrix (5) be <u>equivalent</u> to one satisfying (3). Specifically, there must exist a

3×3 matrix Λ of functions of γ and ζ, analytic for $|\zeta| \leqslant 1$ and for all γ including $\gamma = \zeta^{-1}$, such that if we multiply (5) on the right by Λ, we obtain a matrix g satisfying (3). It is easily seen that g and $g\Lambda$ give rise to the same (Φ, A_j), so we are justified in calling two such matrices equivalent.

In the SU(2) case, the procedure for splitting upper triangular matrices has been known for some time [11]. It was used in the construction of SU(2) monopoles [4 - 7], and Hitchin subsequently proved that all SU(2) monopoles arise from upper triangular matrices [8]. The same sort of thing can be done for SU(3) (indeed, for any gauge group), although the analysis becomes somewhat more complicated. I shall illustrate the method by using it to investigate the fact mentioned in the previous section, that the number of zero-modes of an SU(2) imbedding depends on the details of the imbedding.

The upper triangular g-matrix which gives rise to the SU(2) 1-monopole [1] is

$$\begin{bmatrix} \zeta e^{\gamma} & \gamma^{-1}(e^{\gamma} - e^{-\gamma}) \\ 0 & \zeta^{-1}e^{-\gamma} \end{bmatrix} .$$

This has charge n = 1, so when we imbed it, we will obtain an SU(3) solution which has either 4n-1 = 3 zero-modes or 8n-2 = 6 zero-modes. The imbedding with 6 zero-modes is

$$\begin{bmatrix} \zeta e^{\gamma} & 0 & \gamma^{-1}(e^{\gamma}-e^{-\gamma}) \\ 0 & 1 & 0 \\ 0 & 0 & \zeta^{-1}e^{-\gamma} \end{bmatrix} \tag{6}$$

and one expects that these zero-modes will correspond to matrices of the form

$$\begin{bmatrix} \zeta e^{\gamma} & \Gamma_1 & \Gamma_2 \\ 0 & 1 & \Gamma_3 \\ 0 & 0 & \zeta^{-1}e^{-\gamma} \end{bmatrix} ,$$

where the Γ_i are suitable functions of γ and ζ. The criteria of smoothness (the Γ_i must be entire in γ) and of reality (in the sense discussed in the previous section) are now applied, and one finds that they determine a 6-parameter family of suitable Γ_i's. The procedure of section 3, leading to explicit expressions for Φ and A_j, can now be carried out. The two requirements that are not guaranteed a priori (namely the "splittability" of g and the boundary condition on Φ) are found to be satisfied.

I shall not display the details of all this here; these will appear elsewhere. However, it is worth making the following qualitative remark. The 6-parameter family of solutions are axially symmetric, so five of the six parameters correspond to the position and orientation of the monopole in $\mathbb{R}^3$. The meaning of the remaining parameter has not yet been understood. One particular value of this parameter corresponds to the (spherically symmetric) imbedding of the SU(2) 1-monopole in SU(3), namely (6).

All this has been for the imbedding with the higher number of zero-modes. What about the other two imbeddings? Here, if one follows the same sort of argument, one finds (as expected) that there are only a 3-parameter family of upper triangular matrices that have all the necessary properties. (The three parameters are, of course, just those in the original spherically symmetric SU(2) monopole, denoting its position in $\mathbb{R}^3$.)

5. Conclusion

We have seen that the twistor method enables one to construct some SU(3) monopole solutions of which zero-mode arguments had predicted the existence. These solutions can be displayed as explicitly as one desires. Clearly the potential of the method has only begun to be realized, and much more can be achieved by using it.

It is perhaps worth mentioning that the case where the residual symmetry group is U(1) × SU(2) can also be dealt with by the twistor method. In this case, the imbedding of an SU(2) n-monopole in SU(3) has topological charge n and magnetic weight 0. A preliminary result has been to produce a 6-parameter family of solutions with topological charge 2 and magnetic weight 1 [12]. This family includes the spherically symmetric solution found by Bais and Weldon [3].

In the cases that have so far been investigated it has turned out that the expressions for monopoles are much simpler in twistor space than in space-time. In particular, the g-matrix is considerably less complicated than the corresponding expressions for Φ and A_j. The lesson of this is that one should try to do one's calculations in the twistor-space picture whenever possible, including, perhaps, perturbation theory about the classical monopole solutions. But as yet very little work has been done on this subject.

References

1. M.K. Prasad & C.M. Sommerfield, Phys. Rev. Letters 35 (1975), 760-762.
2. D. Wilkinson & F.A. Bais, Phys. Rev. D 19 (1979), 2410-2415.
3. F.A. Bais & H.A. Weldon, Phys. Rev. Letters 41 (1978), 601-604.
4. R.S. Ward, Commun. Math. Phys. 79 (1981), 317 - 325.
5. M.K. Prasad, Commun. Math. Phys. 80 (1981), 137-149.
6. R.S. Ward, Phys. Letters B 102 (1981), 136-138.
7. E. Corrigan & P. Goddard, Commun. Math. Phys. 80 (1981), 575-587.
8. N.J. Hitchin, Monopoles and Geodesics. Preprint, Oxford, 1981.
9. E.J. Weinberg, Nucl. Phys. B 167 (1980), 500-524.
10 P. Goddard, J. Nuyts & D. Olive, Nucl. Phys. B 125 (1977), 1-28.
11. E.F. Corrigan, D.B. Fairlie , R.G. Yates and P. Goddard, Commun. Math. Phys. 58 (1978), 223-240.
R.S. Ward, Commun. Math. Phys. 80 (1981), 563-574.
12. R.S. Ward, Phys. Letters B, 107 (1981), 281 - 284.

Monopoles in the Atiyah-Ward formalism

E.Corrigan

Department of Mathematics
University of Durham
U.K.

1. Introduction

Over the last year and a half there has been much progress in understanding the exact multimonopole solutions to the classical field equations of spontaneously broken gauge theory. Many people, working from different points of view, have shed light on the problem and the organisers have asked me to talk about one particular approach. Horvath(1) and Nahm(2) in their talks supply the details of other approaches.

The ingredients of the theory to be discussed(3) are an SU(2) gauge potential $\underline{A}_\mu$ and an adjoint representation Higgs' field $\underline{\phi}$. The classical equations of notion of the theory are

$$D_\mu \underline{F}^{\mu\nu} = - \underline{\phi} \wedge D^\nu \underline{\phi}$$

$$D^2 \underline{\phi} = - \lambda \underline{\phi} (\underline{\phi}^2 - a^2) \qquad \mu\nu = 0,1,2,3. \tag{1.1}$$

$$D_\mu {}^* \underline{F}^{\mu\nu} = 0$$

where D_μ, $F_{\mu\nu}$, ${}^*F_{\mu\nu}$ are the usual covariant derivative, field strength tensor and the dual of the field strength tensor, respectively. The problem is to find all solutions to eq(1.1) such that the total energy

$$\mathcal{E} = \int d^3\underline{x} \, \{ \tfrac{1}{2} (\underline{E}_i \underline{E}_i + \underline{H}_i \underline{H}_i + D_o \underline{\phi} \cdot D_o \underline{\phi} + D_i \underline{\phi} \cdot D_i \underline{\phi}) + \tfrac{1}{4} \lambda (\underline{\phi}^2 - a^2)^2 \} \tag{1.2}$$

is finite, ($\underline{E}_i$,$\underline{H}_i$ are the electric and magnetic parts, respectively. of the field strength tensor), and the Higgs' field $\underline{\phi}$ is asymptotically of length a and constitutes a mapping from the two dimensional sphere at spatial infinity to the sphere of radius a belonging to the homotopy class N. The integer N is the magnetic change of the solution, defined in the usual way (and in appropriate units implied by eqs(1.1)). For N fixed we have[4]

$$\mathcal{E} > 4\pi Na. \tag{1.3}$$

This problem is too difficult. However, deleting the right hand side of the $\underline{\phi}$ equation (that is, keeping the lowest order term in the expansion of the solution in terms of λ)[5] and, introducing a euclidean variable x_4 on which nothing depends we may set

$$\underline{\phi} = \underline{A}_4 \tag{1.4}$$

and write equation (1.1) in the form

$$D_\alpha \underline{F}^{\alpha\beta} = 0 \tag{1.5a}$$

$$D_\alpha \underline{f}^{\alpha\beta\gamma} = 0, \quad f^{\alpha\beta\gamma} = \tfrac{1}{2}\,{}^{\alpha\beta\gamma\rho\sigma}\,\underline{F}_{\rho\sigma}. \tag{1.5b}$$

In eqs(1.5) the indices run from zero to four. Special solutions of eqs(1.5) may be obtained, noting that eq(1.5b) is an identity, by insisting that

$$\underline{F}^{\alpha\beta} = \underline{f}^{\alpha\beta\gamma}\, V_\gamma, \tag{1.6}$$

where V_γ is a constant vector.

When $V_\gamma = (1,0,0,0,0)$ we find[4,6]

$$\mathcal{E} = 4\pi Na$$

and

$$F^{\alpha\beta} = *F^{\alpha\beta}, \; F^{o\beta} = 0 \qquad \alpha,\beta = 1,2,3,4, \tag{1.7}$$

or,

$$\underline{H}_i = \underline{D}_i\phi, \ \underline{E}_i = 0, \ \ D_o\underline{\phi} = 0$$

so that in the gauge $\underline{A}_o = 0$ all the fields are static.

The recent progress has come from a concentrated study of eqs(1.7) in the gauge $\underline{A}_o = 0$. The second equation is trivial, the first expresses self duality and contains the information about the monopoles.

2. Analysis of the self-dual equation I

In this and subsequent sections we shall elaborate the study of the first of eqs(1.7). To begin with let us follow Yang[7], complexify the euclidean variable x_μ and write

$$x = x_4 + i\underline{x}\cdot\underline{\sigma} = \begin{pmatrix} y & -\bar{z} \\ x & \bar{y} \end{pmatrix} = \begin{pmatrix} x_4+ix_3 & x_2+ix_1 \\ -x_2+ix_1 & x_4-ix_3 \end{pmatrix} \tag{2.1}$$

Then, the self duality equation (1.7) reduces to

$$\underline{F}_{yz} = 0, \ \ \underline{F}_{\bar{y}\bar{z}} = 0 \tag{2.2}$$

$$\underline{F}_{y\bar{y}} + \underline{F}_{z\bar{z}} = 0 \tag{2.3}$$

Eqns(2.2) are integrable and, defining for convenience

$$A_\mu = -\frac{i}{2}\,\underline{\sigma}\cdot\underline{A}_\mu \qquad \mu = 1,2,3,4,$$

we find

$$\begin{pmatrix} A_y \\ A_z \end{pmatrix} = D^{-1}\begin{pmatrix} \partial_y \\ \partial_z \end{pmatrix} D\,, \qquad \begin{pmatrix} A_{\bar{y}} \\ A_{\bar{z}} \end{pmatrix} = \bar{D}^{-1}\begin{pmatrix} \partial_{\bar{y}} \\ \partial_{\bar{z}} \end{pmatrix}\bar{D} \tag{2.4}$$

where D and $\bar{D}$ are SL(2,c) matrices. They satisfy

$$D^+ = (\bar{D})^{-1}, \tag{2.5}$$

if x_μ is real, in order to guarantee that the vector potential $\underline{A}_\mu$ has real components. In terms of the gauge invariant quantity J:

$$J = D\bar{D}^{-1},$$

equation (2.3) reads[8]

$$\partial_y (\partial_y J\, J^{-1}) + \partial_z (\partial_z J\, J^{-1}) = 0 \tag{2.6}$$

and bears some resemblance to σ model equations, much studied in two dimensions.

A convenient choice of gauge and $D, \bar{D}$ is to take

$$J = \frac{1}{\sqrt{\phi}} \begin{pmatrix} 1 & 0 \\ -\varepsilon & \phi \end{pmatrix} \frac{1}{\sqrt{\phi}} \begin{pmatrix} 1 & \gamma \\ 0 & \phi \end{pmatrix} , \tag{2.7}$$

in which case eqs(2.6) reduce to a set of three coupled equations for ε, ϕ and γ. These equations were studied some time ago and may be linearised by one of the following infinite set of transformations (or ansatze)[9].

Let[10]

$$\begin{pmatrix} \varepsilon_\ell & \phi_\ell \\ \phi_\ell & \gamma_\ell \end{pmatrix}^{-1} = \text{corner entries of} \begin{pmatrix} D_{\ell+1} \end{pmatrix}^{-1} \tag{2.8}$$

where $\ell = 1,2,\ldots$ and the matrix $D_{\ell+1}$ is defined as follows:

$$(D_{\ell+1})_{ij} = \Delta_{-\ell+i+j} \qquad i,j = 0,1,\ldots\, \ell \tag{2.9}$$

the functions Δ_r each satisfying Laplace's equation,

$$\partial^2 \Delta_r = 0 . \tag{2.10}$$

In addition, the Δ's obey the stronger conditions (reminiscent of Cauchy-Rieman equations)

$$\begin{pmatrix} \partial_y \\ \partial_z \end{pmatrix} \Delta_r = \begin{pmatrix} -\partial_{\bar{z}} \\ \partial_{\bar{y}} \end{pmatrix} \Delta_{r+1} \tag{2.11}$$

For $\ell = 0$, $\varepsilon_o = \phi_o = \gamma_o = \Delta_o$, $\partial^2\Delta_o = 0$.

The deeper significance of the function Δ_r will become clearer later, in the next section, but for the moment we shall be content to remark that for a suitable choice of them we can compute ε_ℓ, ϕ_ℓ, γ_ℓ, and hence $D, \bar{D}$ and the vector potentials via equations (2.7) and (2.4). Eq(2.6) is satisfied automatically.

The significance of these ansatze[11] for the monopoles becomes apparent with two further remarks. The vector potentials defined via eq(2.4) are required to be independent of x_4. Taking each of the functions Δ_r to have the same dependence on x_4 guarantees this and leads to

$$\Delta_r = e^{i\alpha x_4}\, \tilde{\Delta}_r(x_1, x_2, x_3). \tag{2.12}$$

in which case (2.10) reads

$$\nabla^2 \tilde{\Delta}_r = \alpha^2 \tilde{\Delta}_r \,. \tag{2.13}$$

Moreover, we also have with this dependence on x_4 the elegant result[12]

$$|\underline{A}_4|^2 = |\underline{\phi}|^2 = \alpha^2 - \nabla^2 \ln \det \tilde{D}_\ell \tag{2.14}$$

where $\tilde{D}_\ell$ is defined via (2.9), but in terms of $\tilde{\Delta}_r$ instead of Δ_r. Noting that for large $|\underline{x}|$ the dominant part of the solution to eq(2.13) is

$$\tilde{\Delta}_r \sim \frac{e^{\alpha|\underline{x}|}}{|\underline{x}|}\, Cr, \tag{2.15}$$

where Cr is independent of $|\underline{x}|$, we can compute the asymptotic value of the Higgs' field. It is

$$|\underline{\phi}|^2 \sim \alpha^2 - \frac{2\alpha\ell}{|\underline{x}|} \tag{2.16}$$

from which we deduce that the constant α is actually a, the asymptotic length of the Higgs' field. Further, provided all the potentials are regular everywhere, the energy is

$$\mathcal{E} = \int d^3x \ \tfrac{1}{2}\Delta^2|\underline{\phi}|^2 = 4\pi\ell a \tag{2.17}$$

In other words, the magnetic charge of the solution is identical with the ansatz label. To describe a monopole of charge N we need to choose appropriately a set of 2N+1 Δ's, linked by equations (2.11).

It is perhaps worth remarking that eq(2.14) is reminiscent of the multi-soliton formulae that occur in the solution of two dimensional field theories such as the Sine-Gordon theory or the Korteweg-de Vries equation. At least in so far as the monopoles are obtained by putting together non-linearly solutions to linear equations. Whether or not the monopole is a soliton in the strict sense remains to be seen via an eventual study of time dependent solutions to the equations (1.5) or (1.1). One might speculate that in the situation $\lambda = 0$ the time dependent solutions describe non-radiating monopoles (true solitons) which radiate when the 'true' coupling constant of the classical theory, λ, is switched on again. Whether this, or one of several other alternatives, is actually the case is yet to be investigated.

Because of a theorem of Taubes[(13)] we expect to be able to find a solution for the vector potentials which is regular everywhere, but not uniquely. In fact, up to gauge transformation the solutions of magnetic charge N should depend on a set of 4N-1 independent parameters, as predicted earlier by Weinberg[(14)]. Indeed, this is possible but to learn how to pick the functions Δ_r we have to acquire a better understanding of their significance.

3. Analysis of the self-dual equations II[15]

In the four dimensional complex space with coordinates y, $\bar{y}$, z and $\bar{z}$ there are two collections of null planes. One set of null planes is self-dual the other anti-dual, (in the sense that the antisymmetric tensor constructed from any pair of independent directions in the plane is self-dual or anti-dual, respectively). Null planes can be characterised by pairs of complex two-vectors (twistors), up to multiplication by a non zero complex number. That is, x (of eq(2.1)) lies on the null plane described by (π,ω) provided

$$x\pi = \omega. \tag{3.1}$$

Clearly, pairs of points satisfying the same equation (3.1) are null separated and, for any non zero λ, the pair $(\lambda\pi,\lambda\omega)$ describes the same plane as the pair (π,ω).

Null planes may be labelled in many ways but a convenient way is to pick a particular point on each one. For example, possible choices are[10]

$$x_1 = \begin{pmatrix} \omega_1/\pi_1 & 0 \\ \omega_2/\pi_1 & 0 \end{pmatrix} \tag{3.2a}$$

$$x_2 = \begin{pmatrix} 0 & \omega_1/\pi_2 \\ 0 & \omega_2/\pi_2 \end{pmatrix} \tag{3.2b}$$

x_1 is a good label except for those planes for which $\pi_1 = 0$, x_2 is good for those planes for which $\pi_2 \neq 0$. For most of the time they are equally good (but different) labels for null planes. However when $\pi_2 = 0$ we may use x_1 and, when $\pi_1 = 0$ we may use x_2. Together, (3.2a,b) describe all null planes.

It is convenient to define the complex variables μ,ν,ζ as follows

$$\zeta = \pi_1/\pi_2$$

$$\mu = i\omega_1/\pi_1 = x_3 + ix_4 - (x_1 + ix_2)\zeta = i(\bar{y} + z\zeta)$$

$$\nu = i\omega_2/\pi_2 = -x_3 + ix_4 - (x_1 - ix_2)/\zeta = i(y - \bar{z}/\zeta) \qquad (3.3)$$

since these are the combinations of the components of x,ω and π that appear in eqs(3.2).

A self-dual field strength is trivial when restricted to an anti dual plane and hence the parallel transport associated with the vector potential A_μ is independent of the path taken along the null plane. In particular, the SL(2,c) matrix g_{12} defined by,

$$g_{12} = \mathrm{Pexp} \int_{x_1}^{x_2} A_\mu dx^\mu, \qquad (3.4)$$

where x_1 and x_2 are the two special points defined above, is dependent only upon x_1 and x_2, (and hence μ,ν and ζ), and not upon the path taken through the null plane from x_1 to x_2.

Ward[15] pointed out that these facts allow an alternative description of the vector potential for a self dual gauge field as follows. The coordinate patches x_1 and x_2, corresponding to null planes for which $\zeta \neq 0$ or ∞, respectively, and the patching matrix g_{12} of eq(3.4) constitute a description of a 2 dimensional holomorphic vector bundle over CP^3 (the set of all anti-dual null planes). What is more, given any such vector bundle the vector potential at a real point x can be recovered from the corresponding patching matrix. This because

$$g_{12} = g_1 g_2^{-1} = \mathrm{Pexp} \int_{x_1}^{x} A_\mu dx^\mu \quad \mathrm{Pexp} \int_{x}^{x_2} A_\mu dx^\mu \qquad (3.5)$$

along any contour from x_1 to x_2 via x (wholly in the null plane) and so,

$$\begin{pmatrix} A_y \\ A_z \end{pmatrix} - \zeta \begin{pmatrix} -A_{\bar z} \\ A_{\bar y} \end{pmatrix} = g_1^{-1}\left(\begin{pmatrix} \partial_y \\ \partial_z \end{pmatrix} - \zeta \begin{pmatrix} -\partial_{\bar z} \\ \partial_{\bar y} \end{pmatrix}\right) g_1$$

$$= g_2^{-1}\left(\begin{pmatrix} \partial_y \\ \partial_z \end{pmatrix} - \zeta \begin{pmatrix} -\partial_{\bar z} \\ \partial_{\bar y} \end{pmatrix}\right) g_2 \tag{3.6}$$

The equality of the two expressions in eq(3.6) is guaranteed because we must have

$$\left(\begin{pmatrix} \partial_y \\ \partial_z \end{pmatrix} - \zeta \begin{pmatrix} -\partial_{\bar z} \\ \partial_{\bar y} \end{pmatrix}\right) g_{12} \equiv 0 \tag{3.7}$$

as a consequence of the special dependence of the patching matrix g_{12} on $y, \bar y$, z and $\bar z$ through the variables μ, ν and ζ. Note also that g_1 is analytic at $\zeta = \infty$ and g_2 is analytic at $\zeta = 0$ and, a direct comparison with Yang's formalism leads to the identification

$$g_2(\zeta = 0) = D, \quad g_1(\zeta = \infty) = \bar D. \tag{3.8}$$

If we were to take[9]

$$g_{12} = \begin{pmatrix} \zeta^{\ell} & \rho \\ 0 & \zeta^{-\ell} \end{pmatrix} \tag{3.9}$$

then the laurent coefficients of ρ

$$\rho = \sum_{-\infty}^{\infty} \Delta_{-r}\, \zeta^r \tag{3.10}$$

would, as a consequence of eq(3.7), satisfy eqs(2.11) automatically. Performing the splitting indicated in eq(3.5)[10] leads to the same expression for $D, \bar D$ (and hence $\varepsilon_\ell, \phi_\ell, \gamma_\ell$) as we obtained previously eq(2.8) by linearising Yang's equation. That g_{12} is a sufficient choice was argued by Atiyah and Ward[9] for the instanton case and, more recently by Hitchin[16] for the monopole situation. For the latter, as we saw earlier, the integer ℓ actually corresponds to the magnetic charge of the solution.

(In the instanton case ℓ had no obvious physical significance).

The importance from our point of view of this alternative description of the Δ's lies in the extra constraints arising from the holomorphy of the vector bundle and the special dependence of the function ρ upon the variables μ,ν and ζ. We can think of ρ as a sort of monopole generating function and the constraints limit it to just $4\ell-1$ degrees of freedom (aside from changes of gauge).

4. The monopole patching function[17,18]

(i) We shall say that the patching matrix g_{12} is equivalent to g'_{12} and write

$$g_{12} \sim g'_{12}, \tag{4.1}$$

if there are two SL(2,c) matrices a and A regular in the neighbourhood of $\zeta = \infty$, 0 respectively, such that

$$g_{12} = a\, g'_{12} A. \tag{4.2}$$

The matrices a,A depend only on μ,ν,ζ and, splitting g'_{12} yields a gauge potential A'_μ gauge equivalent to the potential A_μ, obtained by splitting g_{12}.

(ii) We shall demand that the patching matrix g_{12} that we can split (i.e. of the upper triangular form eq(3.9)) is equivalent to a patching matrix independent of x_4.

(iii) The upper triangular form useful for calculations will not lead to real potentials $\underline{A}_\mu$ even when the coordinates x_μ are real. The condition on a patching matrix guaranteeing reality of the vector potentials is

$$[g_{12}(x_\mu,\zeta)]^+ = g_{12}(x_\mu, -1/\zeta^*). \qquad (4.3)$$

obviously the upper triangular form does not satisfy this so we demand that it be equivalent to a patching matrix that does.

(iv) In order to maintain an unambiguous definition of the laurent coefficients of ρ it seems that we need to ensure that the patching matrix has no singularities in the ζ plane which move with $\underline{x}$. Otherwise, the contours used to define the laurent coefficients,

$$\Delta_r = \frac{1}{2\pi i} \oint_C \frac{d\zeta}{\zeta} \zeta^r \rho, \qquad (4.4)$$

may be pinched by singularities and yield singular expressions for the Δ's. This by itself, however, will not guarantee that the vector potential is regular everywhere since, combinations of the functions Δ_r such as det D_ℓ, (eq(2.14)), may vanish and cause a singularity to develop in the potentials A_μ. The precise condition on ρ that prevents this kind of singularity occurring is not known at present, though the analysis of an alternative view by Nahm[2] (this volume) may well provide the answer.

The content of the constraints (ii), (iii) and (iv) has been analysed by Ward[17], and by Corrigan and Goddard[18]; here we shall study each in turn.

Constraint (ii) is reminiscent of something we said earlier, in eq(2.12), about the dependence of the functions Δ_r on the coordinate x_4:

$$\Delta_r = e^{ix_4 a} \tilde{\Delta}_r(x_1,x_2,x_3).$$

However, as a condition on ρ this cannot be because ρ depends upon μ,ν,ζ only not on x_4 alone. In fact we must have

$$\rho = e^{a(\mu+\nu)/2} \quad \tilde{\rho}(\tfrac{(\mu-\nu)}{2}, \zeta) \tag{4.5}$$

and ρ becomes a function of just two complex variables γ, ζ, where ζ is given by,

$$\gamma = (\mu, \nu)/2, \tag{4.6}$$

and is independent of x_4.

Hence,

$$g_{12} \sim \begin{pmatrix} \zeta^{\ell} e^{a\gamma} & \tilde{\rho}(\gamma,\zeta) \\ 0 & \zeta^{-\ell} e^{-a\gamma} \end{pmatrix} \tag{4.7}$$

in the sense of (i) above. The right hand side of eq(4.7) is independent of x_4 altogether. From now on we take (4.7) to be the typical patching matrix.

As a preliminary to examining constraint (iii) we may remark that γ obeys

$$\gamma(-1/\zeta^*) = (\gamma(\zeta))^* \tag{4.8}$$

when the components of $\underline{x}$ are real. It is also easy to see that (4.7) need only be multiplied by an SL(2,c) matrix on the right (or the left) in order to bring it to a form in which eq(4.3) is satisfied. In other words, we demand that

$$g'_{12} = \begin{pmatrix} \zeta^{\ell} e^{a\gamma} & \tilde{\rho} \\ 0 & \zeta^{-\ell} e^{-a\gamma} \end{pmatrix} \begin{pmatrix} \alpha & \beta \\ \delta & \varepsilon \end{pmatrix}, \quad \alpha\varepsilon - \beta\delta = 1, \tag{4.9}$$

is a patching matrix that satisfies eq(4.3). Each of the functions $\alpha, \beta, \delta, \varepsilon$ is dependent only upon γ, ζ (so that the x_4 independence of g'_{12} is maintained) and regular in a neighbourhood of $\zeta = 0$.

Evaluating the content of eq(4.3) leads to

$$\zeta^{-\ell} e^{-a\gamma} \delta = (-)^{\ell} \zeta^{\ell} e^{-a\gamma} (\delta(-1/\zeta^*))^* \tag{4.10}$$

and to

$$\tilde{\rho} = \frac{(-)^{\ell} (\varepsilon(-1/\zeta^*))^* e^{-a\gamma} - \beta(\zeta) e^{a\gamma}}{\delta \zeta^{-\ell}} . \tag{4.11}$$

There are no other constraints provided the determinant conditions $\alpha\varepsilon - \beta\delta = 1$ is maintained. Eq(4.10) is easily solved for the function δ and we can write a polynomial in γ;

$$\delta = \zeta^{\ell} \sum_{o}^{\ell} a_r(\zeta) \gamma^r, \equiv a_{\ell} \zeta^{\ell} \prod^{\ell} (\gamma - \gamma_i(\zeta)) \tag{4.12}$$

whose coefficients are polynomials in ζ and $1/\zeta$ of degree $\ell - r$. These coefficient polynomials also satisfy

$$(a_r(\zeta))^* = a_r(-1/\zeta^*). \tag{4.13}$$

The first coefficient, a_{ℓ}, which is just a real constant, is removeable by an equivalence of type (i); all the rest share a total of $\ell(\ell+2)$ real parameters which are not removeable. The alternative expression for δ in factored form contains functions $\gamma_i(\zeta)$ $i = 1,\ldots\ell$, which will not, except in special circumstances, be polynomials. It is, however, useful to consider such a factorisation for examining the condition (iv), to which we now turn.

At first sight eq(4.11) does not appear to be anything more than an alternative rendering of $\tilde{\rho}$, since ε and β are arbitrary. However, the denominator of the expression for $\tilde{\rho}$ has zeros whenever ζ takes a value such that γ coincides with one of the γ_i, $i = 1,\ldots\ell$. Moreover, these zeros are $\underline{x}$ dependent, and lead to moving singularities in $\tilde{\rho}$ in violation of condition (iv), unless the numerator contains a compensating zero at the same value of ζ. It is this more subtle requirement that leads to a set of constraints on the $\ell(\ell+2)$ free parameters mentioned above. To see how it works we shall make the following ansatz. Set the function α identically zero and the determinant condition requires

$$\epsilon = -1/\beta \equiv \exp f(\gamma,\zeta), \tag{4.14}$$

the last definition supposing also that ϵ has no zeros in the neighbourhood of the origin in which we are working (otherwise β would have a pole). Then, when $\gamma = \gamma_i$ we require

$$\exp\left(f(\gamma,\zeta) + (f(\gamma,-1/\zeta^*))^* - 2a\gamma\right) = (-)^{\ell+1} \tag{4.15}$$

in order to provide the needed zero. This can be arranged by defining f to satisfy:

$$f(\gamma,\zeta) + (f(\gamma,-1/\zeta^*))^* - 2a\gamma = \sum_{i=1}^{\ell} n_i \prod_{j\neq i}\left(\frac{\gamma-\gamma_j}{\gamma_i-\gamma_j}\right) \tag{4.16}$$

where

$$e^{n_i} = (-)^{\ell+1} \qquad i = 1,\ldots,\ell. \tag{4.17}$$

For a clue to the choice of the complex numbers n_i we compare with some special cases considered previously. In fact the axisymmetric[11,19] monopoles strongly suggest making the choice for which n_i is $2i\pi$ times the weights of a spin $s = \frac{\ell-1}{2}$ representation of SU(2). Thus

$$\begin{aligned} &\ell = 1, \quad s = 0 \qquad n_1 = 0 \\ &\ell = 2, \quad s = \tfrac{1}{2} \qquad n_1 = i\pi,\ n_2 = -i\pi \\ &\ell = 3, \quad s = 1 \qquad n_1 = 2i\pi,\ n_2 = 0,\ n_3 = -2i\pi, \end{aligned} \tag{4.18}$$

etc.

the assignments being made in such a way that the right hand side of eq(4.16) is actually real, (because of condition (4.10) and the definition (4.12) each γ_i always occurs with its complex conjugate, conjugation in the sense of (4.13) with regard to the ζ dependence). That this is always the correct choice to make is also suggested by Nahm's approach in which these representations of SU(2) occur naturally. (Note also that if we examine the rotation properties of the patching matrix we find that the central $2\ell-1$ set of Δ's in ρ transform as a spin $\ell-1$ representation of the rotation group, but this does not seem to have any connection with the deeper remarks leading to eqs(4.18).)

Equation (4.16) looked at in detail is rather a mess because the γ_i functions will generally contain quite complicated branch points. However, these branches will move with the parameter set defining δ. We suppose that all the cuts in the ζ plane can be drawn in such a way that we continuously deform away from the axisymmetric situation and the expressions in eq(4.16) have a laurent expansion in the neighbourhood of the unit circle. Then we can write

$$f(\gamma,\zeta) + (f(\zeta,-1/\zeta^*))^* - 2a\gamma = \sum_{o}^{\ell-1} Cr(\zeta)\, \gamma^r, \qquad (4.19)$$

where,

$$Cr(\zeta) = \sum_{-\infty}^{\infty} Cr^{(p)}\, \zeta^{-p} . \qquad (4.20)$$

However, this is not enough because f and f* depend on ζ and $1/\zeta$ respectively, (since f is analytic at $\zeta = 0$), which means that the right hand side of eq(4.19) must be split in order to isolate the function f. In order to make this splitting without also splitting γ (which is not allowed because of x_4 independence) and, in order not to upset the asymptotic value of $|\underline{A}_4|$ we have to impose:

$$c_1^{(o)} = -2a$$

$$c_2^{(-1)} = c_2^{(o)} = c_2^{(+1)} = 0$$

$$\vdots \qquad (4.21)$$

$$c_{\ell-1}^{(-\ell+2)} = c_{\ell-1}^{(-\ell+3)} = \ldots = c_{\ell-1}^{(o)} = \ldots = c_{\ell-1}^{(\ell-2)} = 0,$$

a set of $(\ell-1)$ constraints. The coefficients $C_r^{(p)}$ depend implicitly (and transcendentally) on the parameters we isolated before, and the constraints mean that only

$$\ell(\ell+2) - (\ell-1)^2 = 4\ell-1$$

of them are actually effective. The transcendental nature of the constraints precludes any obvious interpretation of these parameters as positions of the monopole although, presumably, some combinations of them will have such an interpretation. At least it has proved possible to show for small values of the parameters[17,20] and in situations with special symmetry[21] that the Higgs' field has separated zeros. The fact that the number of effective degrees of freedom is $4\ell-1$ encourages us to think that the ansatz outlined above is actually the whole story but it remains to be proved that the construction leads to regular vector potentials and is complete. The alternative approaches of Hitchin and Nahm throw more light on these questions.

To conclude we look at the case of two monopoles in a little more detail. For the function δ we may write,

$$(\gamma-\gamma_1)(\gamma-\gamma_2) = \gamma^2 -(\gamma_1+\gamma_2)\gamma + \gamma_1\gamma_2$$

$$\equiv \gamma^2 + A\zeta + B - A/\zeta, \qquad (4.22)$$

the last line following from a special choice of origin and rotation frame - to force six of the eight parameters to zero. A and B are both real. For this special choice we also have

$$f + f^* - 2a\gamma = \frac{2i\pi}{\gamma_2-\gamma_1}\gamma$$

and thus a constraint

$$-2a = \frac{1}{2i\pi} 2i\pi \oint_C \frac{d\zeta}{\zeta} \frac{1}{\gamma_2-\gamma_1}$$

$$\text{or,} \qquad a = \frac{1}{4}\oint_C \frac{d\zeta}{\zeta} \frac{1}{(A/\zeta-B-A\zeta)^{\frac{1}{2}}} \quad . \tag{4.22}$$

The contour C encircles one of the two cuts in the ζ plane (in this example it is easy to see that this is always possible). Eq(4.23) constitutes a constraint between A and B, and is a complete elliptic integral of the first kind. When A is zero B takes the special value required by the axisymmetric solution, namely $B = \pi^2/4a^2$. It is perhaps worth remarking at this point that any other choice for B in the axisymmetric case ($B = 9\pi^2/4a^2$ for example) might be just as good as far as satisfying (iv) is concerned but will lead to singular vector potentials.[11]

5. Conclusion

It is encouraging that so much progress has been made in the classical multimonopole solutions albeit in this rather special static situation. However, the deeper questions concerning the time-dependent scattering solutions, the soliton nature of the monopole and the role to be played by the monopoles in quantum field theory remain to be examined. Even within the context of the theory presented above there remain many details to be clarified. In particular, the relationship between the various approaches needs to be explored and may provide a better understanding of the 4N-1 free parameters on which the multipmonopoles depend. Finally, all the work reported here is for the gauge group SU(2) it remains to be seen if the multimonopole solution for larger gauge groups can be treated in a similarly comprehensive way[22,2].

I am grateful to Chris Athorne, David Fairlie and Peter Goddard for many discussions about monopoles.

References

1. P.Forgacs, Z.Horvath and L.Palla, Phys.Lett.B102 (1981) 131, this volume and KFKI preprint 1981-82.
2. W.Nahm, this volume and a talk given at the International Summer Institute on Theoretical Physics, Freiburg, 1981 CERN TH 3172.
3. P.Goddard and D.Olive, Rep.Prog.Physics 41 (1978) 1357.
4. E.B.Bogomolny, Sov.J.Nucl.Phys.24 (1976) 449.
5. M.K.Prasad and C.M.Sommerfield, Phys.Rev.Lett. 35 (1975) 760.
6. S.Coleman, S.Parke, A.Neveu and C.M.Sommerfield, Phys.Rev.D15 (1977) 554.
7. C.N.Yang, Phys.Rev.Lett 39 (1977) 1377.
8. Y.Brihaye, D.B.Fairlie, J.Nuyts and R.G.Yates, J.M.P.19 (1978) 2528.
9. M.F.Atiyah and R.S.Ward, Comm.Math.Phys.55 (1977) 117.
10. E.Corrigan, D.B.Fairlie, P.Goddard and R.G.Yates, Comm.Math.Phys.58 (1978) 223.
11. Realised by R.S.Ward, Comm.Math.Phys.79 (1981) 317.
12. M.K.Prasad, Comm.Math.Phys.80 (1981) 137.
13. A.Jaffe and C.Taubes, Vortices and Monopoles (Birkhauser, Boston) 1980.
14. E.Weinberg, Phs.Rev.D20 (1979) 936.
15. R.S.Ward, Phys.Letts.61A (1977) 81.
16. N.J.Hitchin, Monopoles and geodesics, Oxford preprint, 1981.
17. R.S.Ward, Phys.Lett.B102 (1981) 136, Comm.Math.Phys.80 (1981) 563.
18. E.Corrigan and P. Goddard, Comm.Math.Phys.80 (1981) 575.
19. M.K.Prasad, A.Sinha and L.L.Chau-Wang, Phys.Rev.D23 (1981) 2321.
    M.K.Prasad and P.Rossi, Phys.Rev.Lett.46 (1981) 806.
20. S.A.Brown, M.K.Prasad and P.Rossi, Phys.Rev.D24 (1981) 2217.
    A.Soper, Multimonopoles close together, Cambridge preprint DAMTP 81/85.
21. L.O'Raifeartaigh and S.Rouhani, Rings of monopoles with discrete axial symmetry: explicit solution for N = 3, Dublin preprint DAS-STP-81-31.
22. J.Burzlaff, Transition matrices of self-dual solutions to SU(N) gauge theory, Kaiserslauten preprint 1981.
    R.S.Ward, Phys.Lett.B107 (1981) 281
    C.Athorne, private communication, Cylindrically and Spherically Symmetric Monopoles in SU(3) Gauge Theory, Durham preprint, 1981.

THE CONSTRUCTION OF ALL SELF-DUAL MULTIMONOPOLES BY THE ADHM METHOD

W. Nahm
CERN, Geneva, Switzerland.

From geometry it is well known that a Riemannian manifold can either be described internally, giving e.g. the metric as a function of suitable co-ordinates, or by an embedding into some Euclidean space. Embeddings often allow a better visualization of the manifold. On the other hand, they tend to be rather arbitrary. However, this inconvenience does not exist, if one has a canonical embedding, like that of S^n into R^{n+1}. If one walks along a closed path in a Riemannian space S, trying to conserve one's orientation at every step, the orientation at the end of the path nevertheless will be different from the one at the start. This phenomenon is described by the Riemannian connection, which can also be visualized with the help of an embedding of S into some R^k. In fact, the latter also yields an embedding of the fibre bundle of tangent planes into $S \times R^k$. The way to conserve one's orientation as one moves along some path in S consists in successive projections to neighbouring tangent planes, as they succeed each other along the path.

Gauge fields are also described by a fibre bundle and a connection. They only differ in that the fibres are not related to the geometry of the base space. However, the connection can be described by an embedding, too. We shall consider gauge fields on a Euclidean base space R^4, which takes the role of S. The fibre bundle will be described by an embedding into $R^4 \times M$, where M is a Hermitean vector space. For each point $X \in R^4$, the fibre is a subspace $M(X)$ of M. For a gauge group $G = SU(n)$, the $M(X)$ are n-dimensional complex spaces, for $G = SO(n)$ they are real, i.e. Euclidean, and for $G = USp(2n)$, they are quaternionic. The connection is given by the projections from M to the $M(X)$. Let these projections have the form

$$P(X) = \sum_{i=1}^{n} v_i(X)\, v_i^+(X) \quad , \tag{1}$$

with an orthonormal basis $v_i(X)$, $i = 1,\ldots,n$, of $M(X)$. Then the connection is given by

$$A_\mu^{ij} = v^{i+}\, \partial_\mu\, v^j \quad . \tag{2}$$

Different choices of the basis correspond to gauge transformations of the connection.

As for the Riemannian spaces, embeddings of gauge fields are most useful, when they are canonical. For self-dual gauge fields of finite action the ADHM construction [1] yields such an embedding. For it the space M can be described[2),3)] as the solution space of a system of partial differential equations in R^4, namely of

$$D^{A'A}\chi = \tilde{\Omega}^{A'A} + \Omega^{A'}_{\ B'}\, x^{B'A} \tag{3}$$

and the Dirac equations

$$D^{+}_{AA'}\,\tilde{\Omega}^{A'B} = D^{+}_{AA'}\,\Omega^{A'}_{\ B'} = 0 \quad . \tag{4}$$

Here χ, Ω and $\tilde{\Omega}$ are in the fundamental representation of G. Both the co-ordinates x and the covariant derivatives D are written in quaternionic notation:

$$x = e_\mu\, x^\mu \tag{5}$$

$$D = e^\mu\, \mathcal{D}_\mu \quad . \tag{6}$$

The quaternions e_μ are represented by 2×2 matrices with indices A, A'. Both $\tilde{\Omega}$ and Ω must be square integrable, whereas χ only is required to approach a covariant constant at infinity. The solutions of Eqs.(3) and (4) can be described by the pair $\Omega^{A'}_{\ B'}$, $B' = 1,2$, of square integrable solutions of the Dirac equation, and by a solution σ of the equation

$$\mathcal{D}^2\sigma = 0 \tag{7}$$

which approaches a covariant constant at infinity. One has

$$\chi = \sigma + 2(\mathcal{D}^2)^{-1}\,\Omega^{B'}_{\ B'} \tag{8}$$

$$\tilde{\Omega}^{A'A} = -\Omega^{A'}_{\ B'}\, x^{B'A} + D^{A'A}\sigma + 2D^{A'A}(\mathcal{D}^2)^{-1}\,\Omega^{B'}_{\ B'} \quad , \tag{9}$$

where the inverse of $\mathcal{D}^2$ is taken with image in the space of square integrable functions.

The scalar product in M is given by

$$\left\langle \bar{\Omega},\bar{\sigma} \middle| \Omega,\sigma \right\rangle = \frac{1}{4\pi^2}\int \bar{\Omega}^{+B'}_{\ A'}\,\Omega^{A'}_{\ B'}\, d^4x + (\bar{\sigma}^{+}\sigma)(\infty) \quad , \tag{10}$$

where the latter term may be calculated at any point of the sphere at infinity.

The subspaces $M(X)$ are defined by

$$\tilde{\Omega}^{A'A} + \Omega^{A'}{}_{B'} \, X^{B'A} = 0 \quad . \tag{11}$$

For $G = SU(n)$ and instanton number k, there are k linearly independent solutions of the Dirac equation and n independent solutions of Eq.(7), such that the dimension of M is $n+2k$. The left-hand side of Eq.(11) is a pair of solutions of the Dirac equation. Thus $M(X)$ is obtained from M by imposing $2k$ conditions and has dimension n, as it should be.

By Eq.(11), $\dot{\Omega}$ has to vanish for elements of $M(\infty)$, such that only $\chi(\infty)$ enters into the scalar product of Eq.(10). More generally, one can prove that for elements of $M(X)$

$$\langle \bar{\Omega},\bar{\sigma} | \Omega,\sigma \rangle = (\bar{\chi}^{+}\chi)\,(X) \tag{12}$$

Thus an element of $M(X)$ is determined uniquely by $\chi(X)$. If we write

$$\Omega^{A'}{}_{B'} = v^{+r}{}_{B'} \, \rho^{A'}{}_{r} \quad , \tag{13}$$

$$\sigma = v^{+k} \, \sigma_k \tag{14}$$

with orthonormal bases ρ_r, σ_k of the solution spaces of the Dirac equation and Eq.(7), Eq.(11) takes the form

$$v^{+r}{}_{B'}(X^{B'A} + e_{\mu}^{B'A} \, a^{\mu s}_{r}) + v^{+k} \, a^{sA}_{k} = 0 \quad , \tag{15}$$

where

$$a^{\mu s}_{r} = \int \rho^{+s}_{A'} \, x^{\mu} \, \rho^{A'}_{r} \, d^4x \tag{16}$$

$$a^{sA}_{k} = \int \rho^{+s}_{A'} \, D^{A'A} \, \sigma_k \, d^4k \tag{17}$$

Here we shall not discuss how one proves the ADHM construction, i.e. how one proves that the bundle given by Eq.(15) is the same as the bundle obtained directly from the gauge potential. Basically one lifts both bundles from the base space R^4 to the base space CP^3 of anti-self-dual planes in the complexification of R^4. The new bundles have natural holomorphic structures, and it is not too difficult to see that they are isomorphic. Eq.(15) usually is written in the compact form

$$v^{+}\Delta = 0 \quad . \tag{18}$$

Starting from Eq.(11) it is easy to see that $\Delta^+\Delta$ commutes with the quaternions, and it is not hard to show that it is invertible. Inversely, these two conditions are sufficient to show that Eq.(15) yields a self-dual field.[4)] So far for the instantons. Self-dual multimonopoles also can be written as pure self-dual gauge field configurations in R^4, with a connection whose component A_0 is the Higgs field. Thus formally we have a great similarity to instanton configurations. However, the boundary conditons are quite different. Instead of finite action we have finite energy, and invariance with respect to shifts in Euclidean time. Nevertheless, most of the formalism developed above can be taken over without any change. The only major difference is that M now becomes a Hilbert space. To explore the details of the construction, we have to consider the behaviour of the Higgs field at infinity, which up to gauge transformations can be written in the form

$$iA_0^{ij} = \delta^{ij}\,(z_i - k_i/2r) + o(1/r^2)\ , \tag{19}$$

where the k_i are integers and $z_i \leqslant z_{i+1}$.

The number of independent solutions of Eq.(7) which approach covariant constants at spatial infinity is now equal to the number of vanishing k_i . Put

$$I = \{i|k_i = 0\}\ . \tag{20}$$

Then σ can be written in the form

$$\sigma = \sum_{i\in I} s_i^+ \exp(ix_0 z_i)\ \sigma_i(\vec{x}) \tag{21}$$

with orthonormal $\sigma_i(\infty)$. Similarly, Ω now can be written as

$$\Omega^{A'}{}_{B'}(x) = \int \sum_{r=1}^{k(z)} \rho_r^{A'}(x,z)\ v_{A'}^{+r}(z)\ dz\ . \tag{22}$$

Here the

$$\rho_r^{A'}(x,z) = \exp(ix_0 z)\ \rho_r^{A'}(\vec{x},z) \tag{23}$$

form a basis of the solution space of the Dirac equation and are normalized to

$$\int \rho_{A'}^{+s}(\vec{x},z)\ \rho_r^{A'}(\vec{x},z)\ d^3x = 2\pi\delta_r^s\ . \tag{24}$$

At fixed z their number is given by

$$k(z) = \sum_i k_i \,\vartheta(z_i - z) \quad . \tag{25}$$

The scalar product in M may be written as

$$\langle \bar{v},\bar{s}|v,s\rangle = \int \sum_{r=1}^{k(z)} \bar{v}^{+r}_{A'} v^{A'}_r \, dz + \sum_{i\in I} \bar{s}^{+}_i s_i \quad . \tag{26}$$

Eq.(24) still leaves the freedom of z-dependent unitary transformations, which we use to achieve

$$\int \rho^{+s}_{A'}(\vec{x},z) \frac{\partial}{\partial z} \rho^{A'}_r(\vec{x},z) \, d^3x = 0 \quad . \tag{27}$$

This yields

$$v^{+r}_{B'}(z) \left(X^{B,A} \delta^s_r + i\, e^{B'A}_m T^{ms}_r(z) - i \frac{\overleftarrow{d}}{dz} \right)$$

$$+ \sum_{i\in I} s^{+}_i a^{SA}_i \delta(z - z_i) = 0 \quad , \tag{28}$$

where

$$2\pi\, T^{ms}_r(z) = i \int \rho^{+s}_{A'}(\vec{x},z)\, x^m \rho^{A'}_r(\vec{x},z) \, d^3x \tag{29}$$

$$a^{SA}_i = \int \rho^{s}_{A'}(\vec{x},z_i) \left(e^{A'A}_m \mathcal{D}^m + i\delta^{A'A} z_i \right) \sigma_i(\vec{x}) \, d^3x \quad . \tag{30}$$

If Eq.(28) is written in the form of Eq.(18), the product $\Delta^+\Delta$ again commutes with the quaternions, such that

$$\frac{dT^1}{dz} = [T^2 T^3] + \sum_{i\in I} \alpha^1_i \delta(z - z_i) \text{ and cyclic} \quad , \tag{31}$$

where

$$\alpha^\mu_i e_\mu = a^A_i a^{+}_{iA} \quad . \tag{32}$$

Eq.(31) takes care of the singularities associated to z_i with $k_i = 0$. What happens for non-zero k_i has been described elsewhere [5),6)]. Here we will just summarize the results. According to Eq.(25), $v(z)$ acquires $2|k_i|$ new components, as one moves across z_i. At z_i, Eq.(31) is only valid for the matrix elements of $e_m T^m$ between the old components. The matrix elements between old and new components stay finite at z_i, in fact one can show that they are of order $(z-z_i)^{(|k_i| - 1)/2}$. The matrix elements between the new components have simple poles. The residues are given by the representation matrices of the $|k_i|$-dimensional representation of SU(2). In particular, for $k_i = \pm 1$ the pole is absent. If k_i, k_j have the same sign and belong to coinciding $z_i = z_j$, the matrix elements of mixed type between new components stay finite, whereas the $k_i \times k_i$ and $k_j \times k_j$ submatrices have poles of the kind described before.

It has been noted [7),8)] that for an equation of the form (31) the spectrum of $y_m T^m(z)$ does not depend on z, if $y \in C^3$ is a null vector. In our case the spectrum only changes if one passes a z_i. If one parametrizes y as

$$y = \left(\frac{1+\zeta^2}{2}, \frac{1-\zeta^2}{2i}, i\zeta\right) \quad \zeta \in \mathbb{C} \tag{33}$$

the spectrum is given by the roots of the polynomial

$$\det(\eta - y_m T^m(z)) = p(\eta,\zeta) \tag{34}$$

which also appears in different treatments of the self-dual multimonopoles[9)-11)], and for G = SU(2) determines the complete configuration. But note that for n distinct z_i we have $(n-1)$ polynomials. Using this spectrum one can integrate Eq.(28) explicitly. We first solve the related equation

$$\left(i\frac{d}{dz} + i\, e_m T^m(z) + X\right) w(z) = 0 \tag{35}$$

by the ansatz

$$w(z) = (1 + i\, e_m u^m)\, f(z) \quad . \tag{36}$$

Let (u,ξ,ξ') be a right-handed system of orthonormal vectors in R^3. Then we find

$$\left(i\frac{d}{dz} + X_0 + (T_m - iX_m)\, u^m\right) f = 0 \tag{37}$$

$$(\xi_m + i\xi'_m)\,(T^m - iX^m)\, f = 0 \quad . \tag{38}$$

Eqs.(37) and (38) are compatible because of Eq.(31). For X in a general position, the equation

$$\det \left[(\xi_m + i\xi'_m)\,(T^m - iX^m) \right] = 0 \qquad (39)$$

yields $2k(z)$ distinct vectors u. The corresponding f belongs to one-dimensional spaces, which can be determined algebraically. Insertion into Eq.(37) yields a differential equation for a single function, which can be integrated explicitly. Thus we obtain $2k(z)$ linearly independent solutions of Eq.(35). For each of them and for any solution of Eq.(28) one has

$$v^+(z)\, w(z) = v^+(z_0)\, w(z_0) \qquad (40)$$

such that the general solution of Eq.(28) now can be determined by linear algebra.

The main problem which remains is the determination of the general solution of Eq.(31). There are indications that this equation can be solved in terms of Abelian integrals, but for the moment the problem is open.

Let us, however, discuss which $T^m(z)$ yield orthogonal and symplectic gauge groups. Taking the complex conjugate of the Dirac equation, one obtains

$$\varepsilon^{B'A'}\, C\, \rho^*_{rA'}(\vec{x},z) = c_{rs}\, \rho^{sB'}(\vec{x},-z) \quad , \qquad (41)$$

where C is the charge conjugation matrix for the fundamental representation of G. Because of Eq.(27), the c_{rs} are independent of z, up to the obvious changes at the z_i . For orthogonal G one finds

$$cc^* = -1 \qquad (42)$$

and for symplectic G

$$cc^* = 1 \qquad (43)$$

In the latter case, a convenient choice of the basis transforms c to the unit matrix. In any case, Eq.(29) yields

$$T^{*m}(-z) = -\, c\, T^m(z)\, c^+ \quad . \qquad (44)$$

For $G = SU(2) = USp(2)$, one obtains a well-known condition [9)-11)] for the polynomial of Eq.(33).

Other features of the ADHM construction for self-dual monopoles have been discussed elsewhere [5),6),12),13)]. In particular, it is easy to show that every solution of Eq.(31) for which the T^m are non-singular between the z_i yields a regular multimonopole. Moreover, the solution of Eq.(28) immediately yields the Green function for the operator $\mathcal{D}^2$ in the fundamental representation. For tensor products or for the solutions of the Dirac equation one needs in addition the inverse of $\Delta^+\Delta$. The latter is not hard to get, as the solutions of Eq.(35) which are of the form (36) yield $2k(z)$ solutions of the homogeneous equation

$$\left[\left(i \frac{d}{dz} + X_0\right)^2 + \sum_m (T_m(z) - iX_m)^+ (T^m(z) - iX^m)\right] f(z) = 0 \quad (45)$$

and as usual one can fit together two solutions to obtain instead a delta function on the right-hand side.

I would like to thank M. Atiyah and N. Hitchin for enlighting discussions of their related work.

REFERENCES

1) M. Atiyah, V. Drinfeld, N. Hitchin and Yu. Manin, Phys. Letters 65A, 185 (1978).

2) E. Witten in Complex Manifold Techniques in Theoretical Physics, Eds. D. Lerner and P. Sommers (Pitman 1979).

3) H. Osborn, CERN preprint TH-3120 (1981).

4) E. Corrigan, D. Fairlie, P. Goddard and S. Templeton, Nucl. Phys. B140, 31 (1978).

5) W. Nahm, CERN preprint TH-3172 (1981).

6) W. Nahm, ICTP, Trieste, preprint IC/81/238.

7) D. Olive in Current Topics in Elementary Particle Physics, Eds. K.H. Mütter and K. Schilling (Plenum Press, 1981), p.199.

8) N. Hitchin, prive communication.

9) E. Corrigan and P. Goddard, Commun. Math. Phys. 80, 575 (1981).

10) P. Forgács, Z. Horváth and L. Palla, Budapest preprint KFKI-1981-92.

11) N. Hitchin, "Monopoles and geodesics", preprint (1981).

12) W. Nahm, Phys. Letters 90B, 413 (1980).

13) W. Nahm, Phys. Letters 93B, 42 (1980).

N.S. Manton

Institute for Theoretical Physics, University of California, Santa Barbara, CA 93106, USA.

I. INTRODUCTION

A number of physicists and mathematicians have recently made striking progress in constructing exact solutions of the Bogomolny equations, $F_{ij} = \varepsilon_{ijk} D_k \Phi$. (See several of the articles in this volume.) These solutions represent static configurations of several monopoles of the Prasad-Sommerfield type, either superposed at one point or separated in space. Since the equations are nonlinear, the superposition is non-trivial.

Ultimately we are interested in the quantum dynamics of these monopoles. Presumably the quantum field theory has a state corresponding to a single monopole with a definite momentum, and also asymptotic states corresponding to several well separated monopoles with definite momenta [1]. We would like to know the scattering amplitudes. The monopoles are probably not solitons in the strict sense, so in the scattering process, perturbative quanta (including photons) will be produced. Also monopole, antimonopole pairs could appear, or be annihilated.

Without a complete understanding of quantized Yang-Mills theory, it is not feasible to try to completely solve the quantum dynamics of monopoles (where the Yang-Mills field is coupled to a Higgs field). In this lecture we shall describe some special situations where it can be argued that only a finite number of degrees of freedom are relevant, and not the infinite number of the complete theory. Then the physics is more tractable.

In section II we show that the classical motion of BPS monopoles, in the limit of small relative velocities, can be described by a geodesic motion in the parameter space of the exact static solutions. This illustrates the general principle that at low energies, only the collective coordinates corresponding to zero modes of the soliton configuration are excited, and their number is finite. Although we have not studied them carefully, there are gapless non-zero modes too, which may be important. Physically, we expect them to be the long wavelength photons, or long wavelength excitations of the massless Higgs field. Some problems which appear when this classical picture is extended to a quantum mechanical one are briefly mentioned.

In section III we discuss monopole, antimonopole pair production in a weak magnetic field, a process where the total number of solitons changes. We can view it as a tunnelling phenomenon whose rate is controlled by an "instanton", a solution of the Euclidean field equations. For weak fields this instanton has the form of a single closed monopole world line in the background field, and we can show that the only important degrees of freedom of the full quantum field theory are the collective coordinates of the monopole. From the 4-D Euclidean point of view there is therefore no change of soliton number. The dynamics reduces to that of a single point particle, with one internal degree of freedom associated with electric charge. Here the effect of the gapless modes can be computed exactly. They are responsible for the Coulombic interaction between the monopole, antimonopole pair during the production process.

The material of sections II and III is a condensed version of three earlier papers, to which the reader is referred for more details [2,3,4]. Finally I would like to thank Professor Salam and the organizers for inviting me to participate in this most worthwhile conference, and for their hospitality in Trieste.

I would like to thank Mike Lowe for some helpful comments.

II. SCATTERING OF BPS MONOPOLES

In a gauge theory, the field equations are not the canonical equations corresponding to an ordinary Hamiltonian. This is because A_o (the time component of the gauge potential) is not a true dynamical variable. A Hamiltonian formulation is possible only if Gauss's Law is imposed as a constraint on the initial data.

However, there has recently been developed, by Singer and others [5], a new approach in which the field equations can be reinterpreted as canonical equations, and moreover all gauge freedom is eliminated. For most purposes it would be hard to work with this new formalism, as it is not very explicit. Nevertheless, our result on the scattering of monopoles is very easy to obtain from this point of view.

Let A be the set of all smooth field configurations $\{A_i(x), \Phi(x)\}$ in $\mathbb{R}^3$. Here A_i are the spatial components of the gauge potential and Φ the adjoint Higgs field. Let G be the group of gauge transformations, that is, the set of differentiable functions from $\mathbb{R}^3$ into the gauge group G. The true configuration space is the quotient $C = A/G$, obtained by identifying gauge equivalent fields. Note that time, and the A_o component of the gauge potential, play no role here.

$\mathcal{C}$ is naturally a Riemannian manifold. Let c and c + δc be neighbouring configurations in $\mathcal{C}$, with representatives (A_i, Φ) and $(A_i + \delta A_i, \Phi + \delta\Phi)$. Part of $(\delta A_i, \delta\Phi)$ is generally an infinitesimal gauge transformation of (A_i, Φ), but this is not a true displacement, so we project it out. In other words, fix $(\delta A_i, \delta\Phi)$ to be in background gauge

$$D_i \delta A_i + [\Phi, \delta\Phi] = 0 \quad , \tag{2.1}$$

hence orthogonal to infinitesimal gauge transformations. Then

$$h_c(\delta c, \delta c) = \int d^3x (\mathrm{tr}(\delta A_i \delta A_i) + \mathrm{tr}(\delta\Phi\delta\Phi)) \tag{2.2}$$

defines the metric on $\mathcal{C}$, and it is independent of the choice of representatives of c and c + δc.

There is also a potential function U defined on $\mathcal{C}$. It is the usual non-kinetic part of the Bogomolny Lagrangian,

$$U(c) = \int d^3x \left(\frac{1}{4}\, \mathrm{tr}(F_{ij} F_{ij}) + \frac{1}{2}\, \mathrm{tr}(D_i\Phi\ D_i\Phi) \right) \tag{2.3}$$

The integral is gauge invariant, so well-defined on $\mathcal{C}$.

Having a metric h on $\mathcal{C}$ and a potential U, we can define a dynamical system with action

$$I(c) = \int dt\left(\frac{1}{2} h_c(\dot{c}, \dot{c}) - U(c)\right) \tag{2.4}$$

This is the immediate generalization of the action defined by Babelon and Viallet [5] for a pure Yang-Mills theory. Since the Lagrangian is nonsingular, the conversion to a Hamiltonian formalism is quite simple. One can check that the motion of the point c on the infinite-dimensional curved manifold $\mathcal{C}$ with potential U, corresponds to the evolution of a representative of the configuration c according to the Yang-Mills-Higgs equations.

Let us assume now that the gauge group is SU(2). In the sector with monopole charge n, the minimum of U is attained for all fields satisfying the Bogomolny equation $F_{ij} = \varepsilon_{ijk} D_k \Phi$, and these form a 4n-1 dimensional manifold, which we call min(U). Suppose that c(t) is a trajectory on $\mathcal{C}$ initially tangent to min(U), and for which $\dot{c}(t)$ is initially small. Well separated monopoles in slow relative motion provide an example. By energy conservation, the trajectory c(t) must remain close to min(U). Moreover, as $\dot{c}(t)$ tends to zero, the motion is constrained to lie on min(U), but on that manifold is free. c(t) is therefore a geodesic motion, with the metric on min(U) being that induced from the metric already defined on $\mathcal{C}$. This is what we wished to demonstrate. Unfortunately, we have no explicit information about the metric or its geodesics, so have no concrete results on monopole scattering.

Although the above argument can be made rigorous for motion in simple potential wells, for the monopole problem there may be difficulties because there are an infinite number of normal modes orthogonal to the tangent space of min(U), and their spectrum, though positive, has no gap. However, the only modes whose frequencies are small are the photons and Higgs field excitations of long wavelength. We could probably deal with them explicitly by using classical radiation theory for point charges.

One intriguing possibility is that when monopoles scatter they can turn into dyons, monopoles with electric charge. Asymptotically, we expect the 4n-1 dimensional parameter space to be the product of $\mathbb{R}^{3n}$ (the positions of n monopoles) with n-1 circles, representing relative U(1) phases. Motion at constant speed on these circles yields dyons (with total electric charge zero). Unless the metric on min(U) has some special symmetries, translational motion could be partly converted to motion on these circles in the scattering process, since the simple product structure of the parameter space is not maintained when the monopoles are close together.

We would like to extend our classical picture of monopole dynamics at low energy to a quantum mechanical one, by deriving from the quantum field theory an effective action for paths on min(U), from which quantum mechanical scattering amplitudes could be obtained using Feynman's path integral. The effective action is not just the classical action (the path length), because it must take account of the field fluctuations away from min(U), at least in the Gaussian approximation. In addition to renormalizing the monopole and dyon masses, there could be a real finite effect due to these (as in the Casimir effect). Because min(U) is not just the set of all monopole positions, but includes internal coordinates in a non-trivial way, the quantum mechanics may have surprising features, perhaps strange statistics.

III. MONOPOLE PAIR PRODUCTION IN A WEAK MAGNETIC FIELD

It is well known that an electric field polarizes the vacuum, and if it extends over a sufficiently large region real charged pairs appear, which tend to neutralize the source of the field. Schwinger showed that the rate of e^+e^- pair production in a field E is [6]

$$\Gamma_{QED} = \frac{e^2E^2}{4\pi^3} \sum_{n=1}^{\infty} \frac{1}{n^2} \exp\left(-\frac{n\pi m^2}{eE}\right) \quad (1 + O(e^2)) \tag{3.1}$$

where m is the electron mass. For spinless charged particles

$$\Gamma_{\substack{\text{Scalar}\\ \text{QED}}} = \frac{e^2E^2}{8\pi^3} \sum_{n=1}^{\infty} \frac{(-1)^{n+1}}{n^2} \left[\exp\left(-\frac{n\pi m^2}{eE}\right)\right](1 + O(e^2)) \tag{3.2}$$

We are interested in a Yang-Mills-Higgs model, with only an electromagnetic U(1) gauge group unbroken, and having a spherically symmetric monopole of the 't Hooft, Polyakov type. We shall mention only in passing the Prasad-Sommerfield limit, where the Higgs field becomes massless.

In such a model, a magnetic field is unstable, since its energy can be reduced by adding a sufficiently well separated monopole, antimonopole pair. Monopole pair production can be treated semi-classically as a tunnelling phenomenon provided the field is weak. (For monopoles of mass $\sim 10^2$ GeV, "weak" here means $B \lesssim 10^{20}$ Gauss!) This method is superficially quite different from Schwinger's, and based on vacuum decay calculations pioneered by Langer [7] and extended to relativistic field theory by Stone, Coleman and Callan [8]. In a weak field the tunnelling involves essentially just one monopole pair, which corresponds to a single monopole world line, since magnetic charge is conserved.

Multi-monopole processes are suppressed exponentially by factors of $\exp(-\pi M^2/gB)$, where M is the monopole mass, B the magnetic field and g the magnetic charge of the monopole. This evades the strong coupling problem, for it is the interaction between monopole world lines mediated by photons which is proportional to $g^2 \sim 137$. Pair production in a weak field is therefore similar to the quantum dynamics of non-relativistic solitons, interacting with relativistic conventional particles. There, a perturbative treatment is possible because each soliton retains its identity [1].

We find the pair production rate for monopoles is

$$\Gamma_M = \frac{g^2 B^2}{8\pi^3} \left[\exp\left(- \frac{\pi M^2}{gB} + \frac{g^2}{4}\right)\right] \left(1 + O\left(\frac{g^5 B}{M^2}\right) + O(e^2)\right) \qquad (3.3)$$

The analogy between this result and (3.2) suggests that a magnetic version of scalar QED could be an effective Lagrangian for describing monopoles, thus realizing the approximate duality between electric and magnetic phenomena which was Dirac's original motivation for contemplating monopoles. Note that monopoles of the simplest type do appear to be spinless. We shall comment further on the similarities and differences between (3.3) and (3.2) below.

To compute the pair production rate, we first identify it with the decay rate of the "vacuum" (here, the constant magnetic field), which in turn is twice the imaginary part of the vacuum energy, E_o. E_o is given by the Euclidean functional integral (schematically)

$$\exp(-E_o T) = \lim_{T\to\infty} \int (d\phi)\exp(-S(\phi)) \tag{3.4}$$

and in perturbation theory E_o is real. However there is also an instanton stationary point of the action S, when the boundary conditions correspond to a constant magnetic field at large distances. This represents a monopole loop in the background field. Among the small oscillation normal modes about the instanton, one is always negative, leading to an imaginary part for the vacuum energy. Since the instanton is localized, a space-time volume factor VT is associated with it. In the field-theoretic WKB approximation, one must allow an arbitrary number of these instantons, which form a non-interacting dilute gas. One obtains

$$\exp(-E_o T) = C \sum_{n=0}^{\infty} \frac{(-iK[\exp(-S_{c\ell})]VT)^n}{n!} \tag{3.5}$$

where $S_{c\ell}$ is the one-instanton action, K is a ratio of one-loop determinants in the presence and absence of the monopole loop, and C is a real normalization constant, in fact, the perturbative term. Thus, the pair production rate per unit volume is

$$\Gamma_M = 2K \exp(-S_{c\ell}) \tag{3.6}$$

That there is a monopole loop instanton can be seen by considering the Euclidean space equation of motion for a point magnetic charge g in an electromagnetic field

$$M\ddot{x}_\mu = - g\tilde{f}_{\mu\nu}\dot{x}_\nu \tag{3.7}$$

Here $\tilde{f}_{\mu\nu}$ is the dual of the Maxwell tensor, and the derivatives are w.r.t. proper time. For a constant magnetic field B along the z-axis, the monopole world line is a circle of radius $R_{c\ell} = M/gB$ in the z-t plane. Physically, the monopole, antimonopole pair tunnel to a separation $2R_{c\ell}$ and then propagate classically in Minkowski space, separating further along hyperbolic world lines.

The action of the instanton (leaving the radius R momentarily free) is

$$S = 2\pi MR - gB\pi R^2 - \frac{1}{4} g^2 + O(\frac{1}{R}) \tag{3.8}$$

The first three terms are those for a point monopole, but since in a weak field R will be much greater than the monopole size, the corrections are small. The finite part of the Coulomb contribution to the action is $-g^2/4$. There is also a divergent part (for a point monopole) but this is absorbed in the Coulomb contribution to the monopole mass. It seems that in the Prasad-Sommerfield limit, (3.8) remains correct. The coefficient of the Coulomb term might have changed, since for a static monopole, antimonopole pair the Coulomb energy is doubled due to the long-range Higgs field [9]. With the circular geometry here, however, there is no finite part of the scalar Coulombic action.

S is stationary w.r.t. R when $R = R_{c\ell} = M/gB$, so the actual action of the instanton is $S_{c\ell} = \pi M^2/gB - g^2/4$. Note that d^2S/dR^2 is negative. Variation of the radius of the monopole loop therefore provides the imaginary eigenvalue in the one-loop determinant.

Eq. (3.8), which was derived fairly rigorously in ref. [2], illustrates the general principle that in the infrared limit the action can be defined just in terms of the collective coordinates; and notice that the gapless modes do contribute. The generalization of the first two terms of (3.8) to non-circular trajectories is $ML - gBA$, where L is the total proper time, and A is the area of the projection of the trajectory onto the z-t plane. The Coulomb term must be g^2 times some dimensionless function of the shape of the trajectory, but we do not know what this function is in general.

The one-loop determinant factor K is not computable in closed form for arbitrary values of B. However, for small B the fluctuations of the fields about the instanton can be divided into two types:

those that simply alter the shape of the monopole world line, whose effect can be computed exactly, and those that involve an internal excitation of the monopole. The eigenvalues for small oscillations of a static monopole are unknown, but we know their effect is to renormalize the monopole mass. Similarly, for a monopole loop of large radius, we can show that the internal excitations of the monopole lead solely to a mass renormalization.

With the latter modes integrated out, the functional integral reduces to a path integral, so the pair production can be treated as a quantum mechanical problem involving just one point particle. Unlike in section II, we have here an effective action, because for a single monopole there is no essential difference between the classical and one-loop actions.

We find that

$$K = \frac{M^2}{16\,\pi^3 R_{c\ell}^2} \exp(-2\pi\, R_{c\ell} \Delta M) \tag{3.9}$$

where ΔM is the one-loop quantum correction to the monopole mass. It fcllows that the pair production rate is

$$\Gamma_M = \frac{g^2 B^2}{8\pi^3} \exp\left(-\frac{\pi M_R^2}{gB} + \frac{g^2}{4}\right) \tag{3.10}$$

where M_R is the renormalized monopole mass.

Unfortunately we have missed charge renormalization effects in this semi-classical calculation, although some idea of why this is, is discussed in ref. [3]. This is a pity, since there is still considerable debate about whether the renormalization constants for e and g are identical [10] or reciprocal [11]. (See, for example, some of the other contributions in this volume.)

So far we have tacitly assumed that the instanton is unique. However, we know that the parameter space for a single monopole is $\mathbb{R}^3 \times S^1$, giving a position and a phase angle. (This extra phase is left out of the 4n-1 count, because it is associated with a pure gauge transformation. It has a physical significance though, because it is non-vanishing at infinity.) When the phase varies linearly with time, the static monopole becomes a dyon. In a background magnetic field, too, there can be a varying phase accompanying the motion in space. To satisfy the field equations the angular frequency should be constant, and quantized, so that the phase is single-valued around the monopole loop. This quantization is not the same as the electric charge quantization of the physical dyons, although there is a connection.

We must sum over instantons with all allowed values of the angular frequency, and this leads to the rate of production of monopoles and dyons

$$\Gamma = \sum_N \frac{g^2 B^2}{8\pi^3} \exp\left(-\frac{\pi M_N^2}{gB} + \frac{g^2}{4}\right) \qquad (3.11)$$

Here M_N is the semi-classical mass of the dyon of charge $Q=Ne$, so (3.11) is just the naive extension of (3.10) with dyons included, although our derivation is somewhat different.

Let us return now to the approximate duality between (3.3) and (3.2). Schwinger's result holds for weak coupling but arbitrary field strength E. If E were small, only the first term in the sum would contribute, and the analogy would be complete if the $(1 + O(e^2))$ corrections could be summed up to give $\exp(e^2/4)$. These corrections are photon exchange diagrams which must be added to Schwinger's basic diagram, a single closed electron propagator in the background field. In fact it is shown in ref. [3] that for small E, but arbitrary coupling, including these Coulomb effects gives precisely $\exp(e^2/4)$. We do not expect the duality to be exact in the higher order corrections to (3.3), because these terms depend on the internal structure of the monopole, which is not universal.

IV. SUMMARY

We have shown that in the low energy limit, multi-monopole dynamics can be reduced to a finite dimensional problem, since only the collective coordinates are excited. There still remain many problems, however. What is the quantum mechanics of interacting BPS monopoles, and what is the physical significance of the "internal" degrees of freedom? Is there a consistent way of renormalizing the magnetic charge? Can we get some explicit information about the geodesics on the 4n-1 dimensional parameter space? Have we really understood the role of the gapless modes?

ACKNOWLEDGMENT

This research was supported in part through the National Science Foundation under grant PHY77-27084.

REFERENCES

1. J. Goldstone and R. Jackiw, Phys. Rev. D11, 1486 (1975); J.-L. Gervais, A. Jevicki and B. Sakita, Phys. Rev. D12, 1038 (1975); L.D. Fadeev and V.E. Korepin, Phys. Rep. 42C, 1 (1978).
2. I.K. Affleck and N.S. Manton, Nucl. Phys. B194, 38 (1981).
3. I.K. Affleck, O. Alvarez and N.S. Manton, Princeton University preprint (1981), to appear in Nucl. Phys. B.
4. N.S. Manton, ITP Santa Barbara preprint 81-116.
5. I.M. Singer, Comm. Math. Phys. 60, 7 (1978);and lecture in "Perspectives in Modern Field Theories" (Stockholm, 1980); M.S. Narasimhan and T.R. Ramadas, Comm. Math. Phys. 67, 121 (1979); O. Babelon and C.M. Viallet, Comm. Math. Phys. 81, 515 (1981); P.K. Mitter and C.M. Viallet, Comm. Math. Phys. 79, 457 (1981).
6. J. Schwinger, Phys. Rev. 82, 664 (1951).
7. J.S. Langer, Ann. Phys. (N.Y.) 41, 108 (1967).
8. M. Stone, Phys. Rev. D14, 3568 (1976); S. Coleman, Phys. Rev. D15, 2929 (1977); C.G. Callan and S. Coleman, Phys. Rev. D16, 1762 (1977).
9. N.S. Manton, Nucl. Phys. B126, 525 (1977).
10. J. Schwinger, Phys. Rev. 151, 1048 (1966); 1055.
11. S. Coleman (unpublished).

REFERENCES

1. J. Goldstone and R. Jackiw, Phys. Rev. D11, 1486 (1975).

2. [illegible] Phys. Rev. D12 [illegible]

[illegible] Phys. Rept. 23C, [illegible] (1976).

[illegible] (1981).

[illegible]

[illegible] (1983).

[illegible]

[illegible] Phys. Rev. D [illegible]

[illegible] (1981).

[illegible] Phys. Rev. [illegible]

7. [illegible] Phys. [illegible]

[illegible]

[illegible]

[illegible] 629 (1971).

[illegible]

AXIALLY SYMMETRIC MONOPOLES FOR AN ARBITRARY GAUGE GROUP*

F.A. BAIS

Institute for Theoretical Physics
Princetonplein 5, P.O.Box 80006
3508 TA Utrecht, The Netherlands

Abstract

An algebraic method for the construction of the axially symmetric BPS-Monopole configurations in theories with arbitrary gauge group is presented.

*Status report on work in collaboration with R. Sasaki.

1. Introduction

Recently important progress towards the complete multimonopole solution for the SU(2) Yang-Mills-Higgs theory has been made. The generalization to the case with an arbitrary compact semisimple gauge group G poses an interesting problem. We present as a first step in that direction an algebraic procedure for constructing the axially symmetric monopole configurations for arbitrary G [1)2)3)]. It is an integration method which can be used for all static axially symmetric self-dual gauge fields, as well as for a large class of two dimensional nonlinear sigma models.

2. Ansatz

The ansatz for the static axially symmetric gauge potentials [1)] involves 2d (d=dim. G) functions of the coordinates ρ and z. The components A_ν in cylindrical coordinates read:

$$\begin{aligned}
A_t &= \sum_\alpha X_1^\alpha\, T_1^\alpha + \sum_\gamma \sigma_1^\gamma\, T_3^\gamma \\
A_z &= \sum_\alpha W_1^\alpha\, T_2^\alpha \\
A_\rho &= \sum_\alpha W_2^\alpha\, T_2^\alpha \\
-A_\phi &= \sum_\alpha X_2^\alpha\, T_1^\alpha + \sum_\gamma \sigma_2^\gamma\, T_3^\gamma
\end{aligned} \tag{2.1}$$

where the summation over $\alpha(\gamma)$ runs over positive (simple) roots. The generators T_i^α (i=1,2,3) span the regular SU(2) sub-algebra's associated with the positive roots.

3. Reduced system

Before we give the explicit form of the self-duality condition (Bogomolny-equations) we introduce some convenient variables:

$$y = \rho + iz \qquad \bar{y} = \rho - iz \tag{3.1}$$

and define

$$A_y = \tfrac{1}{2}(A_\rho - iA_z) \qquad A_{\bar{y}} = \tfrac{1}{2}(A_\rho + iA_z) \ ,$$

$$\Phi_y = -\tfrac{1}{2}(A_\phi + iA_t) \qquad \Phi_{\bar{y}} = -\tfrac{1}{2}(A_\phi - iA_t). \tag{3.2}$$

The equations now take the following form:

$$\partial_y A_{\bar{y}} - \partial_{\bar{y}} A_y - i\left[A_y, A_{\bar{y}}\right] - i\left[\Phi_y, \Phi_{\bar{y}}\right] = 0, \tag{3.3a}$$

$$\partial_y \Phi_{\bar{y}} - i\left[A_y, \Phi_{\bar{y}}\right] + \tfrac{1}{2}\left(M_y \Phi_{\bar{y}} + M_{\bar{y}} \Phi_y\right) = 0, \tag{3.3b}$$

$$\partial_{\bar{y}} \Phi_y - i\left[A_{\bar{y}}, \Phi_y\right] + \tfrac{1}{2}\left(M_y \Phi_{\bar{y}} + M_{\bar{y}} \Phi_y\right) = 0, \tag{3-3c}$$

In order to enlarge the symmetry properties of the system we have introduced the additional variables M_y and $M_{\bar{y}}$ which obey

$$\partial_y M_{\bar{y}} = \partial_{\bar{y}} M_y = - M_y M_{\bar{y}} \tag{3.4}$$

and thus are expressable in terms of a harmonic function $V(\rho,z)$ as

$$M_y = V^{-1} \partial_y V \qquad M_{\bar{y}} = V^{-1} \partial_{\bar{y}} V;$$

$$\partial_y \partial_{\bar{y}} V = 0 \tag{3.5}$$

For the choice $V \equiv \rho$ (i.e. $M_y = M_{\bar{y}} = \frac{1}{2}\rho$) eq. (3.3) reduces to the one we want to solve.

4. Nonlinear Sigma Models

To point out the equivalence of our system of equations with an interesting class of nonlinear sigma models [1] we define the variables

$$a_y = A_y + i\Phi_y \qquad , \; a_{\bar{y}} = A_{\bar{y}} + i\Phi_{\bar{y}} \tag{4.1}$$

These are "gauge potentials" which take a real value in the Lie Algebra of some non compact form G^* of the gauge group G. The algebra consists of $L_2 \oplus i(L_1 \oplus L_3)$ where $L_i = \left\{T_i^\alpha ;\ \forall\alpha\right\}$. Two of the three equations express the fact that $f_{y\bar{y}} = 0$, i.e. that a_μ is a pure gauge

$$a_\mu = -i(\partial_\mu g)g^{-1} \tag{4.2}$$

where $g \in G^*$. The remaining equation reduces to

$$\vec{\nabla}\,(V(\vec{\nabla}\mu))\mu^{-1} = 0 \tag{4.3}$$

$$\vec{\nabla} = (\partial_\rho, \partial_z) \text{ and}$$

$$\mu = g^\dagger g \tag{4.4}$$

Because of (4.4) it is evident that μ is in fact an element of the coset space G^*/K where K is the maximal compact subgroup of G^* generated by L_2. We note that the system (4.3) has an additional global G^* symmetry corresponding to multiplying g from the right.

5. Relation to (Super) Gravity

It is known for some time that for G = SU(2) the system (4.3) is equivalent to the Ernst equation of General Relativity,[4] which describes static axially symmetric solutions to pure Einstein gravity. Choosing the particular (triangular) parametrisation,

$$g = \begin{bmatrix} f^{-\frac{1}{2}} & f^{-\frac{1}{2}}\psi \\ 0 & f^{\frac{1}{2}} \end{bmatrix} \tag{5.1}$$

and defining the complex function $\varepsilon = f + i\psi$ one recovers the Ernst equation in its usual form:

$$(\mathrm{Re}\,\varepsilon)\, \vec{\nabla}(\rho\vec{\nabla}\varepsilon) - \rho(\nabla\varepsilon)^2 = 0 \tag{5.2}$$

This equation has been studied extensively in the context of classical relativity and an algebraic proliferation of solutions is possible[5].

The nonlinear sigma models (4.3) constitute an interesting generalisation of the Ernst equation and we will devise the algebraic machinery necessary to solve them in the remainder of this paper.[2][3]

There is an alternative way in which one may envisage to generalise the Ernst equation, namely, by considering pure (N=1) supergravity in higher dimensions and imposing axial symmetry in the (X_1,X_2) plane as well as independence of $(X_3,X_4,\ldots,X_n)$. It is well known that extended supergravity theories are obtained in a similar way by dimensional reduction from some appropriate n to 4 dimensions. Julia[6] has argued that imposing static axial symmetry on these models and limiting oneself to the bosonic sector of the theory one obtains sigma models, which presumably correspond to a subclass of the models (4.3). In fact for the sequence of N-extended supergravity (N=7,6,5,...,0) one obtains the sigma models corresponding to $G = E_8, E_7, E_6$, SO(10), SU(4), SU(3) ⊕ SU(2), SU(2) ⊕ U(1) (?) and SU(2) respectively.

6. K - Invariance

There is a subgroup K of the original group G which preserves the structure of the ansatz (2.1)[1] namely ($\mu = y,\bar{y}$)

$$A_\mu \to A'_\mu = \Omega A_\mu \Omega^{-1} - i\,(\partial_\mu \Omega)\Omega^{-1} \tag{6.1a}$$

$$\Phi_\mu \to \Phi'_\mu = \Omega \Phi_\mu \Omega$$

where

$$\Omega = \exp\left[i \sum_\alpha f^\alpha(\rho, z)\, T_2^\alpha\right] \in K$$

The residual gauge group $K \subset G$ is generated by L_2 and has dimension $\frac{1}{2}(d-\ell)$, $d = \dim G$, $\ell = \operatorname{rank} G$.

7. Triangularity (R-gauge).

The residual K-invariance can be used to choose a convenient parametrisation of the group element $g \in G^*$ which determines the gauge fields through eq's (4.1) and (4.2.). A theorem of group theory called the Iwasawa decomposition states that G^* factorizes into two parts

$$G^* = KT, \tag{7.1}$$

in which the Lie algebra of T consists of abelian and nilpotent parts only. For example, if G = SU(N) than $G^* = SL(N,R)$, K = SO(N) and T the group of NxN upper triangular matrices with determinant unity. Therefore, by an appropriate K - transformation one can always make a_μ triangular:

$$a_\mu = -i(\partial_\mu \tau)\tau^{-1}, \quad \tau \in T \tag{7.2}$$

8. $\sum$-Invariance

There is a rather subtle discrete symmetry group of the system (3.3) which we denote by $\sum$ - invariance [2), where

$$A_y \to \tilde{A}_y = RA_yR^{-1}, \quad A_{\bar{y}} \to \tilde{A}_{\bar{y}} = SA_{\bar{y}}S^{-1}$$

$$\Phi_y \to \tilde{\Phi}_y = S\Phi_yS^{-1} + M_y\eta, \quad M_y \to M_y \tag{8.1}$$

$$\Phi_{\bar{y}} \to \tilde{\Phi}_{\bar{y}} = R\Phi_{\bar{y}}R^{-1} + M_{\bar{y}}\eta, \quad M_{\bar{y}} \to M_{\bar{y}}$$

The transformations R form a subgroup of the Weyl group of G. S and η are uniquely determined for a given R. In the case of G = SU(N) there are N different choices of R (including the identity transformation $R_o = S_o = 1$, $\eta_o = o$). They are

$$R_1 = \begin{bmatrix} 01 & & \\ & 01 & \\ & & 01 \\ (-1)^{N-1} & & 0 \end{bmatrix} \quad , \; R_1^{-1} S_1 = \text{diag.}\ (-1,+1,\ldots,+1), \quad , \; \eta_1 = \text{diag}\ \frac{1}{N}(1,1,\ldots 1,\ 1-N);$$

$$R_2 = R_1^2 \quad , \quad R_2^{-1} S_2 = \text{diag.}\ (-1,-1,+1,\ldots,+1), \quad \eta_2 = \text{diag.}\ \frac{1}{N}(2,\ldots,2,2-N,2-N), \tag{8.2}$$

We note that though the $\sum$- transformation preserves triangularity, it changes the reality structure of the solution. In order to avoid introducing various Lie algebras at intermediate stages we keep the Lie algebras of G, G^* and K intact and complexify the coefficient functions. As will become clear shortly, the $\sum$-invariance appears naturally as a symmetry of the linear scattering problem associated with the system (3.3).

9. Γ-Invariance

The third symmetry of the system (3.3) is a direct consequence of the generalization of the problem through the introduction of the variables M_μ[2]. This so called Γ-transformation acts as follows

$$A'_\mu = A_\mu \ , \ \Phi'_y = \gamma^{\frac{1}{2}}\Phi_y \quad , \quad \Phi'_{\bar{y}} = \gamma^{-\frac{1}{2}}\Phi_{\bar{y}} \ ,$$

$$M'_y = \gamma M_y \ , \ M'_{\bar{y}} = \gamma^{-1} M_{\bar{y}}. \tag{9.1}$$

The scalar function γ is a solution of the completely integrable Riccati equation

$$d\gamma = (\gamma - 1)\left[\gamma M_y dy + M_{\bar{y}} d\bar{y}\right] \tag{9.2}$$

A one-parameter (k) solution is given by

$$\gamma = \frac{k-i(V-iZ)}{k+i(V+iZ)}, \quad \partial_{\bar{y}} (V+iZ) = 0$$

where Z is by definition the conjugate harmonic function of V. We note that the Γ-transformation destroys triangularity, and to be able to combine various Γ-and ∑-transformations one has to restore the triangularity by a suitable K-transformation. This K-transformation can be determined algebraically with the help of the linear scattering problem.

10. The linear Scattering Problem (LSP).

The G^* gauge potentials a_μ defined in (4.1) which are pure gauges (4.2) remain so after we apply a Γ-transformation :

$$a'_\mu \equiv \omega_\mu = -i\,(\partial_\mu \mathcal{R})\mathcal{R}^{-1} \qquad (10.1)$$

Equivalently we may write the linear system

$$\partial_\mu \mathcal{R} = i\omega_\mu\, \mathcal{R}. \qquad (10.2)$$

The integrability condition of this linear system is just that the gauge fields satisfy our nonlinear system (3.3) assuming γ is a solution to (9.2).

The LSP of (10.2) has a local gauge symmetry

$$\mathcal{R} \to \Lambda\mathcal{R}$$

$$\omega_\mu \to \Lambda\omega_\mu\,\Lambda^{-1} - i\,(\partial_\mu\Lambda)\Lambda^{-1}\,, \quad \Lambda \in G^* \qquad (10.3)$$

which we denote as Λ-invariance to avoid confusion with the original gauge invariance. There are two important remarks to be made. In the first place it should be noted that the K-transformation of the G gauge fields (A_μ, Φ_μ) are realized on the G^*-gauge fields ω_μ as Λ transformations. Secondly, and this is more remarkable, also the Σ-transformations correspond to Λ-transformations. For example the Σ-transformation $R = R_1$ of (8.2) corresponds to

$$\Lambda = D_1 = \begin{bmatrix} o & \omega^{1/N} & & & \\ & o & \omega^{1/N} & & \\ & & \cdot & & \\ & & & \cdot & \\ & & & \cdot & o\ \omega^{1/N} \\ (-1)^{N-1}\omega^{-1+\frac{1}{N}} & & & & o \end{bmatrix} , \tag{10.4}$$

$$\omega = \frac{1+\gamma^{\frac{1}{2}}}{1-\gamma^{+\frac{1}{2}}} .$$

11. Triangularity Restoration

We have mentioned before how a Γ-transformation on a triangular solution $A = (A_\mu, \Phi_\mu, M_\mu)$ of the non linear system (3.3) yields a new solution which is not triangular. Triangularity can however be restored by an appropriate K-transformation, which we show to obtain from the solution of the LSP.[3]

Suppose we have a solution $\{\mathcal{R}, \gamma\}$ of the LSP for the field A. We can make the Iwasawa decomposition of $\mathcal{R}$ (being an element of G^*):

$$\mathcal{R} = \Omega^{-1}\tau \quad \Omega \in K \quad \tau \in T \tag{11.1}$$

Substituting this back into the LSP (10.2) we obtain

$$\Omega\, \omega_\mu \Omega^{-1} - i(\partial_\mu \Omega)\Omega^{-1} = -i(\partial_\mu \tau)\tau^{-1} \tag{11.2}$$

which states that Ω is exactly the gauge function that brings $\Gamma(\gamma)$ A into the triangular gauge appearing at the right hand side. Thus by applying the K-transformation Ω on a solution ΓA we get the triangular solution $K(\Omega)\Gamma(\gamma)A$, which we denote symbolically as $I(\Omega,\gamma)A$. Extracting $\Omega(\Omega^{-1})$ from a given $\mathcal{R}$ is a purely algebraic procedure. For the case of G=SU(N) i.e. $\mathcal{R} \in SL(N,R)$ the orthonormal column vectors of Ω^{-1} are obtained from the column vectors of $\mathcal{R}$ by means of the Schmidt orthonormalisation procedure.

In the previous section we noted that the linear system for $\sum A$ is Λ-gauge equivalent to the linear system for A. Therefore, the LSP for $\sum A$ has a solution

$$\left\{\mathcal{R}' = D\mathcal{R} = D\Omega^{-1}\tau \;,\gamma\right\} \tag{11.3}$$

Again we can decompose $\mathcal{R}'$ into its K and T factors which we denote by

$$\mathcal{R}' = (\Omega)_\Delta^{-1}\tau' \tag{11.4}$$

Observe that $(\Omega)_\Delta$ is determined algebraically from D and Ω in eq. (11.4), and does not depend on τ. If we are to apply a Γ-transformation on the solution $\sum A$ yielding the non-triangular solution $\Gamma(\gamma)\sum A$, than triangularity is restored by applying the K-transformation $(\Omega)_\Delta$ on $\Gamma(\gamma)\sum A$.

12. Algebraic Proliferation

We have exhibited all technical ingredients necessary to generate a hierarchy of solutions by algebraic manipulation.[3)] The basic pattern consists of two steps.

As an initial condition we assume a zero-th generation ("seed") solution $\overset{o}{A} = (\overset{o}{A}_\mu, \overset{o}{\Phi}_\mu, \overset{o}{M}_\mu)$ to the original equation (3.3) as well as solutions to the LSP (10.2) for $\overset{o}{A}$, in particular the K-part of $\mathcal{R}$ (the triangularity restoring functions) Ω and the corresponding γ's:

$$\left(\overset{o}{\Omega}_1, \overset{o}{\gamma}_1\right), \ldots, \left(\overset{o}{\Omega}_m, \overset{o}{\gamma}_m\right), \ldots \qquad (12.1)$$

The index denotes a particular choice of integration constants, i.e. the set (12.1) may be infinite.

The seed solution may be a "trivial" vacuum solution for which the LSP can be solved casily. For the monopole problem the vacuum is typically characterized by the ℓ(= rank G) parameters which characterize the vacuum expectation value of the Higgs field in the adjoint representation.

The first step is to apply the I = KΓ transformation to $\overset{o}{A}$ by picking any pair $(\overset{o}{\Omega}, \overset{o}{\gamma})$ from (12.1). To be definite let us construct $I(\overset{o}{\Omega}_1, \overset{o}{\gamma}_1)\,\overset{o}{A}$. The γ-equation (9.2) for $I(\overset{o}{\Omega}_1, \overset{o}{\gamma}_1)\overset{o}{A}$ has solutions $\gamma = \overset{o}{\gamma}_m / \overset{o}{\gamma}_1$ for all m. Furthermore, because of the group property of K we have that $\overset{o}{\Omega}_m \overset{o}{\Omega}_1^{-1}$ is the triangularity restoring function for the solution $\Gamma(\overset{o}{\gamma}_m/\overset{o}{\gamma}_1)\; I\; (\overset{o}{\Omega}_1, \overset{o}{\gamma}_1)\overset{o}{A}$. Hence we have a solution $I(\overset{o}{\Omega}_1 \overset{o}{\gamma}_1)\overset{o}{A}$ and its triangularity restoring functions with corresponding γ's:

$$(\overset{o}{\Omega}_m \overset{o}{\Omega}_1^{-1}, \overset{o}{\gamma}_m / \overset{o}{\gamma}_1) \quad , \quad \forall_m \tag{12.2}$$

A similar situation as in (12.1).

As a second step we apply a $\sum$-transformation on $I(\overset{o}{\Omega}_1,\overset{o}{\gamma}_1)\overset{o}{A}$ yielding $\sum I(\overset{o}{\Omega}_1,\overset{o}{\gamma}_1)\overset{o}{A}$, which we call a first generation solution:

$$\overset{1}{A} \equiv \sum I(\overset{o}{\Omega}_1,\overset{o}{\gamma}_1)\overset{o}{A} \tag{12.3}$$

As explained in the previous section (11.3-4) the triangularity restoring functions with corresponding γ's for $\overset{1}{A}$ are given by

$$(\overset{1}{\Omega}_m \equiv (\overset{o}{\Omega}_m \overset{o}{\Omega}_1^{-1})_\Delta, \overset{1}{\gamma}_m \equiv \overset{o}{\gamma}_m / \overset{o}{\gamma}_1), \ \forall m \tag{12.4}$$

We are back to the same situation as in the beginning. We can repeat the sequence of both steps ad libitum. The higher the generation the more parameters are included in the solution.

In order that the obtained solutions do indeed satisfy the original equation, one has to adjust the parameters in γ such that indeed $M_y = M_{\bar{y}} = 1/2\rho$ for the final (and initial) solution. Also the reality condition has to be imposed. However, for each group there are characteristic "block" transformations $H = (\sum I)^n$ where for certain n both conditions are satisfied automatically.[7] As a matter of fact the $\sum$-transformation changes the asymptotic behaviour of the fields corresponding to the introduction of a (typically Dirac type) magnetic charge of $\frac{eg}{4\pi} = \eta$ (see eq. (8.1)). One does therefore expect that in general several $\sum$-transformations have to be applied in order to have a magnetic change which is compatible with a regular solution.

Conclusions

We have discussed an algebraic scheme for the generation of axially symmetric BPS monopoles in arbitrary gauge theories. The method is ready for use. It has the virtue that the transformations are applied to the gauge potentials directly. We have not discussed the following issues which are presently under unvestigation and/or will be published seperatedly.

1) The construction of explicit solutions

2) The details of the full infinite dimensional symmetry groups, which allow us to generate new solutions. They are generated by certain Kac-Moody algebra's characterized by the extended Dynkin diagrams of the group G (G^*). The Σ-transformations correspond to the discrete symmetries of the rootlattices of these Kac- Moody algebra's.

3) The generalisation of this method to higher dimensions so that the limitations to axial symmetry can be dropped.

References

1. F.A. Bais and R. Sasaki, Nucl. Phys. B. (to be published).
2. F.A. Bais and R. Sasaki, "Solution generating techniques for static axially symmetric self-dual gauge field equations", University of Utrecht preprint.
3. F.A. Bais and R. Sasaki, "Algebraic construction of static axially symmetric self-dual gauge fields for an arbitrary group", University of Utrecht preprint.
4. F.J. Ernst, Phys. Rev. 167 (1968) 1175.

   L. Witten, Phys. Rev. D19 (1979) 718.

   P. Forgacs, Z. Horvath and L. Palla, Phys. Rev. Lett 45 (1980) 505.
5. D. Maison, Phys.Rev.Lett. 41 (1978) 521, B.K. Harrison, Phys. Rev. Lett. 41 (1978) 1179.

   G. Neugebauer, I. Phys. A12 (1979) L67.
6. B. Julia, "Super space and Supergravity" (S.W. Hawking and M.Rocek, editors) Cambridge University Press, Cambridge (1981); "Infinite Lie Algebra's in Physics", Preprint LPTENS 81/44.
7. F.A. Bais and R. Sasaki, (in preparation).

HARRISON-NEUGEBAUER TYPE TRANSFORMATIONS FOR INSTANTONS: MULTICHARGED MONOPOLES AS LIMITS

A. Chakrabarti

Centre de Physique Théorique de l'Ecole Polytechnique,
Plateau de Palaiseau, 91128 Palaiseau, Cedex, France.

I generalise the formalism of FHP [1] for <u>axially symmetric</u> monopoles to the case of finite action instantons [2]. The monopoles are then obtained as a simple limit of our instanton sequences. I give here only some indications concerning the nature and scope of the results. Details are in [2]. Our technique [3] is interesting for studying instantons and also for the impact it has on the on the study of monopoles. It selects out particular classes of instantons with remarkable properties worth exploiting. For the simplest sequence ("1-chain", with charge -1 PS monopole as limit) we demonstrated this by obtaining among other things, explicit corrections to the "dilute gas approximations" for <u>arbitrary index</u> [3(b)]. We obtained as by products, immediately and effortlessly, Green's functions for a PS background [3(a)]. In [2] I generalise our construction to instanton sequences having axi-symmetric, multicharged monopoles as static limits. In the process I, presumably, obtain explicit instanton sequences in successive Atiyah-Ward classes. One starts as bollows.

Take flat Euclidean space, $ds^2 = dt^2 + dr^2 + r^2(d\theta^2 + \sin^2\theta\, d\varphi^2)$. Define $(t+ir) = \tanh[(\tau+i\rho)/2]$ when for $-\infty < t < \infty$, $0 \leq r < \infty$; $-\pi \leq \tau \leq \pi$, $0 \leq \rho < \infty$. Construct selfdual, finite action solutions depending on (ρ, θ) only. My ansatz is ($f = f(\rho,\theta)$ etc), $A_\tau = f^{-1}\left(\partial_\rho f \frac{\sigma_3}{2} + \partial_\rho \psi \frac{\sigma_1}{2}\right)$, $A_\rho = f^{-1}\partial_\rho \psi \frac{\sigma_2}{2}$, $A_\theta = f^{-1}\partial_\theta \psi \frac{\sigma_2}{2}$, $(\sin\theta)^{-1} A_\varphi = f^{-1}\left(\partial_\theta f \frac{\sigma_3}{2} + \partial_\theta \psi \frac{\sigma_1}{2}\right)$. Selfuality constraints reduce to the <u>modified</u> Ernest equation ($\epsilon = f + i\psi$, $\bar{\epsilon} = f - i\psi$)

$$(\epsilon + \bar{\epsilon})\left[\partial_\rho^2 \epsilon + (\sinh\rho)^{-2}(\partial_\theta^2 \epsilon + \cot\theta\, \partial_\theta \epsilon)\right] = (\partial_\rho \epsilon)^2 + (\sinh\rho)^{-2}(\partial_\theta \epsilon)^2 .$$

I now construct suitably <u>modified Harrison's (H) and Neugebauer (I) transformations</u> in terms of $M_1 = (2f)^{-1}\partial_+\epsilon$, $M_2 = (2f)^{-1}\partial_+\bar{\epsilon}$, $N_1 = (2f)^{-1}\partial_-\bar{\epsilon}$, $N_2 = (2f)^{-1}\partial_-\epsilon$ $\left(2\partial_\pm = \partial_\rho \pm (\sinh\rho)^{-1}\partial_\theta\right)$

The complete formmalism is in [2]. Here I quote some results. I start from the vacuum $f = e^{\alpha\rho}, \Psi = 0$. Applying a suitable B=IH, I get the "1-chain" mentioned before with $f = (\alpha \sinh\rho)^{-1} \sinh\alpha\rho \sin\theta , \; \Psi = \cos\theta$. Regularity at $\rho \to \infty$ restricts α (=2,3,...). The index is P=(α-1). In the 't Hooft picture this corresponds [3(a), (b)] to the generating function $\Lambda = 1 + \sum_{k=1}^{\alpha-1} \lambda_k^2 \left[(t-c_k)^2 + r^2 \right]$ where $\lambda_k = \left[\cos\left(\frac{k\pi}{\alpha}\right) \right]^{-1}, \; c_k = \cot \frac{k\pi}{\alpha}$. Setting $\tau = t'/\alpha$ and $\rho = r'/\alpha$ one obtains the PS monopole with $f' = (\sin\theta \sinh r')/r' , \; \Psi' = \cos\theta$, a major simplification compared to the corresponding FHP result. Iterating n times our transformations with suitably chosen parameters one obtains the "n-chain" with P = n(α-n), α=n+1, n+2, ... The monopole limit is obtained as before with magnetic charge $= \lim_{\alpha \to \infty} P/\alpha = n$. Various possible further developments are indicated in [2] , such as the construction of instanton sequences having as limits (4n-1) parameter separable monopole solutions of charge n.

References : 1. P. Forgacs, Z. Horvath and L. Palla. See Horvath's talk in this meeting and the sources quoted there.

2. A. Chakrabarti, preprint (Dec. 81) of same title as here.

3.(a) H. Boutaleb-Joutei, A. Chakrabarti and A. Comtet, Phys. Rev. D23, 1781(1981).

(b) A. Chakrabarti and A. Comtet, Phys. Rev. D (to appear soon).

(c) A. Chakrabarti, A. Comtet and F. Koukiou, preprint (Sept.81).

In [2] I use many results found in (c) in the context of complex, selfdual solutions. In (a, b, c) references can be found to further applications of our technique to (d) merons, (e) nonselfdual complex solutions and (f) SU(3) instantons.

INTERNAL STRUCTURE OF THE NON-ABELIAN MONOPOLES

Y.M. Cho

Lab. de Physique Theorique et Hautes Energies, 4 place Jussieu, 75230 Paris, Cedex 05, France.

So far the discussions on the non-Abelian monopoles in the literature have been by and large on their topological nature or on the corresponding classical (finite energy) solutions. Now I wish to discuss about an overall relative symmetry [1] which exists among the monopoles themselves. As a motivation let us recall that there exist, for instance, six monopoles which have the unit strength $4\pi/g$ in SU(3). One may then ask oneself whether there is a symmetry among these, and if so, what is it precisely, and how the monopoles transform among themselves under the symmetry. The issue cannot be avoided if one wishes to represent the non-Abelian monopoles with field operators in one's theory, and could also become relevant for the monopoles in grand unfieid theories. Now I will argue that, first of all, there is indeed a symmetry called <u>the color reflection</u> [2], a finite subgroup of the original gauge symmetry, among the monopoles of a given magnetic strength. Secondly, the monopoles form linear, sometimes disjoint, multiplets [1] under the color reflection.

The existence of the symmetry may have a deep physical implication in a theory like QCD where the gauge symmetry is supposed to be unbroken, because under this circumstance not all the monopoles but <u>only the reflection invariant combination of them may be accepted in the physical sector</u>. Actually this is precisely what one may need if the confinement in QCD is to be achieved by the condensation of the monopoles. This is so since the dual Meissner effect [2,3] will guarantee only the confinement of the colored flux, but not the color itself (e.g., for $q\bar{q}$ it will admit all three $r\bar{r}$, $b\bar{b}$ and $y\bar{y}$ (r, b, y = red, blue, yellow), none of them color singlet). To complete the color confinement one clearly needs the color reflection invariance.

At first thought the existence of the symmetry may appear difficult to understand because the magnetic charge is defined as a homotopic charge of a certain field configuration which may not be changed by a regular gauge transformation. If so, one may conclude that it must be impossible to transform one monopole to another of a different magnetic charge. However, this objection is only partly valid, due to the fact that the homotopy can determine the magnetic charge uniquely <u>only up to</u> the Weyl degrees of freedom [2,4].

To prove our claim let us first recall that the non-Abelian magnetic charge may best be defined by imposing <u>the magnetic symmetry,</u> [5,6] or by introducing an internal Killing vector $\hat{m}$ which has Cartan's subgroup H as its little group, into the potential. The magnetic symmetry will restrict the holonomy group of the potential to be precisely the Cartan's subgroup, and make the dual structure of the restricted potential explicit. Then the magnetic charge may be defined by the second homotopy of the coset space G/H fixed by the Killing vector $\hat{m}$. More importantly all the point-like monopoles may explicitly be written in terms of the Killing vector [5,7]. Once this is understood, the Weyl degrees of the non-Abelian magnetic charges can easily be derived from the reflection degrees of the Killing vector, and the existence of the reflection symmetry may be demonstrated explicitly [1].

As an illustration one may easily show that the 2-dimensional monopole-antimonopole pair of SU(2) forms a regular representation Γ_R which is reducible to $\Gamma_1 \oplus \Gamma_2$. Consequently only the Γ_1 which forms the invariant subspace may be accepted in the physical sector. As for SU(3) the monopoles with $|Q_m| = 4\pi/g$ form a 6-dimensional regular representation $\Gamma_R = \Gamma_1 \oplus \Gamma_2 \oplus 2\,\Gamma_3$, and again only the invariant subspace Γ_1 may be admitted to the physical sector. However, the monopoles with $|Q_m| = 4\pi\sqrt{3}/g$ form two disjoint triplets $\Gamma_t = \Gamma_1 \oplus \Gamma_3$ so that out of the six monopoles two will now appear in the physical sector, which may be related to each other by the charge conjugation. This kind of intriguing complexity is a general feature of more complicated groups.

REFERENCES

1 - Y.M. Cho, LPTHE Report 81/12, to be published.

2 - Y.M. Cho, Phys. Rev. Lett. 46, 302 (1981) ; Phys. Rev. D 23, 2415 (1981).

3 - S. Mandelstam, Talk at this Conference. Compare this with Ref. 2.

4 - P. Goddard, J. Nuyts and D. Olive, Nucl. Phys. B 125, 1 (1977).

5 - Y.M. Cho, Phys. Rev. Lett. 44, 1115 (1980) ; Phys. Rev. D 21, 1080 (1980).

6 - It must be emphasized that the non-Abelian magnetic charge is -and should be-defined without reference to any Higgs field. It is unfortunate to remark that so far the role of the Higgs field has been overemphasized in defining the magnetic charge, and as a result there has been quite a misunderstanding and confusion on this matter in the literature, and even at this conference. The Higgs field becomes necessary only if one wants to have a finite energy solution.

7 - In this approach the topological nature of the Wu-Yang monopole becomes evident in terms of the Killing vector $\hat{m}$. Compare our approach with Ref. 3.

STIEFEL-SKYRME-HIGGS MODELS, THEIR CLASSICAL STATIC SOLUTIONS AND YANG-MILLS-HIGGS MONOPOLES

V.K.DOBREV (INRNE, Sofia, BULGARIA)

A new series of models is introduced by adding Higgs fields to the earlier proposed euclidean four-dimensional Skyrme-like models with Yang-Mills composite fields constructed from Stiefel manifold fields:

$$S = -\frac{1}{2}\int d^4x \, tr\, \tilde{F}_{\mu\nu}\tilde{F}_{\mu\nu} - \int d^4x \, tr(\nabla_\mu \varphi)(\nabla_\mu \varphi) + \lambda \int d^4x \, V(\varphi), \tag{1a}$$

$$\tilde{F}_{\mu\nu} \equiv \partial_\mu \tilde{A}_\nu - \partial_\nu \tilde{A}_\mu + [\tilde{A}_\mu, \tilde{A}_\nu]; \quad \tilde{A}_\mu \equiv \phi^+ \partial_\mu \phi; \quad \nabla_\mu \varphi \equiv \partial_\mu \varphi + [\tilde{A}_\mu, \varphi]; \tag{1b}$$

where the elements of the Stiefel manifold $\mathbb{F}V_{Nn}$ $(N > n)$ are described by $N \times n$ matrices ϕ with real ($\mathbb{F}=\mathbb{R}$), complex ($\mathbb{F}=\mathbb{C}$) or quaternionic ($\mathbb{F}=\mathbb{H}$) entries, obeying $\phi^+\phi = 1_n$. The action (1a) is H_n(=O(n), U(n),Sp(n)) - gauge invarint (for $\mathbb{F} = \mathbb{R}, \mathbb{C}, \mathbb{H}$,resp.) under $\phi(x) \mapsto \phi(x)h(x)$, $\varphi(x) \mapsto h(x)^+\varphi(x)h(x)$, $h(x) \in H_n$.(That restricts the possible choices of V.) The equations of motion for (1a) are:

$$(D_\nu \phi)(\nabla_\mu \tilde{F}_{\mu\nu} + [\nabla_\nu \varphi, \varphi]) = 0, \quad D_\nu \phi \equiv \partial_\nu \phi - \phi \tilde{A}_\nu, \tag{2a}$$

$$\nabla_\nu \nabla_\nu \varphi + \frac{\lambda}{2}\frac{\partial V}{\partial \varphi} = 0. \tag{2b}$$

We start the investigation of the models (1a) with their classical <u>static</u> versions, i.e. with fields independent of x_4. For static fields the action functional (1a) is replaced by the energy functional:

$$E = -\frac{1}{2}\int d^3x \, tr\, \tilde{F}_{kl}\tilde{F}_{kl} - \int d^3x \, tr(\nabla_k \varphi)(\nabla_k \varphi) + \lambda \int d^3x \, V(\varphi). \tag{3}$$

We introduce the composite "magnetic" field $\tilde{B}_k = \frac{1}{2}\varepsilon_{klm}\tilde{F}_{lm}$ and the basis X_a of the Lie algebra h_n with $tr(X_a X_b) = -\frac{1}{2}\delta_{ab}, \tilde{B}_k = X_a \tilde{B}^a_k$, etc. Then in analogy with the noncomposite YMH case we have:

$$E - \lambda \int d^3x V(\varphi) = \frac{1}{2}\int d^3x \left[\tilde{B}^a_k \tilde{B}^a_k + (\nabla_k \varphi)^a (\nabla_k \varphi)^a\right] \geq \left| \int d^3x \, \tilde{B}^a_k (\nabla_k \varphi)^a \right|. \tag{4}$$

In (4) we assume that E is finite which implies as usually

$$\tilde{A}_k(x) \xrightarrow[r\to\infty]{} h(\hat{x})^+ \partial_k h(\hat{x}), \quad r \equiv \sqrt{x_1^2 + x_2^2 + x_3^2}, \quad \hat{x} \equiv x/r, \quad h(\hat{x}) \in H_n, \tag{5a}$$

$$\varphi^a(x) \xrightarrow[r\to\infty]{} -C\hat{\varphi}^a(x), \quad C \in \mathbb{R}, \quad \hat{\varphi}^a(\hat{x})\hat{\varphi}^a(\hat{x}) = 1. \tag{5b}$$

A simple way to realize (5a) for our models is to require

$$\phi_{A\alpha} \xrightarrow[r\to\infty]{} 0 \quad , \; A=n+1,\ldots,N; \quad \alpha=1,\ldots,n \,. \tag{6}$$

Following the analogy with the monopole case we conclude that in the "Prasad-Sommerfield" limit $\lambda\to 0$ the energy of a configuration fulfilling (5) is minimal provided the "Bogomol'ny" equation

$$\tilde{B}_k = \pm \nabla_k \varphi \tag{7}$$

is satisfied. Combining (7) with the Bianchi identities

$$\varepsilon_{jk\ell}\nabla_\ell \nabla_k \varphi + [\tilde{B}_j, \varphi] \equiv 0 \,, \qquad \nabla_k \tilde{B}_k \equiv 0 \tag{8}$$

we can see that every solution of (7) satisfies not only the static version of (2a,b) with $\lambda = 0$, but also the more restrictive static noncomposite Yang-Mills-Higgs equation

$$\nabla_j \tilde{F}_{jk} + [\nabla_k\varphi, \varphi] = 0. \tag{9}$$

An important conclusion now follows that for fixed gauge group H_n every minimal energy solution of (7) with <u>arbitrary N</u> provides a minimal energy solution of the YMH model with the same gauge group H_n. One may speculate that this arbitrariness could be appropriate to accommodate the multimonopole solutions. This approach seems a possible adaptation of the ADHM multi-instanton construction to the monopole case. We shall illustrate our program by reproducing the BPS monopole. For that it is enough to consider HV_{21} and take

$$\phi(x) = \begin{pmatrix} \hat{x}_k q_k \left(1 - Cr/\sinh C(r+r_0)\right)^{1/2} \\ \left(Cr/\sinh C(r+r_0)\right)^{1/2} \end{pmatrix} , \; \varphi^a(x) = \hat{x}_a \left(\frac{1}{r} - C\coth C(r+r_0)\right), \tag{10}$$

where C is the constant in (5b) and r_0 is an arbitrary positive constant; q_k are the quaternionic units.

This result was published in a Trieste preprint, IC/81/88. After that we learned that several authors found explicit multimonopole solutions using the Atiyah-Ward ansatz or Bäcklund transformations. The connection of these results to our program is not yet clear and is related to the long standing problem of connecting the Atiyah-Ward ansatz and the ADHM construction for instantons. This problem is not solved also for Nahm's (Hilbert space) adaptation of the ADHM construction to the monopole case (cf. W.Nahm, these proceedings).

SPHERICALLY SYMMETRIC MONOPOLES

N. Ganoulis

Fachbereich Physik, Universität Wuppertal, Federal Republic of Germany.

Our starting point are the "self-dual" (or Bogomolny) equations for an arbitrary gauge group G, with the additional requirement that the solutions are spherically symmetric. This problem has been studied by a number of authors[1-5] and we now know quite a lot about the solutions. A brief presentation of results is given below (without derivations). For more details see ref. (5).

We require rotational invariance only up to gauge transformations; the generator is

$$\vec{J} = -i \ \vec{r} \times \vec{\nabla} + \vec{T}$$

$\vec{T}$ satisfy angular momentum commutation relations. The problem is then reduced to one with a single variable, the radial distance r.

It is possible to show that by defining new fields $L^{\pm}$ (associated with the transverse gauge fields) and L_3 (associated with the Higgs field) the "self - dual" equation $\vec{B} = \vec{D}\Phi$ becomes:

$$\frac{d}{dr} L_3 = \frac{1}{2} \left[L^+, L^- \right]$$
$$\frac{d}{dr} L^\pm = \pm \left[L_3, L^\pm \right] \qquad (1)$$

with constraints

$$\left[T_3, L^\pm\right] = \pm L^\pm$$
$$\left[T_3, L_3\right] = 0 \qquad (2)$$

The next step is to make a definite choice for $\vec{T}$, one that corresponds to the principal SU(2) subgroup of G and is defined by the property

$$\left[T_3, E_{\pm\alpha}\right] = \pm E_{\pm\alpha} \quad \text{when } \alpha \text{ is a simple root.}$$

E_α ($E_{-\alpha}$) are raising (lowering) generators of G.
The diagonal generators are denoted H_α.

With this choice for T_3 and because of the constraints (2), the unknown fields $L^\pm$, L_3 are expanded along generators that correspond to simple roots only. After substitutions and change of variables it is possible to show(4,5) that (1) results to a system of non - linear equations

$$\frac{d^2}{dr^2}\,\Theta\alpha = \exp\,(\sum_\beta K_{\alpha\beta}\,\Theta_\beta) \qquad (3)$$

α, β run over simple roots. $K_{\alpha\beta}$ is the Cartan matrix; it has integer entries only and is known for every group G. Equation (3) is related to the Toda lattice(6), a system much studied (7,8) for its integrability and special properties.

In order to integrate (3) one uses a Lax pair constructed out of L_3, $L^\pm$. The final answer can be written in a simple form

$$e^{-\Theta\alpha} = \langle\Lambda_\alpha \mid e^{X^*}\, e^{C+QR}\, e^{X} \mid \Lambda_\alpha\rangle \qquad (4)$$

C and Q are constant diagonal matrices (i.e. can be expanded along H_α). X is lower triangular (expanded along $E_{-\alpha}$) and satisfies

$$e^{-X}\, Q\, e^{X} = Q + \sum_{\alpha\ \text{simple}} E_{-\alpha}$$

X^* is upper triangular: $X^* = (-1)^{T_3}\, X^+\, (-1)^{T_3}$

The matrix element is taken between highest eigenvectors $|\Lambda_\alpha\rangle$ of the fundamental representations of G. The vectors $|\Lambda_\alpha\rangle$ are annihilated by any raising generator.

We see that there are two lots of integration constants contained in C and Q. It turns out that Q is proportional to the Higgs field at infinity. C must be chosen so that the magnetic field is regular at the origin. This condition can be satisfied[5] with

$$e^{C} = \prod_{\alpha>0} (Q.\alpha)^{-H_\alpha}$$

Therefore the solutions (4) are functions of Q only and by inserting a complete set of states $|\mu\rangle\langle\mu|$ can be written as sums of exponents of the radius r:

$$e^{-\Theta_\alpha} = \sum_{|\mu\rangle} A_\mu(Q) \; e^{(Q.\mu)r}$$

There is a systematic way for writing the constants A_μ for a given group. Using these results one can also calculate the magnetic charge matrix g and it turns out that for the physically interesting choices of Q (the asymptotic Higgs field), the magnetic charge is "abelian" (i.e. Q and g are parallel).

References

(1) D. Wilkinson and A. Goldhaber Phys. Rev. D16, 1221 (1977)
(2) F.A. Bais and D. Wilkinson Phys. Rev. D19, 2410 (1979)
(3) A.N. Leznov and M.V. Saveliev Comm. Math. 74, III (1980)
(4) D. Olive Bad Honnef lecture, September 1980 ICTP/80'81-1
(5) N. Ganoulis, P. Goddard and D. Olive Preprint ICTP/81/82-4
(6) M. Toda Prog. Theor. Phys. (Suppl.) 45, 174 (1970)
(7) See references in M. Toda Phys. Rep. 18, I (1975)
(8) B. Konstant Adv. Math. 34, 195 (1979)

THE BOGOMOLNY INEQUALITY FOR EINSTEIN-MAXWELL THEORY

G.W. Gibbons

DAMTP, University of Cambridge, Silver Street, Cambridge CB3 9EW, UK.

In flat space the Maxwell field

$$F = \frac{Q}{r^2}\, dr \wedge dt + P \sin\theta \, d\theta \wedge d\phi \tag{1}$$

represents a singular electric monopole, strength Q, and magnetic monopole of strength P. In curved space this singularity can be hidden inside the event horizon of the Reissner-Nordstrom metric

$$ds^2 = -\left(1 - \frac{2M}{r} + \frac{Q^2+P^2}{r^2}\right)dt^2 + \left(1 - \frac{2M}{r} + \frac{Q^2+P^2}{r^2}\right)^{-1} dr^2 + r^2\left(d\theta^2 + \sin^2\theta\, d\phi^2\right) \tag{2}$$

provided

$$M \geqslant \sqrt{Q^2+P^2} \tag{3}$$

Inequality (3) is the appropriate analogue of the Bogomolny bound in Einstein-Maxwell Theory. One also has explicit multi-black hole solutions for which (3) is saturated due to Papapetrou and Majumdar. They have the form

$$ds^2 = -H^{-2}\, dt^2 + H^2\, d\underline{x}^2 \tag{4}$$

$$F = \frac{dH}{H^2} \wedge dt \tag{5}$$

or any duality rotation of it, where H is any harmonic function on $\mathbb{R}^3$. Choosing H to be 1 plus a sum of N poles gives N extreme black holes. Physically their existence is due to the balance between gravitational attraction and electrostatic repulsion - an obvious analogue of the multi-monopole situation.

In the supersymmetric version of the theory (O(2) supergravity) Q & P are <u>central charges</u> and (3) is a consequence of the algebra. Just as Witten has given a rigorous proof of the positive mass theorem motivated by simple supergravity C. Hull and I have recently given proof of (3) under suitable assumptions. The basic idear is to replace the covariant derivative in Witten's proof by the supercovariant derivative $\left(\hat{\nabla}_\alpha \epsilon = \nabla_\alpha \epsilon - \frac{1}{4} F_{\sigma\beta} \gamma^\sigma \gamma^\beta \gamma_\alpha \epsilon \right.$ where ∇_α is the standard covariant derivative) which occurs in the supersymmetry transformations:

$$\delta \psi_\mu = \hat{\nabla}_\mu \epsilon \qquad \delta A_\mu = \bar{\epsilon} \psi_\mu$$

$$\delta e^c_\mu = \sqrt{2}\, \bar{\epsilon} \gamma^c \psi_\mu$$

of a background purely bosonic field where $\psi_\mu = 0$. One finds that (3) can only be saturated if there exist supercovariantly constant spinors and this happens essentially only if the metric has Papapetrou-Majumdar form. In this case the background posesses a superinvariance.

These metrics admit 4 fermion zero modes which are solutions of the Dirac equation which tend to a constant at infinity. Two of these are supercovariantly constant. The result is that extreme black holes fit into 4-fold multiplets rather than the 16-fold multiplets one might have expected. This is because of the central charges and is again familiar from monopole theory.

In the quantum theory this extreme case is also picked out because, if there are no charged fields present, the extreme holes have zero temperature and are thus stable against thermal evaporation via the Hawking effect. They thus qualify, in some sense, as the solitons of the Einstein-Maxwell theory. In fact zero temperature is essential if supersymmetry is not to be broken since it is broken at finite temperature.

One can identify a candidate effective field theory for the solitons - it is the (0,½) hypermultiplet which was coupled to N=2 supergravity by Zachos with the consequent gauging of the central charge. This theory is clearly not self dual. The situation for the N = 8 theory is less clear.

References

1. G.W. Gibbons in Heisenberg Memorial Symposium ed. H.P. Durr and P. Breitenlohner to be published by Springe in the Lecture notes in Physics Series.

2. G.W. Gibbons & C.M. Hull DAMTP preprint to appear in Phys Letts B.

Lax pairs for Bogomolny equations concerning vortices and monopoles

D. Maison
Max-Planck-Institut für Physik und Astrophysik
Munich, Fed.Rep.Germany

Bogomolny equations are extremal cases of inequalities [1] for the action of solutions of certain elliptic field theories in terms of a topological charge. They include the self-duality equations of 4-dim. Yang-Mills-theory and the 2-dim. CP^n-model. Important consequences of the 1st-oder Bogomolny equations are:

i) they entail the 2nd order Euler-Lagrange equations,
ii) the action of the solution is proportional to the topological charge,
iii) the stress-tensor vanishes point-wise,

and hence they describe the interaction free superposition of localized topological charge. A unifying viewpoint for the various instances of Bogomolny equations is to consider them as holomorphy conditions in superspace with regard to a supersymmetric extension of these theories [2].

Lax pairs for Bogomolny equations are desirable because they may be used to

i) find infinitely many conservation laws
ii) prove complete integrability
iii) construct explicit solutions via the ISM (Inverse Scattering Method)
iv) quantize the theory via the Q(antum) ISM.

For the BPS-monopole a Lax pair can be found using the fact that the Bogomolny equation is nothing but the self-duality equation for a SU(2)-Y.M. vector potential independent of x^4 whose 4th component is the Higgs field. The s.d. equation can be represented by the condition $F_{\lambda\bar{z}+y,\lambda\bar{y}-z} = 0$ for arbitrary $\lambda \in \mathbb{C}$ where $y = x^1+ix^2$ and $z = x^3+ix^4$.

This is the integrability condition for the linear system [3] $(D_{\bar z} - \lambda D_y)\psi = 0, (D_{\bar y} + \lambda D_z)\psi = 0$ for some $SL(2,\mathbb{C})$ spinor ψ. From this system it is possible to derive the Lax pairs used by Forgacs, Horvath and Palla [4] for the construction of separated BPS-monopoles and by Belinsky and Zakharov [5] for the Ernst equation resulting in the case of axial symmetry.

For the vortices of the Abelian Higgs model the Bogomolny equations are $(\partial_- - iA_-)\varphi = 0$ and $F_{12} = \frac{1}{2}(1-|\varphi|^2)$ with φ the Higgs-field, $A_{\mp} = A_1 \pm iA_2$ and $x_\pm = \frac{1}{2}(x_1 \pm ix_2)$. From $A_- = -i\partial_- \ln\varphi$ one gets the equ. $\partial_+\partial_- u = e^u - 1$ for $u = \ln|\varphi|^2$. A Lax pair for this equation is obtained by putting $\mathcal{D}_\pm = \partial_\pm \mp \frac{1}{2}(\partial_\pm u)\cdot h - e^{u/2} e_\pm + f_\pm$, where h, $e_\pm$ and $f_\pm$ are generating elements of the abstract Lie algebra with the defining relations $[h,e_\pm] = \pm e_\pm$, $[h,f_\pm] = 0$, $[e_+, e_-] = [f_-,f_+] = h$, $[e_+,f_-] = [e_-,f_+] = 0$. From these one finds $[\mathcal{D}_+,\mathcal{D}_-] = [\partial_+\partial_- u - e^u + 1]h$.

A Fock-like representation of the Lie algebra can be constructed chosing a vacuum vector v obeying $e_- v = f_- v = hv - v = 0$. All the non-trivial representations of the algebra turn out to be infinite dimensional. The algebra resembles the Kac-Moody algebra $A_1^{\ 1}$, but it is not clear if it allows also for a representation of the form $x_i \otimes \lambda$, where λ is a 'spectral parameter' and the x_i are finite matrices.

References:

[1] E.B. Bogomolny, Sov. J. Nucl. Phys. 24 (1976) 449.

[2] D. Maison, in preparation.

[3] A.A. Belavin, V.E. Zakharov, Phys.Lett. 73B (1978) 53.

[4] P. Forgâcs, Z. Horvath, Z. Palla, contribution to this conference.

[5] V.A. Belinski, V.E. Zakharov, Sov. Phys. JETP 50 (1979) 1.

The full text of the talk presented at the Conference is available as a preprint MPI-PAE/PTh 3/82.

AXIALLY SYMMETRIC MULTIMONOPOLE SOLUTIONS

Paolo Rossi

Scuola Normale Superiore, Pisa, Italy.

The construction of multimonopole solutions can be explicitly accomplished in the SU(2) gauge theory with vanishing Higgs potential when the monopole charges are superimposed in a single location. This configuration enjoys axial and mirror symmetry.

We employ a complexified version of the theory, with complex coordinates

$$\sqrt{2}p = x_1 + ix_2 \qquad \sqrt{2}\bar{p} = x_1 - ix_2$$

$$\sqrt{2}q = x_3 - ix_4 \qquad \sqrt{2}\bar{q} = x_3 + ix_4$$

Yang's R gauge:

$$A_u = \begin{pmatrix} -\frac{\varphi_u}{2\varphi} & 0 \\ \frac{\rho_u}{\varphi} & \frac{\varphi_u}{2\varphi} \end{pmatrix} \qquad A_{\bar{u}} = \begin{pmatrix} \frac{\varphi_{\bar{u}}}{2\varphi} & -\frac{\bar{\rho}_{\bar{u}}}{\varphi} \\ 0 & -\frac{\varphi_{\bar{u}}}{2\varphi} \end{pmatrix}$$

and Corrigan et al.'s construction of self-dual solutions

$$_n\varphi = \frac{H^{n\times n}_{k-\ell}}{H^{n-1\times n-1}_{k-\ell}} \qquad _n\rho = (-1)^n \frac{H^{n\times n}_{k-\ell-1}}{H^{n-1\times n-1}_{k-\ell}} \qquad _n\bar{\rho} = (-1)^{n+1} \frac{H^{n\times n}_{k-\ell+1}}{H^{n-1\times n-1}_{k-\ell}}$$

where $H^{j\times j}_{k-\ell+m} = \det \begin{vmatrix} \tilde{\Delta}_m & \cdots & \tilde{\Delta}_{n-j+1} \\ \vdots & & \vdots \\ \tilde{\Delta}_{m+j-1} & \cdots & \tilde{\Delta}_m \end{vmatrix}$ is a Toeplitz determinant and

$$\tilde{\Delta}_\ell = \frac{1}{2\pi i} \oint \Omega^{(n)}(\omega_1,\omega_2)\ \zeta^\ell \frac{d\zeta}{\zeta}$$

where $\sqrt{2}\omega_1 = (\bar{q}-p\zeta)$ $\sqrt{2}\omega_2 = -(q+\bar{p}\zeta^{-1})$.

We show that the staticity condition requires

$$\Omega^{(n)}(\omega_1,\omega_2) = \exp(\omega_1+\omega_2)\ F(\omega_1-\omega_2,\zeta)$$

axial symmetry requires no explicit dependence of F on ζ and finally the reality condition dictates

$$F(\omega) = \frac{e^\omega+(-1)^n e^{-\omega}}{q_0 \prod_{k=1}^n (\omega-q_k)}$$

where $\omega = \omega_1-\omega_2$ and the denominator must be a real polynomial.

In order to insure regularity the function $F(\omega)$ must necessarily be entire, which restricts q_k to the roots of the numerator.

Further analysis shows that for the n-monopole configuration one has to choose

$$q_k = z_k \equiv [(\frac{n+1}{2})-k]\ i\pi$$

the following fundamental relation holds for all $|\ell| < n$

$$\tilde{\Delta}_\ell = \Delta_\ell$$

$$\Delta_\ell = \exp ix_k \sum_{k=1}^n \alpha_k \left\{ \left(\frac{x_3-z_k-r_k}{\sqrt{2p}}\right)^\ell \frac{\exp r_k}{2\ r_k} - \left(\frac{x_3-z_k+r_n}{\sqrt{2p}}\right)^\ell \frac{\exp(-r_k)}{2r_k} \right\}$$

where $\alpha_k = \frac{(n-1)!}{(k-1)!(n-k)!}$ and $r_k^2 = x_1^2 + x_2^2 + (x_3 - z_k)^2$.

By employing the following formula for the square of the Higgs field:

$$h^2 = A_u^a A_u^n = 1 - \nabla^2 \ln H_{k-\ell}^{n\times n}$$

and by deleting all exponentially damped terms one obtains the asymptotic behavior

$$h = 1 - \sum_{k=1}^{n} \frac{1}{r_k} + O(\exp -2r_k)$$

corresponding to n monopole charges, as expected. The behavior of the Higgs field on the x_3 axis can be explicitly determined

$$h(x_1 = x_2 = 0) = \left| (\tanh x_3)^{(-1)^n} - \sum \frac{1}{x_3 - z_k} \right| .$$

Further results concern:

a) existence of real singular solutions corresponding to arbitrary locations q_k in the expression for Δ_ℓ (replacing $\tilde{\Delta}_\ell$ in H) and satisfying the relationships among determinants:

$$H_{k-\ell}^{n+m\times n+m} = \left| \frac{-2^{2n-2}\ell^{2ix_n}}{(2p\bar{p})^n} \right|^m H_{k-\ell}^{n-m\times n-m}$$

b) existence of a nonequivalent time dependent representation for the single monopole relating it to the multinstanton solutions:

$$\Omega = \sum_{n=-\infty}^{\infty} \frac{1}{2\omega_1 - 2i\pi n} \frac{1}{2\omega_2 - 2i\pi n} = \frac{\mathrm{sh}\omega}{\omega} \frac{1}{\mathrm{ch}\omega - \mathrm{ch}(\omega_1 + \omega_2)}$$

c) representation of Toeplitz determinants in terms of group integrals for the groups U(j):

$H^{j\times j}_{k-\ell} = I(\det \tilde{\Omega}^{(n)}(U))$ where ζ is replaced by the group element $U \in U(j)$ in all formulae and $I(F) = \int du\, F(U)$ is the invariant integral

d) use of this result in order to compute the large n limit of axi-symmetric monopoles by relating it to a special case of lattice QCD_2 with fermions and to show that the resulting system is translation invariant along the x_3 axis and has and infinite core where the Higgs field vanishes.

References

M.K. Prasad and P. Rossi, Phys.Rev. D24 (1981) 2182-2199

M.K. Prasad and P. Rossi, Phys.Rev. D23 (1981) 1795-1799

S.A. Brown, M.K. Prasad and P. Rossi, Phys.Rev. D24 (1981) 2217-2224 and references therein.

EXPLICIT SOLUTION OF THE CORRIGAN-GODDARD CONDITIONS FOR N MONOPOLES FOR SMALL VALUES OF THE PARAMETERS

L. O'Raifeartaigh, S. Rouhani and L.P. Singh *
Dublin Institute for Advanced Studies, Dublin 4, Ireland.

For an introduction I refer you to the articles by Professor Atiyah[(1)] and Dr. Corrigan[(1)] in this volume and here only state what we mean by the Corrigan-Goddard (C.G.) conditions. The general solution of the Bogomolny's equation as proposed by C.G. is given in terms of a transition matrix,

$$\begin{pmatrix} \dfrac{e^{i\frac{\eta}{2}K_+} + (-1)^n e^{i\frac{\eta}{2}K_-}}{H(\gamma,\zeta)} & -\zeta^n \\ \zeta^{-n} & 0 \end{pmatrix} \tag{1}$$

where

$$H(\gamma,\zeta) = \gamma^n + \sum_{k=1}^{n}\sum_{m=-k}^{k} a_k^{(m)} \zeta^m \gamma^{n-k} = \prod_{r=1}^{n} (\gamma - \omega_r)$$

$$K = K_+ - K_- = \sum_{r=1}^{n} n_r \prod_{s\neq r} \left(\frac{\gamma - \omega_s}{\omega_r - \omega_s}\right)$$

$$n_r = 0, \pm 2, \ldots, \pm(n-1) \quad \text{for n odd} \tag{2}$$

$$n_r = \pm 1, \pm 3, \ldots, \pm(n-1) \quad \text{for n even}$$

There are n(n+2) parameters in $H(\gamma, \zeta)$. The functions K+ and K− are the positive and negative parts of the Laurent expansion of K, such that each is a function of the variables $\mu = it+z+x\zeta$ and $\nu = it-z+x\zeta^{-1}$. For such an expansion to exist K must satisfy certain constraints (4), which we refer to as C.G. conditions[(1)]. Let us expand K as a power series in γ,

$$K = b_{-1} + b_0\gamma + b_1\gamma^2 + \cdots + b_{n-2}\gamma^{n-1} \tag{3}$$

then the n(n−2) C.G. conditions read,

$$\oint \frac{d\zeta}{\zeta}\, \zeta^m b_r = 0 \qquad |m| \leq r \quad , \quad r \geq 1 \tag{4}$$

In fact the C.G. conditions replace the Bogomolny equation. However although the C.G. conditions are easier to solve than the Bogomolny's equation they are

* Permanent address: Physics Department, Utkal University, Bhubenswar 751 004, India.

nevertheless very difficult to solve in general, furthermore the n(n+2) parameters in H are related to each other via the n(n-2) conditions (4) thus making it difficult to see the physical interpretation of the parameters entering via H. Here we shall consider a solution of (4) for small values of the parameters.

Let us consider small variations about the axi-symmetric limit, ($\omega_r = in_r$)

$$H(\gamma,\zeta) = H^s(\gamma) + \sum_{k,m} \delta a_k^{(m)} \zeta^m \gamma^{n-k} , \qquad (5)$$

the small variations in the parameters , $\delta a_k^{(m)}$ can be arranged in a table as in fig.1. there is one parameter to each box, a total of n(n+2) parameters but they are not all independent. For small variations the C.G. conditions are solved and give;

$$\delta a_k^{(m)} = \rho_{kj}^{-1} \, \varepsilon_j^{(m)} , \qquad 1 \le k, j, r \le n \; ; \; |m| \le k$$

$$\rho_{kj} = \sum_r \frac{(in_r)^{k+j-2}}{\sigma_r^2} , \qquad \sigma_r = \prod_{s\neq r} (n_r - n_s) \qquad (6)$$

The n(n-2) conditions (4) reduce the n(n+2) parameters in H to 4n free parameters $\varepsilon_\lambda^{(m)}$, which may be chosen as in fig. 2. (λ only takes the two end values in each column and the rest of $\varepsilon_j^{(m)}$ are zero). Clearly the C.G. conditions admit a solution to first order and all the independent parameters appear. Furthermore (6) does not mix $\varepsilon_\lambda^{(m)}$ of different m, and since the integers n_r come in equal and opposite pairs (6) breaks into two parts of negative and positive parity,

$$\delta a_{2j}^{(m)} = \rho_{2j\,\tau_+}^{-1} \, \varepsilon_+^{(m)}$$

$$\delta a_{2j+1}^{(m)} = \rho_{2j,\tau_-}^{-1} \, \varepsilon_-^{(m)} \qquad (7)$$

where $\tau_\pm$ are even and odd integers taking one value each (see fig.2.). This gives rise to only one free parameter for a fixed m and parity. (In fig.2. the independent parameter of each parity has been circled in one of the columns). Consequently the 4n independent parameters can be identified with the 1-dimensional irreducible representations of the extended rotation group around the z-axis, C_∞ x Parity.

To obtain an interpretation of the parameters one can set all $\varepsilon_j^{(m)}$ equal to zero retaining but one set $\varepsilon_\lambda^{(k)} \neq 0$ such an ansatz is clearly C_k symmetric and describes a ring of k-monopoles sitting on a plane orthogonal to the z-axis, and the rest of the monopoles should be situated on the z-axis.[3]

References.

(1) M.F. Atiyah and R. Ward Comm. Math. Phys. 55, 117 (1977)
R. Ward, Phys. Lett. 61A, 81 (1977) Comm. Math. Phys. 79, 317 (1981)
E. Corrigan and P. Goddard Comm. Math. Phys. 80, 575 (1981)
for a review see; L. O'Raifeartaigh and S. Rouhani New Developments in Mathematical Physics ed. H. Miller and L. Pittner, (Springer Verlag, Wien 1981)

(2) E. Weinberg, Phys. Rev. D20, 936 (1979)
(we have allowed the length of the Higgs field at infinity to be a variable and thus have an extra parameter)

(3) L. O'Raifeartaigh and S. Rouhani, DIAS preprint DIAS-STP-81-31.

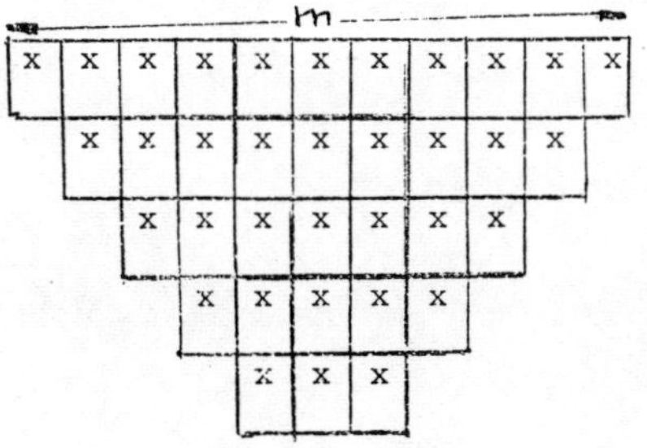

Fig.1. Diagrammatic representation of the n(n+2) parameters which occur in the definition of H in (1). Note that the range of m for each k is $|m| \leq k$.

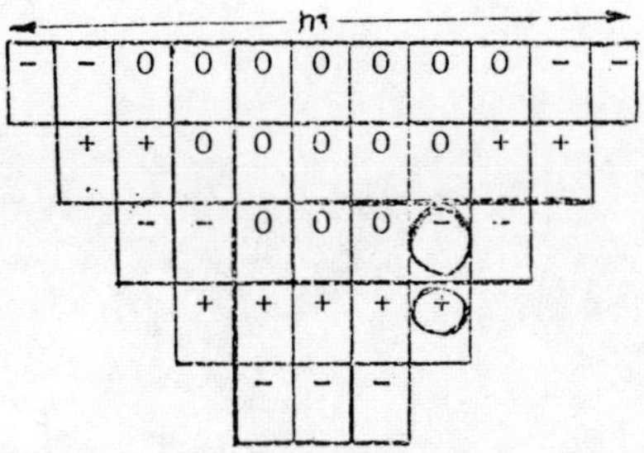

Fig.2. Diagrammatic representation of the 4n independent parameters $\varepsilon_{\pm}^{(m)}$ in (7); where $\delta Q_k^{(m)} = \rho_{kj}^{-1} \varepsilon_j^{(m)}$ and all the $\varepsilon_j^{(m)}$ are zero except the last two in each column. The ρ matrix connects only entries of the same m and the same parity. There is only one independent parameter for each parity and each m, which has been circled for one of the columns.

SU(3) MONOPOLES IN THE YANG R GAUGE

D.H. Tchrakian

St. Patrick's College, Maynooth, Co. Kildare, Ireland,
and
Dublin Institute for Advanced Studies, Ireland.

Abstract: The Bogomolny equations of the SU(3) Yang-mills-Higgs system have been reduced to a set of coupled ordinary differential equations. The SU(2) embedding solutions are recovered.

The Yang R-gauge[1] has played a very important role in the search for multimonopole solutions[2][3] of the SU(2) Yang-Mills-Higgs system subject to the Bogomolny equations. Ward's[2] method draws on the results of Corrigan Fairlie Goddard and Yates[4] which in turn is based on the Yang R-gauge, while Prasad's[3] method is itself a direct application of the R-gauge to the monopole problem.

It is therefore quite natural to enquire whether the R-gauge method can be used successfully in the search for SU(3) self-dual monopoles. We address ourselves to this question in the following, but we would like to remark that the results of such an investigation would throw light on on the problem of finding solutions to the pure Yang-Mills self-duality equations in 4-dimensions as well.

In a previous paper[5], the self-duality equations for the SU(3) Yang-Mills fields were parametrised in the R-gauge, in terms of two scalar functions ϕ_1, ϕ_2 and six complex functions $\rho_1, \rho_2, \rho_3 = \rho\phi_1$; $\bar{\rho}_1, \bar{\rho}_2, \bar{\rho}_3 = \bar{\rho}\phi_1$, of the variables $y, \bar{y} = \frac{1}{\sqrt{2}}(x_1 \pm ix_2)$; $z, \bar{z} = \frac{1}{\sqrt{2}}(x_3 \mp ix_4)$. Here we follow Prasad's method and impose a dimensional reduction procedure to obtain the self-duality equations of the static Yang-Mills-Higgs system in 3-dimensions, with one component of the connection, say A_4, identified with the Higgs field Φ. Then we try to find a one function Ansatz which gives rise to solutions that in addition to satisfying the criteria of reality and regularity, also lead to the correct asymptotic behaviour for the magnitude of the Higgs field, this latter being what decides the topological magnetic charge of the solution.

We start by assuming that all the above mentioned $\phi_a, \rho_i, \bar{\rho}_i$ functions depend on x_μ through the single function $f(x)$, which satisfies

$$f_{y\bar{y}} + f_{z\bar{z}} = 0 . \tag{1}$$

This four dimensional Laplace equation will be converted to a Helmholz equation when the explicit dependence of $f(x)$ on x_4 is determined by a dimensional reduction procedure similar to that of Prasad[3], which makes all the potentials and the Higgs field independent of the time, x_4. The self-duality equations to be solved are equations (10)-(15) of reference(5). To start with, of these equations (10) and (11) are solved directly by the following Ansatz

$$\left.\begin{array}{l} \bar{\rho}_{2\bar{y}} - \bar{\rho}\,\bar{\rho}_{1\bar{y}} = \theta(f) f_z \\ \bar{\rho}_{2\bar{z}} - \bar{\rho}\,\bar{\rho}_{1\bar{z}} = -\theta(f) f_y \end{array}\right\} \quad (2a) \qquad \left.\begin{array}{l} \rho_{2y} - \rho\,\rho_{1y} = \bar{\theta}(f) f_{\bar{z}} \\ \rho_{2z} - \rho\,\rho_{1z} = -\bar{\theta}(f) f_{\bar{y}}, \end{array}\right\} \quad (2b)$$

$$\left.\begin{array}{l} \bar{\rho}_{1\bar{y}} = \psi(f) f_z \\ \bar{\rho}_{1\bar{z}} = -\psi(f) f_y \end{array}\right\} \quad (3a) \qquad \left.\begin{array}{l} \rho_{1y} = \bar{\psi}(f) f_{\bar{z}} \\ \rho_{1z} = -\bar{\psi}(f) f_{\bar{y}}, \end{array}\right\} \quad (3b)$$

which lead, because of (1), to the constancy of ψ and $\bar{\psi}$, as well as to the relations

$$\dot{\theta} + \psi\,\dot{\bar{\rho}} = 0 \quad , \quad \dot{\bar{\theta}} + \bar{\psi}\,\dot{\rho} = 0 \quad ; \quad \dot{\theta} = \frac{d\theta}{df} \text{ etc...} \tag{4}$$

The remaining self-duality equations, (12)-(15) of ref.(5), then give

$$\frac{|\psi|^2}{\phi_1^2} + \frac{1}{2}\frac{|\theta|^2}{\phi_2^2} - \frac{1}{2}\left(\frac{\phi_1}{\phi_2}\right)^2 |\dot{\rho}|^2 + \frac{d^2}{df^2}\ln\phi_1 = 0 \tag{5}$$

$$\frac{1}{2}\frac{|\psi|^2}{\phi_1^2} + \frac{|\theta|^2}{\phi_2^2} + \frac{1}{2}\left(\frac{\phi_1}{\phi_2}\right)^2 |\dot{\rho}|^2 + \frac{d^2}{df^2}\ln\phi_2 = 0 \tag{6}$$

$$\frac{\bar{\psi}}{\phi_1^2}\,\theta + \ddot{\bar{\rho}} + \dot{\bar{\rho}}\,\frac{d}{df}\ln\left(\frac{\phi_1}{\phi_2}\right)^2 = 0 \tag{7}$$

$$\frac{\psi}{\phi_1^2}\,\bar{\theta} + \ddot{\rho} + \dot{\rho}\,\frac{d}{df}\ln\left(\frac{\phi_1}{\phi_2}\right)^2 = 0 . \tag{8}$$

Finding suitable solutions to these equations ammounts to having found a successful Ansatz to the SU(3) self-duality equations. Then it must be checked that they satisfy the correct boundary conditions, e.g. for instantons or for monopoles, and lead to non-singular fields. As we are here interested in monopole solutions, we make one further requirement. This is that the square of the magnitude of the Higgs field be of the following form

$$\|\Phi\|^2 = -\mathrm{tr}\, A_4^2 = \text{const.} - \mu \,\Box \ln f . \tag{9}$$

Then, ascribing the explicit time dependence $e^{i\lambda x_4}$ to each of the functions $\phi_1, \phi_2, \rho_1, \bar{\rho}_1, \rho_2, \bar{\rho}_2, \rho_3, \bar{\rho}_3$, and using the formulae for the potentials given in ref.(5), we compute the value of the constant in (9) to be $\frac{2}{3}\lambda^2$. Evaluating the other terms of $\|\Phi\|^2$ and identifying with (9), we find the two additional conditions

$$\frac{4}{3}\left[\left(\frac{\dot{\phi}_1}{\phi_1}\right)^2+\left(\frac{\dot{\phi}_2}{\phi_2}\right)^2-\left(\frac{\dot{\phi}_1}{\phi_1}\right)\left(\frac{\dot{\phi}_2}{\phi_2}\right)\right]+\left(\frac{\phi_1}{\phi_2}\right)^2\frac{|\dot{\theta}|^2}{|\psi|^2}=\frac{\mu}{\rho^2} \tag{10}$$

$$\frac{|\psi|^2}{\phi_1^2}+\frac{|\theta|^2}{\phi_2^2}=\frac{\mu}{\rho^2}. \tag{11}$$

From (4)(5) and (11) we deduce that $\phi_1\phi_2=\rho^{\frac{3\mu}{2}}$ which then enables us to evaluate the explicit time dependence of ρ, and hence also the coefficient of the term in the asymptotic expansion of (9).

For $\mu=1$, with either $\psi=\bar{\psi}=0$ or $\theta=\bar{\theta}=0$, we recover the two SU(2) embedding solutions. It would be very interesting to find solutions with $\mu\neq 1$.

References

(1) C.N. Yang, Phys. Rev. Letters 38,1377(1977)

(2) R.S. Ward, Commun. Math. Phys. 79, 317 (1981)

(3) M.K. Prasad, Commun. Math. Phys. 80, 137 (1981)

(4) E.F. Corrigan, D.B. Fairlie, P. Goddard and R.G. Yates, Commun. Math. Phys. 58, 223 (1978)

(5) L.P. Singh and D.H. Tchrakian, Phys. Lett. B104, 469 (1981)

Fundamental and Composite Monopoles

Erick J. Weinberg
Physics Department
Columbia University, New York, N.Y. 10027, USA.

One approach to the study of the magnetic monopoles which appear as classical solutions to spontaneously broken gauge theories is to investigate the zero-frequency small oscillation modes. This allows one to determine both the number of parameters which enter an arbitrary static solution and the number of collective coordinates which must be introduced in a quantum treatment of these objects.

In the Prasad-Sommerfield limit index theory methods allow one to relate the number of normalizable zero modes to the topological charges. Thus in the SU_2 theory there are 4n such modes about an arbitrary self-dual solution with n units of magnetic charge.[1] One of these is a gauge mode, so there are only 4n-1 parameters. However, all 4n modes lead to collective coordinates; the variables conjugate to these are the linear momenta and electric charges of n monopoles.

Now consider a theory with a gauge group G of rank r.[2, 3] One can always require that asymptotically in a fixed direction the adjoint Higgs field and the magnetic field be of the form

$$\phi = v \sum_j h_j T_j \qquad B_i = \frac{\hat{r}_i}{4\pi r^2} \sum_j g_j T_j$$

where the T_j are the generators of the Cartan subalgebra. Now suppose that a set of simple roots is chosen which satisfies $\vec{\beta}_i \cdot \vec{h} \geq 0$. Topological arguments then show that $\vec{g}$ must be of the form

$$\vec{g} = 4\pi \sum_i n_i \frac{\vec{\beta}_i}{\vec{\beta}_i^{\,2}}$$

First suppose that the unbroken symmetry is $(U_1)^r$; in this case the n_i are all topologically conserved. There are then r spherically symmetric monopole solutions which may be termed fundamental. The ith such has topological charges $n_a = \delta_{ai}$ and a mass m_i. About it are precisely four zero modes; i.e., it has no internal degrees of freedom. The mass and number of zero modes for any self-dual solution of higher topological charge are

$$M = \sum_i n_i m_i \qquad N = \sum_i 4 n_i$$

The solution should therefore be interpreted as composed of a number of non-interacting fundamental monopoles. Among the solutions which are thus seen to be composite are some which are also spherically symmetric and would appear, without the zero mode analysis, to be very much like the fundamental solutions.

If instead the unbroken group H is non-Abelian, only some of the n_i are topologically conserved; denote these as $\tilde{n}_i$. The index theory methods give the number of zero modes only when $\vec{g}$ is such that the long range magnetic field commutes with all generators of H. In this case there are

$$N = \sum_i k_i \tilde{n}_i$$

zero modes, where the k_i can be determined from the root diagram; in the simplest case, SU_3 broken to $SU_2 \times U_1$, $k = 6$. In almost all cases there is one fundamental solution corresponding to each $\tilde{n}_i$, and the k_i are as expected from naive symmetry arguments. There are however exceptional cases, occurring only in groups with two root lengths, where on the one hand the fundamental solution is not unique, and on the other the k_i are larger than expected.

References

1. E. Weinberg, Phys. Rev. D20, 936 (1979).
2. E. Weinberg, Nucl. Phys. B167, 500 (1980).
3. E. Weinberg, Columbia University preprint CU-TP-224.

This research was supported in part by the U. S. Department of Energy.

II: SUPERSYMMETRIC MONOPOLES AND ELECTRIC-MAGNETIC DUALITY

MAGNETIC MONOPOLES AND ELECTROMAGNETIC DUALITY CONJECTURES

David Olive
Blackett Laboratory, Imperial College, London SW7 2BZ, UK.

1. INTRODUCTION

I am grateful to the organizers of the meeting for this opportunity to review the progress to date concerning the duality conjectures which my collaborators and I made more than four years ago[1,2]. At that time we were well aware of their daring and surprising nature. Nevertheless we thought, and tried to argue, that there was sufficient reason to think about such ideas seriously. Because the ideas refer to the conjectured quantum field theory of the classical monopole solutions arising in spontaneously broken gauge theories in four space time dimensions, the conjectures were difficult to prove (or disprove) and remain so. Despite this, as the years go by, the odds do seem to be shortening that the conjectures are valid, at least in one special sort of spontaneously broken gauge theory, that with N=4 extended supersymmetry, as I shall explain. This enhancement of the plausibility of our speculations is due to a series of unexpected developments, partly triggered by these ideas.

In this talk I shall first set the background to the duality conjectures by reminding you of the revolution in the theory of monopoles initiated by the realization of 'tHooft[3] and Polyakov[4] that magnetic monopoles could arise as classical solutions to spontaneously broken gauge theories.

The eradication of that confusing (though perfectly correct) construct, the Dirac string, facilitated the study of monopoles associated with non-Abelian exact gauge symmetries H (rather than just U(1)), and lead to the discovery of results concerning their classification which applied equally well to the point (singular) monopoles and the soliton monopoles. (We shall call the classical solution monopoles soliton monopoles even though they are not solitons in the precise technical sense used for classical solutions in two space time dimensions). These results will be reviewed in sections 2 and 3.

Physically the most important feature of the soliton monopoles is that they have an internal structure (and hence mass) determined by the overall symmetry group G, and its mode of symmetry breaking. The point monopoles, on the other hand, have a singularity at their centre which is only relieved (presumably) by quantum fluctuations as in QED. In view of the long standing unease with the renormalisation procedure, even in conventional QED, as stressed by Dirac[5] it seems advantageous to have a theory in which the structure occurs in the first approximation.

The possibility of a local field operator creating the monopole state is discussed in section 4, leading to the presentation of the duality conjecture of Goddard, Olive and Nuyts concerning the dual groups H and H^v and their possible interpretation as electric and magnetic gauge symmetry groups. It seems that a spontaneous gauge symmetry breaking by a Higgs field in the adjoint representation with vanishing self coupling is most favourable to the idea. Such theories and their supersymmetric versions are discussed in section 5 where the more specific duality conjecture of Montonen and Olive is introduced. In section 6 the symmetry breaking is studied in more detail by Dynkin diagram techniques and some comments are made about the exact static solutions.

2. TOPOLOGICAL AND MAGNETIC CHARGE CLASSIFICATIONS OF MONOPOLES.

If G is the overall gauge symmetry of the Lagrangian, then, in vacuo, the Higgs field must take values in the coset space manifold G/H. For any finite energy configuration, the Higgs field must, at sufficiently large distances, take values in this manifold and hence furnish a map from the large sphere in space, S_2 , into G/H. These maps fall into disjoint classes, called homotopy classes $\pi_2(G/H)$ which are equivalence classes with respect to deformations. The classes therefore provide gauge invariant conserved quantities and possess natural addition properties. They can therefore be thought of as "quantum numbers" of purely classical origin.

The authors[6,7,8] (two Russian groups, one member of whom, Fateev, is present here, and Sidney Coleman) who independently discovered this topological classification of classical field configurations in terms of the boundary conditions on the Higgs fields, also demonstrated its equivalence to a topological classification in terms of the long range gauge fields (i.e. those pertaining to H) via the beautiful equation

$$\pi_2(G/H) \quad \cong \quad \pi_1(H)_G \qquad (1)$$

On the right hand side the Higgs field has disappeared and so this classification applies equally to point monopoles as indeed had been discovered long before by Lubkin[9]. It was also rediscovered, simultaneously with the work above [6,7,8], by Wu and Yang[10] who formulated it in a particularly clear cut way using the concept of section. The precise meaning of $\pi_1(H)_G$ and the relationship between the different approaches has been reviewed by Goddard and Olive[11].

If H has an invariant U(1) subgroup (so that it is <u>not</u> semisimple) then it can be written locally as a product of the U(1) and the residual subgroup, K:

$$H \;=\; "U(1) \times K" \qquad (2)$$

The inverted commas are to remind us that the direct product structure is valid locally but not necessarily globally. The generator of the U(1) subgroup is a K singlet.

It follows from the structure (2) that

$$\pi_1(H) \supset \pi_1\,(U(1)) \;\cong\; \mathbb{Z}$$

the group of integers. Integer quantum numbers are the kind most familiar to physicists and it turns out that these are proportional to the magnetic charge g defined as the flux of the U(1) magnetic field out of a large sphere. This is the only gauge invariant kind of magnetic flux which can be defined.

Equation (1) does not tell us the allowed numerical values of g. This information is supplied by the generalized Dirac quantization condition[(12)], discovered by Corrigan and myself[(13)], valid whenever the exact symmetry group H has the form (2).

The quantization condition reads

$$\exp(igQ/\hbar) = k \in K \tag{3}$$

Q is the U(1) generator normalized so that its eigenvalues correctly give the U(1) electric charges of the elementary fields.

Acting on K singlet states, the element **k** of K occurring on the right hand side of (3) must, by definition, reduce to unity, so that

$$\exp(igQ_s/\hbar) = 1 \tag{4}$$

and the usual Dirac quantization condition[(12)] is satisfied by the projection, Q_s, of Q onto the K singlet subspace and g.

In general, since the left hand side of equation (3) is a K singlet, the element k of K on the right hand side of equation (3) must also be a K singlet. Mathematically this means that k must be an element of the centre of K, Z(K). Physically, the most interesting case is that K is semisimple. Then Z(K) is a finite group. Furthermore it follows from equation (3) that k must be an element of a fixed cyclic subgroup of Z(K), Z say. If Z has $|Z|$ elements, then, since it is cyclic, it has a generating element of order $|Z|$, and it follows from (3) that

$$\exp(igQ|Z|/h) = 1 \qquad (5)$$

so that for K non-singlet states fractional U(1) charges are allowed (with respect to K singlet states defining the unit), with $|Z|$ denoting the relevant fraction. For example, if $K \equiv SU(3)$, $|Z|$ can equal either 1 or 3 as the centre of SU(3) consists of the cyclic group composed of the three cube roots of unity. Nature apparently chooses $|Z| = 3$ as the quarks, which are SU(3) colour triplets indeed possess fractional charges as compared to colour singlet electrons.

Much more can be concluded from equation (3) concerning the correlation between K transformation properties, U(1) charges, and possible magnetic monopoles[14].

3. MAGNETIC WEIGHTS AND NON-ABELIAN MONOPOLES.

As we said the magnetic charge g defined above in connection with an invariant U(1) subgroup of the exact gauge symmetry H is the only kind of gauge invariant magnetic flux which can be definёd. Nevertheless it is possible to define a generalized gauge covariant magnetic charge associated with classical field configurations exhibiting a generalized inverse square law at large distances:

$$B_i = \frac{G}{4\pi} \frac{r_i}{r^3} \;; \quad D_i G = 0 \;, \quad |r| >> 1 \qquad (6)$$

The significance of such configurations is that all known static finite energy classical solutions to spontaneously broken gauge theories behave thus in their rest frame[1]. It would be interesting to have a counter-example but lacking that we must take the form (6) seriously.

The coefficient G in (6) is a covariantly constant element of the Lie algebra of H and so varies by a gauge transformation on the large sphere surrounding the monopole whose presence G signifies. If H has the form(2) indicating an invariant U(1) subgroup, then the coefficient of Q in G is indeed proportional to g as defined above. Otherwise G provides us with an attribute of the monopole in question which we think of as the non-Abelian magnetic charge.

To understand G we must classify the structure modulo gauge transformations. This requires the introduction of the theory of Lie algebras and their representations. This theory is, willy-nilly, the appropriate mathematical framework for understanding the structure of non-Abelian Lie groups whether or not we are interested in monopoles. It is explained in the famous lectures by Racah[15] and the book by Humphreys[16] and elsewhere and the reader is urged to investigate.

A basis for the Lie algebra of H can be taken to consist of a linearly independent set of mutually commuting generators $T_1,\ldots T_r$, (with r = rank H) together with step operators $E_{\pm\alpha}$ corresponding to the roots $\pm\alpha$

$$\left[T_i, E_{\pm\alpha}\right] = \pm\alpha_i E_{\pm\alpha}$$

Assuming the usual orthonormalization it is found that in any representation of the group H, the possible eigenvalues, λ_i of the T_i ;

$$T_i\ |\lambda_i> \quad = \quad \lambda_i\ |\ \lambda_i> \tag{7}$$

consist of points lying in an r dimensional lattice, called $\Lambda(H)$, the weight lattice of H . For example with the usual conventions, the weight lattice of SO(3) consists of the eigenvalues of the third component of angular momentum and hence the set of integers. The weight lattice of SU(2) includes both integers and half integers because of the extra spinor representations (which are excluded when SO(3) is considered because they are not then single valued).

This lattice structure is a quantized structure arising from the commutation relations of the Lie algebra. The elementary fields of the theory transform according to representations of the gauge group H and can be labelled by certain weights characterising these representations.

It is a mathematical theorem that there exists a gauge transformation h, rotating G into the Cartan subalgebra

$$h\,G\,h^{-1} \;=\; \frac{4\pi}{e}\sum_{1}^{r}\gamma_i T_i \tag{8}$$

These γ_i are called the "magnetic weights" of the monopole[1], in contradistinction to the "electric weights" associated with the representation according to which the elementary fields transform.

The reason they are called weights is that they must satisfy a generalization of the Dirac quantization condition

$$\exp(ieG) \;=\; \exp(4\pi i\sum_{i}^{r}\gamma_i T_i) \;=\; 1 \tag{9}$$

Since this is to be valid for all states in equation (7) it follows that

$$2\,\gamma\cdot\lambda = \text{integer} \quad , \qquad \lambda\varepsilon\Lambda(H)$$

so that γ must also lie on a lattice[17], that reciprocal to $\Lambda(H)$. Thus γ also has a quantized structure, apparently not because of a Lie group structure but because of the Dirac quantization condition. Nevertheless it can be shown that the reciprocal lattice on which γ lies is indeed the weight lattice of a specific Lie group, called H^{v} which has the same dimension and rank as H but differs in terms of an interchange of long and short roots, and in terms of the global structure (if the roots all have the same length). Since $(H^V)^V = H$ we can talk of H and H^V as dual groups[1].

Now it is an important fact that when H is non-Abelian, γ_i is not the unique magnetic weight corresponding to G: any γ_i' obtained from γ_i by a sequence of Weyl reflections (reflections in the hyperplanes perpendicular to the roots) will also do since these reflections can be achieved by gauge transformations. These reflections generate a finite group, called the Weyl group and its effect constitutes the sole ambiguity in the magnetic weights[1]. This ambiguity is a strictly non-Abelian effect since an Abelian group has no Weyl group.

When we consider a single magnetic monopole, the different values of γ related by Weyl reflections are physically equivalent since they are related by gauge transformations. The correspondence with the previous topological classification (1) in terms of $\pi_1(H)$ is easily made. It can be shown[1] that the magnetic weight lattice $\Lambda(H^V)$ splits into disjoint components, cosets $\Lambda(H^V)/\Lambda_r(H^V)$, in one to one correspondence with the elements of $\pi_1(H)$, where $\Lambda_r(H^V)$ denotes the root lattice of H^V obtained by adding its roots.

When we consider a solution describing two magnetic monopoles, it seems that the magnetic charges associated with each must mutually commute, and so be capable of being gauge rotated into the same Cartan subalgebra. Then there is a Weyl group degree of freedom between the two monopoles which is physical since it alters gauge invariant quantities such as the energy density[1].

Thus, although the inverse square law (6) looks distressingly Abelian (since by equation (8) it can be gauge transformed into a Cartan subalgebra), nevertheless there appears to be a real physical difference between Abelian and non-Abelian monopoles in that the non-Abelian monopoles display an extra discrete degree of freedom, characterised by the Weyl group which manifests itself when interactions are considered. To give the simplest example, the response of a non-Abelian monopole to a uniform non-Abelian magnetic field is not uniquely determined without specifying its group orientation with respect to the external field.

It is highly unlikely that all possible points of the magnetic weight lattice should correspond to stable monopoles. In fact in 1979, Brandt and Neri[18] and Coleman[19] independently found that the inverse square law (6) itself was unstable except for special magnetic weights now called "minimal weights"[20]. These minimal weights are those for which the energy in the Coulomb tail is minimized for given topological quantum number and thus correspond to the points of a coset $\Lambda(H^v)/\Lambda_r(H^v)$ closest to the origin. This interpretation was established for SU(N) by Brandt and Neri[18], for the classical groups by Coleman[19] and for all the Lie groups by Goddard and myself[20]. The minimal weights for a given topological quantum number $\pi_1(H)$ are all Weyl group conjugates and this means that, modulo gauge transformations, there is a unique monopole for each topological quantum number.

This result relies on a rapid approach to the inverse square law (6) and is not valid for the self-dual monopoles to be explained below in section 5. E. Weinberg[21,22] has given explicit examples of self-dual, or BPS, (and hence stable) solutions with the same topological quantum number but gauge inequivalent magnetic weights.

This concludes our brief review of the classification theorems. Most of the results apply equally to point monopoles and soliton monopoles. However the impetus leading to their discovery was largely due to the discovery by 'tHooft[3] and Polyakov[4] of the soliton monopoles.

4. THE STRUCTURE OF MONOPOLES AND THE GON DUALITY CONJECTURE.

Having said this, the most interesting and important feature of soliton monopoles is that they have an internal structure. Roughly speaking the inverse square law behaviour (6) extends down to a finite radius inside which the full overall gauge group G is excited. In particular the mass is finite even when the energy of the tail (6) is included. On the other hand point monopoles have zero radius and infinite mass since the energy integral diverges at small radius. In conventional quantum field theory, the smoothing out of a point singularity only occurs after radiative corrections and renormalizations are dealt with. Presumably one must hope for something similar in respect of point monopoles but for soliton monopoles the regularity is guaranteed ab initio.

The structural parameters of the soliton monopole, mass, radius etc., are of course determined by the parameters of the original Lagrangian and therefore have to do with the electrically charged particles. Thus the monopole parameters are not free once the electrical parameters are given. Thus the theory is highly constrained. In recent years, since the advent of the dual string model and supersymmetric theories such constraints have come to be regarded as desirable since they are symptomatic of unexpected symmetry.

An important question remains to be settled: in the full quantum theory are there quantum states corresponding to localised monopoles? If so,can these states be created by local quantum field operators and if so, what equations of motion do these operators satisfy?

The difficulty in answering this question points to our abject ignorance in understanding what are very physical theories, namely spontaneously broken gauge theories. It is unlikely that definitive answers will emerge for some years yet, but as I shall show, discussion of possible answers does lead to interesting consequences which can be checked.

In space-time of 1+1 dimensions there exists an interesting theory in which these questions possess an answer. The Sine-Gordon model has soliton solutions which can be created by a local fermi quantum field operator satisfying the equations of motion of the massive Thirring model[23]. There are thus two alternative Lagrangians (Sine-Gordon and massive Thirring) related by a quantum equivalence. In terms of the Sine-Gordon field ϕ, the Thirring field ψ is a transcendental and non-local function in the sense that ϕ and ψ do not commute at space like intervals. The "quantum number" carried by the soliton is "topological" when the Sine Gordon Lagrangian is considered but "Noether" with respect to the Thirring model. See section 3 of the review[11] by Goddard and myself for more discussion of this.

Let us suppose that the choice of G and its mode of breaking to H is such as to allow quantum field operators for the monopole solitons. The candidate theory for this operator should be consistent with the properties of non-Abelian monopoles discussed in the previous section, 3. Because of the inverse square law tail (6), we expect long range interactions which, in the case of Abelian monopoles, are certainly mediated by gauge fields because of the characteristic possibilities of attraction or repulsion depending on whether the charges are unlike or like. For non-Abelian monopoles we might expect the mediation of particles associated with H^v as a gauge group. The magnetic weights of a monopole soliton would then correspond to the weights of the representation of H^v

according to which its quantum field operator transforms and we would have a new understanding of the quantized structure of the magnetic weights as being due to the Lie algebraic structure of the dual or magnetic group H^v. (see my Lausanne 1979 lecture[24] for more details).

It is a significant fact that the magnetic weights of a stable heavy monopole, being minimal as explained above, constitute the weights of a minimal representation of H^v, and that such representations are the only ones all of whose weights are related by gauge transformations via the Weyl group[20]. Also the ambiguity in the interaction between two monopoles mentioned above can now be interpreted as relating to the fact that the combination of two representations of H^v is ambiguous owing to its decomposition into a Clebsch Gordan series of irreducible representations, all providing alternative possibilities[1].

These then are the sorts of arguments adduced by Goddard, Nuyts and myself[1] when we made the conjecture in 1976 and we still do not know whether the idea is correct or not. I can say that the idea does not disagree with anything obvious and that it does succeed in making coherent the properties of non-Abelian monopoles discussed in section 3.

To illustrate further, it would say that the only absolutely conserved quantity is the centre of H^v. But by the mathematical construction of H^v this precisely equals the topological quantum number $\pi_1(H)$ which is of course conserved. To avoid any confusion it should be realisedthat the conservation of the magnetic weights is not implied. The only constraints on the representation of the monopoles before and after any scattering are those furnished by the Clebsch Gordan rules as GON explained[1].

If the idea is correct that there are two alternative gauge Lagrangians we see that according to the Dirac quantization condition the two gauge coupling constants are inversely related. Each Lagrangian has its own perturbation expansion in powers of the relevant gauge coupling constant with solitons treated non-perturbatively. This is in contra-distinction to the ideas presented at this meeting where an attempt is made to construct a Lagrangian involving both coupling constants simultaneously and to develop series in both simultaneously[25].

In 1+1 dimensional space time not all soliton solutions possess local field operators. For example the ϕ^4 soliton does not, as far as I know, and the Sine Gordan soliton does. Presumably this is connected with another miracle of the Sine Gordan system, the possession of an infinity of conservation laws as a consequence of its compete integrability and its susceptibility to the inverse scattering[26,27] method.

It could well be that the existence of the local monopole field operator holds in certain cases depending on the mode of symmetry breaking, the introduction of extra symmetries and maybe even the choice of gauge groups.

Indeed it was very soon realized, in 1977, by Montonen and myself[2], that the 'tHooft Polyakov example of G= SU(2) broken by a triplet Higgs field to H = U(1), was particularly favourable to the idea when the Prasad Sommerfield limit[28] of vanishing Higgs self-interaction was considered. All subsequent effort has gone in this direction and the rest of my talk will be devoted to this, and to the more general case of an overall gauge symmetry G broken by a Higgs field in the adjoint representation with vanishing self-interaction. As we shall see this precludes the case that the exact gauge symmetry H is semisimple.

This exclusion is a pity because the QCD group, SU(3) is simple and its dual group, $SU(3)^V$ is $SU(3)/Z_3$. This means that the triplets of the two groups are mutually incompatible and this could be an explanation of quark confinement. In fact the inspiration for thinking about duality as far as I was concerned was the fact that the confinement arguments of 'tHooft , Mandelstam[29] and others[30] alternated between magnetic and electric formulations of the theory. Unfortunately these ideas still lack a mathematically precise formulation and it is the promise of an exact theory which encourages us in the direction of adjoint Higgs. (It may not after all be accidental that the exact gauge symmetry group of nature is not semisimple).

I want to close this section by making a phenomenological remark concerning the GON conjecture. We saw that stable massive monopoles had to minimize the energy in the Coulomb tail for given topological quantum number. If the GON conjecture has any validity, this means that they must correspond to special irreducible representations of the magnetic group H^{v} called minimal representations. These are listed for any simple Lie group in [20] and in the table below, labelled by their dimensionalities, they appear for the popular grand unified or unified groups thought to describe the particle interactions[31].

TABLE 1

Candidate gauge group	Minimal representations
E_7	56
E_6	27, $\overline{27}$
E_5 = SO(10)	10, 16, $\overline{16}$
E_4 = SU(5)	5,10, $\overline{10}$, $\overline{5}$
E_3 = SU(3)xSU(2)	(3,1), ($\overline{3}$,1),($\overline{3}$,2),(3,2),(1,2)

The fact is that these are precisely the representations to which the fundamental fermions, quarks and leptons, are assigned (with the exception of the 10 of SO(10))[24,32]. I know of no other explanations of this fact and I think that it is an important feature of monopole theory that it can potentially explain it. The usage of the notation

E_5 for SO(10), etc., will become clearer in section 6 when I have more to say about the Lie groups in Table 1.

5. SELF-DUAL MONOPOLES, DUALITY AND SUPERSYMMETRY.

Montonen and I realized very soon(2) that symmetry breaking by a Higgs field in the adjoint representation and with vanishing self coupling(28) was particularly favourable to the GON conjecture and indeed enabled a more specific conjecture to be formulated. Before explaining this I shall develop some of the remarkable properties of these theories, including some properties whose discovery was provoked by the conjecture.

We consider a Lagrangian density

$$\mathcal{L} = -\frac{1}{4}(F_{\mu\nu})^2 + \frac{1}{2}(D_\mu\phi)^2 \qquad (10)$$

The energy is given by

$$E = \frac{1}{2}\int d^3x \left[E_i^{\,2} + B_i^{\,2} + (D_i\phi)^2 + (D_o\phi)^2\right] \qquad (11)$$

and can assume its minimum value, zero, for any field configuration in which $D_\mu\phi$ and $F_{\mu\nu}$ vanish. This implies that the length of ϕ, denoted a, is constant in vacuo and that if $a \neq 0$, ϕ can be regarded as a Higgs field which breaks the overall gauge symmetry G to a subgroup H discussed below. Since the self interaction of ϕ vanishes, the Lagrangian itself does not specify a, nor the orbit with respect to G of ϕ in vacuo. This information is to be regarded as coming from some sort of boundary condition imposed from outside.

Let us define the generator of G in the direction of ϕ in vacuo as(33)

$$Q = e\hbar\phi.T/a \qquad (12)$$

The generators of the exact gauge symmetry H are precisely those generators of G which commute with Q and therefore consist of Q itself and the set of generators of G commmuting with Q yet orthogonal to Q in the Lie algebra of G. So the group H obtained by exponentiating this structure has precisely the form "U(1)xK" [33] discussed above in connection with equation (2). Further, Q is correctly normalized for the validity of the quantization condition (3). We shall henceforth assume that ϕ in vacuo is such that K is semisimple, leaving until the next section the precise mathematical condition for this. Then as explained above, $\pi_1(H) \supset \mathbb{Z}$ so that topologically non-trivial field configurations carrying magnetic charge are possible.

The second term in the Lagrangian density (10) provides a mass term for those gauge potentials of G not corresponding to H. This is the Higgs-Kibble-Brout-Englert mechanism[34]. Specifically

$$M^2_{\alpha\beta} = e^2\hbar^2\phi\, T_\alpha T_\beta \phi = a^2(Q^2)_{\alpha\beta}$$

Using (12) and the fact that in the adjoint representation the generators T can be represented by totally antisymmetric structure constants. Hence diagonalizing Q and hence M^2 the masses of the gauge particles are given simply as[35,36]

$$M = a\,|q| \qquad (13)$$

where q is the U(1) charge carried by the particle in question. The scalar particles which survive pair off with the H gauge particles and so have q=0. Since there is no mass term in (10) for them they likewise satisfy (13). Thus the mass formula (13) applies to all the quantum excitations of the elementary fields occurring in the Lagrangian (10).

In order to calculate the mass of a classical field configuration consider the form (11) for the energy and rewrite

$$\tfrac{1}{2}\,(B_i^{\,2} + (D_i\phi)^2) = \tfrac{1}{2}\,(B_i \mp D_i\phi)^2 \;\pm B_i D_i\phi$$

The volume integral of the last term may be integrated by parts to yield $\pm$ ag where g is the U(1) magnetic flux defined in section 2. We conclude that the energy E is bounded below by $a|g|$ with equality if and only if

$$B_i = \pm D_i\phi \qquad D_o\phi = E_i = 0 \tag{14}$$

This implies that the configuration is at rest so that its energy equals its mass(37,38)

$$M = a\,|g\,| \tag{15}$$

Configurations satisfying (14), I shall call "self-dual configurations" and the resultant monopoles "self-dual monopoles". To understand why, let me introduce an artificial fifth, space-like dimension and identify the fifth component of gauge potential with the Higgs field (which is possible as it lies in the adjoint representation)(39,36)

$$W_5 = \phi \tag{16}$$

Equation (14) is the self-duality condition in the space of four Euclidean dimensions 1,2,3 and 5. Also the five dimensional gauge Lagrangian density reduces to (10) as ϕ is independent of x_5.

Because of equation (14) the Higgs and magnetic fields contribute equally to the energy density in (11) thereby doubling the usual contribution of the Coulomb tail (6). The self-dual monopoles therefore have, in a certain sense, an infinite radius and differ in this essential way from the monopoles considered before.

The magnetic monopole solutions which are to be interpreted as particles should be stable and hence minimize E for given g. Hence they are self-dual and satisfy the mass formula (15) which bears a remarkable resemblance to the formula (13) for gauge particles. In fact we may also consider dyon solutions[40] with electric and magnetic U(1) charges q and g. When their energy is minimized for fixed charge their mass is given by[37,38]

$$M = a\sqrt{q^2+g^2} \tag{17}$$

Remarkably this is a universal mass formula applicable to all the particle states of the theory, whether scalar or gauge excitations of the original fields or soliton monopoles or dyons[2, 36].

Now let us return to the duality conjecture. The results above hold for any choice of semisimple G but we shall now focus, temporarily, on G = SU(2). Then H = U(1) and it turns out that there are precisely two spherically symmetric monopole solutions with equal and opposite charges of magnitude $4\pi/e$ where e is the gauge coupling constant. The dual group of $U(1)_E$ in the sense of GON is indisputably $U(1)_M$ with magnetic charge as generator. The alternative Lagrangian involving field operators for the monopoles would also involve a $U(1)_M$ gauge field and ought to produce the original heavy gauge particles as soliton solutions with electric charges $\pm e\hbar$.

Montonen and I proposed [2] that the three fields formed an SU(2) gauge triplet with gauge coupling constant $4\pi/e\hbar$, and masses arising from an isotriplet Higgs field with length in vacuo, a' say. In other words the two alternative Lagrangians each have the form (10) with gauge particles and monopoles exchanging roles and respective coupling constants e and $4\pi/e\hbar$. The original Lagrangian yields the mass formula (17) and the second

$$M = a'\sqrt{q^2+g^2}$$

and so there is perfect agreement if a=a'. This agreement requires the vanishing of the Higgs self interaction in both cases.

So the spectrum of states, and their masses agree with the conjecture as does another feature, the non relativistic force exerted between the massive states. Owing to the long range Higgs effect, like magnetic charges exert no forces on each other while unlike charges attract with twice the expected force, as was found by Manton[41] using classical arguments. On the other hand Born diagram calculations show that the massive gauge particles behave similarly owing to the massless scalar particle providing a long range attraction which cancels the Coulomb repulsion and doubles the Coulomb attraction[2].

So far the picture seems attractive but there are snags concerning the quantum theory: gauge particles carry spin h, how can the monopoles do this? Would not quantum corrections vitiate the mass formula (17) which was essential to the working of the idea? We argued that these features were associated with a symmetry and would work if the quantization respected that symmetry. This was essentially correct but in an unexpected way. The necessary symmetry is apparently supersymmetry[42] which can fit the spin 0 and spin 1 fields of (10) into a single supermultiplet if extended supersymmetry with either N=2 or N=4 is considered. These theories[43,44] can be obtained by dimensional reduction from simple theories in D=6 and 10 dimensions respectively where a gauge field couples minimally to a massless chiral spinor field which in D=10 is also Majorana. The four dimensional theory is found by supposing that the fields do not depend on the surplus dimensions. Thus the N=2 theory has two scalar fields and the N=4 theory, six. All fields lie in the adjoint representation of the gauge group and the only interactions amongst the scalars involve the squares of their mutual commutators. Thus the Lagrangian density is indeed given by (10) plus extra terms which vanish when the extra fields vanish. Thus the classical solutions of (10) remain solutions of the extended Lagrangian.

Two approaches have been made to the quantum treatment of these theories. One is the semiclassical approach considering quantum oscillations about the self dual monopole as reviewed by Osborn at this meeting[45]. These oscillations consist of an expansion in terms of zero modes and independent supersymmetric harmonic oscillators each with vanishing zero point energy. Assuming convergence of the sum over oscillators it follows that the first quantum correction to the energy, and hence the mass as given by (15) vanishes[42]. The zero modes have to do with the realization of the supersymmetry algebra and indicate what sort of multiplet the monopoles lie in. The fermionic zero modes are represented by Dirac gamma matrices indicative of the varying spin content[46]. (Yang[47] considered the analogous problem for point monopoles in his talk here).

The second approach is to take the supersymmetry algebra as a statement of exact operatorial identities. This would be true if the theory could be quantized in a supersymmetric way. We shall accept this, which is a basic tenet of supersymmetry theory, even though, as far as I know, it has not been convincingly established for the supersymmetric gauge theories under consideration.

Then it would follow that the monopole states must represent the supersymmetry algebra and contain components with different spins (since the vacuum with zero energy and momentum is the only trivial representation.)

Graphical arguments indicate that there should be no radiative corrections to the self interactions of the scalar fields[48], thereby preserving the Prasad-Sommerfield limit.

The crucial equation in the supersymmetry algebra reads in D=6 or 10 dimensions :

$$\{Q_\alpha \, , \bar{Q}_\beta\} = \Big[(1+ \Gamma_{D+1}) \, \Gamma \cdot \mathbb{P} \Big]_{\alpha\beta} \tag{18}$$

$\mathbb{P}^\mu$ is the D momentum and when the theory is reduced from 6 to 4 dimensions (say) it turns out that its fifth and sixth components become aq and ag respectively[49,36]. Thus the electric and magnetic charges appear on the right hand side of the fundamental anticommutator in four dimensions, thereby playing the role of "central charges" [49]. (Zizzi has investigated this dimensional reduction carefully, paying particular attention to the intermediate stage in five dimensions[50]).

From (18) it follows that

$$2^{D/2} \, \mathbb{P}^0 = \sum_\alpha \{Q_\alpha, Q_\alpha^{\ +}\} \geq 0$$

so that the energy is fundamentally positive (thereby extending the feature of equation (11)). Furthermore[36]

$$2^{D/2} \, \mathbb{P}^2 \, \mathbb{P}^0 = \sum_\alpha \{(\Gamma . \mathbb{P} Q)^{\dagger}_\alpha \, , (\Gamma . \mathbb{P} Q)_\alpha\} \geq 0$$

Hence $\qquad \mathbb{P}^2 \geq 0 \qquad$ (19)

with $\qquad \mathbb{P}^2 = 0 \qquad$ (20)

if and only if $\Gamma . \mathbb{P} . Q = 0$.

Now if we reduce from D=6 to D=4 with the identification of aq and ag as the excess components of momentum we see that equation (20) is precisely the mass formula (17) now rederived from the supersymmetry algebra (18) which is supposed to be exact to all orders in the quantum theory. More precisely (19) is exactly valid and equation (20) holds for states $|s\rangle$ satisfying the subsidiary supersymmetry condition[36]

$$\Gamma.\mathbb{P}\, Q|s\rangle = 0, \quad (\Gamma.\mathbb{P}Q)^{\dagger}|s\rangle = 0 \tag{21}$$

One can check that self-dual configurations satisfy (21) in the sense of being invariant with respect to the indicated supersymmetry transformations.

Thus equations (21) are the quantum analogues of the self-duality equations and undoubtedly will be much studied in the future.

Notice how in this theory supersymmetry unexpectedly unifies the different mass generation mechanisms; the topological mass of the monopole solitons and the Higgs mechanism for the gauge particles, as well as relating them to extra dimensional effects.

There remains the choice between N=2 and N=4 extended supersymmetry, that is, reduction from six or ten dimensions.

The attractive feature of N=2 is that the rotation in the 5-6 plane rotates between q and g and therefore can be thought of as a duality rotation[36]. The duality rotation symmetry of the free Maxwell equations was noted over fifty years ago and has long intrigued physicists. The theory we are considering seems to be the closest to realizing that symmetry with charged matter present.

If, in the N=4 theory, we choose a phase in which, in vacuo, all six scalar fields are parallel in group space (thereby satisfying the condition that their mutual commutators vanish) then the six extra components of momentum give essentially two charges q and g as before so the same mass formulae hold[46].

For N=4 Osborn has emphasized[46] that there is only one possible supermultiplet of spin less than or equal to one and hence capable of satisfying renormalizable field equations. The original gauge particles belong to such a multiplet and we would expect the monopoles to do so too, and this is confirmed by the semiclassical method mentioned above[51] (which also indicates that in the N=2 theory the monopoles lie in a multiplet not containing spin $\hbar$[52]).

One might expect q and g to renormalize similarly (because of duality) yet might want the Dirac quantization condition to be preserved[51,53]. This is difficult to arrange unless the Callan Symanzik beta function vanishes. In fact the N=4 supersymmetric gauge theory is the only known four dimensional theory in which this happens, at least to 3 loops[54] and possibly altogether[55].

The conclusion is that the N=4 supersymmetric extension of the Lagrangian (10) seems to be the only theory for which the Montonen Olive conjecture could be valid and that its validity is reasonably plausible at least for G = SU(2).

6. GENERAL GROUPS.

Many of the statements in the preceding section were true regardless of the choice of gauge group, for example, the universal mass formula (15) and the possibility of rendering the theory N=4 extended supersymmetric, but what is not yet clear beyond the G= SU(2) case is the precise spectrum of elementary monopole solutions. As well as the difficulty of obtaining all the solutions, there is an uncertainty in their interpretation as I shall explain.

According to the generalized Dirac quantization condition (9) the magnetic weights must lie not only on the weight lattice of H^v, but also of G^v. In order that there be no singularity in the solution it must be topologically trivial in G. Hence the magnetic weight must actually lie in the root lattice of G^v (a subset of the weight lattice)[17]. Therefore it is plausible to follow Bais[56] and generalize the conjecture of Montonen and myself from G = SU(2) to any G and expect that in the alternative Lagrangian the elementary monopole states be described by the heavy gauge particles of the group G^v, since these correspond to certain roots of G^v.

There are several differences compared to the G = SU(2) case. One is that some roots of G^v will correspond to roots of K^v and hence massless gauge particles. Another is that the G^v and G spectrum may be dissimilar if G has roots of unequal length so that the very close resemblance between the two alternative Lagrangians in the G= SU(2) case could not hold.

In order to make the discussion more precise I must discuss the symmetry breaking in more detail and introduce yet more Lie algebra technology[15,16] of the type already used in part to discuss the non-Abelian quantization condition (9).

Let $\alpha^{(1)}$, $\alpha^{(2)}$ ...... $\alpha^{(r)}$ be a set of r= rank G simple roots of G. Then any root of G can be expressed either as a sum or minus a sum of these simple roots. The fundamental weights $\lambda^{(1)}$, ... $\lambda^{(r)}$ are defined by

$$2 \lambda^{(i)} \cdot \alpha^{(j)} / (\alpha^{(j)})^2 = \delta^{ij} \qquad (22)$$

It is always possible to choose a gauge so that the Higgs field ϕ lies in the Cartan subalgebra subspace and satisfies

$$\phi \cdot \alpha^{(i)} \geq 0 \qquad i = 1, \ldots r$$

I mentioned in sections (2) and (5) that it was desirable that K be semisimple in order to get charge quantization. In the gauge just chosen the necessary and sufficient condition for K to be semisimple is that the Higgs field ϕ is parallel to one of the r fundamental weights defined by equation (22)[14]. So there are precisely rank G possibilities for K, given G, and there is a simple algorithm for recognizing the possibilities for K from the Dynkin diagram for G, D(G). D(G) encodes the structure of the Lie algebra of G and assigns to each simple root a point with $4(\alpha^{(i)}.\alpha^{(j)})^2 / (\alpha^{(i)})^2 (\alpha^{(j)})^2$ lines joining points i and j, and an arrow pointing from i to j if $(\alpha^{(i)})^2$ exceeds $(\alpha^{(j)})^2$. Then D(K) is obtained from D(G) by deleting that point (and its links) corresponding to the simple root not orthogonal to ϕ.[14]

As we said, any root may be written

$$\alpha = \sum_{1}^{r} n^i \alpha^{(i)} , \tag{23}$$

in terms of simple roots. According to equation (12) the U(1) electric charge of the gauge particle corresponding to this root is

$$q = e\hbar\alpha.\phi/a = e\hbar\, \alpha.\lambda_\phi / \sqrt{(\lambda_\phi)^2} = n_\phi\, e\hbar(\alpha_\phi)^2/2\sqrt{(\lambda_\phi)^2} = n_\phi \bar{q}$$

where n_ϕ is the (integer) coefficient in (23) of the simple root α_ϕ not orthogonal to ϕ. $\bar{q}$ is the smallest unit of U(1) charge carried by the gauge particles. Thus q is indeed quantized when K is semisimple as anticipated before[14].

The mass of this gauge particle is, by equation (13)

$$M = |n_\phi a\bar{q}|$$

Particles with $|n_\phi|>1$ may be regarded as unstable[56] in the sense that they have the same mass and U(1) quantum number as $|n_\phi|$ gauge particles with U(1) charge $n_\phi \bar{q}/|n_\phi|$. This inconvenience may be avoided for certain special choices of symmetry breaking in which n_ϕ can only take the values 0 or $\pm$ 1 as the different roots of G are considered.

The necessary and sufficient condition for this is that $\lambda_\phi/(\alpha_\phi)^2$ be a minimal weight of G^v [20]. If this were a magnetic weight then this is the condition that the corresponding monopole is Brandt-Neri stable in the sense of section 3. Thus the same mathematical condition plays two distinct physical roles. As we shall show later there is a tendency for this special sort of symmetry breaking to occur in nature.

Now we come to the difficulty mentioned at the beginning of this section. We saw that the magnetic weights of classical monopole solutions belonged to the root lattice of G^v and so could be written

$$\alpha^v = \sum_1^r m_i \alpha^{(i)v} \quad ; \quad \alpha^{(i)v} = \alpha^{(i)}/(\alpha^{(i)})^2$$

If the monopole is self-dual, its mass is $|m_\phi \bar{g} a|$ by a similar argument to that above. We would like to be able to prove that if the solution is elementary in the sense that it should correspond to an elementary field then its magnetic weight α^v must actually be a root of G^v. This would fit in with the generalization of the conjecture of Montonen and myself. The quandary is that we can construct spherically symmetric self-dual solutions with $|m_\phi|$ too large for α^v to be a root of G^v. We could discard these solutions by the argument applied to gauge particles and say that this solution is a special case of a multimonopole solution in which the monopoles accidentally overlap[56,21,57]. This is reasonable but then the same argument might discard genuine roots of G^v if they had $|m_\phi| > 1$. The difficulty could be avoided by considering a symmetry breaking such that the roots of G^v have only $m_\phi = 0$ or ± 1. The necessary and sufficient condition for this is that λ_ϕ be a minimal weight of G[20]. If both this and the previous condition are satisfied then G must have roots all of the same length.[20] Further the spectrum of gauge particles and monopoles would precisely correspond as in the G=SU(2) case which is the simplest example of this situation. Further the K^v

weights defined by the $|m_\phi| = 1$ roots of G^v would be minimal[20]. This case, with what we shall henceforth refer to as "special symmetry breaking" promises to be the most favourable for the generalization of the duality conjecture.

It is for the case when G has roots of unequal length that E. Weinberg[21] found explicit self-dual, $|m_\phi| = 1$ (and hence stable) monopole solutions with K^v weights which are not minimal. It was these counter examples to the Brandt-Neri result which we mentioned in section 3 as also providing topologically equivalent monopoles with gauge inequivalent magnetic weights.

Goddard and I[20] listed all possible "special symmetry breakings" of simple gauge groups G and they constitute a fraction of the possibilities yielding semisimple K. In Table 2 I have drawn a sequence of groups which will illustrate some of these possibilities and are of special interest from both a mathematical and physical point of view[32].

Mathematically Table 2 depicts a series of exceptional groups E_8 down to E_3,[58] obtained successively by the amputation of the extreme point of the right leg of the Dynkin diagram. Unlike the four series of classical groups SU(r+1), SO(2r+1), SO(2r) and Sp(2r) which are infinite in extent, this series is finite but like them, its lowest members are common to other series e.g. SO(10) and SU(5) or else not simple (SU(3) x SU(2)). According to the algorithm mentioned above, a Higgs field along the fundamental weight corresponding to the point about to be amputated yields the exact symmetry subgroup which is locally U(1) times the group corresponding to the amputated Dynkin diagram which appears immediately below in the table.

TABLE 2

G	D(G)	Z(G)
E_8	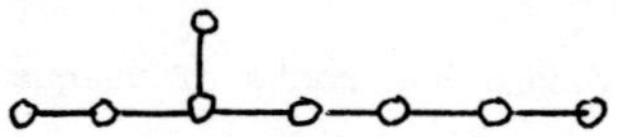	1
E_7	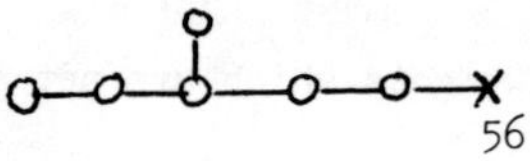	Z_2
E_6	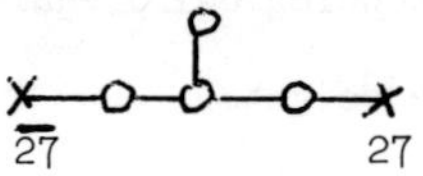	Z_3
E_5=SO(10)	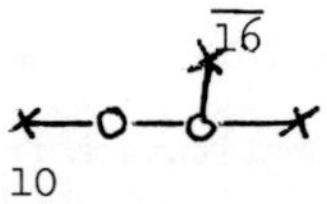	Z_4
E_4= SU(5)	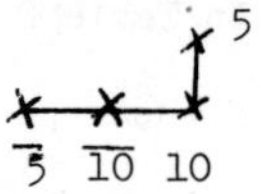	Z_5
E_3= SU(3) x SU(2)	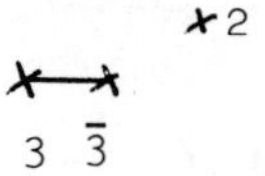	Z_6

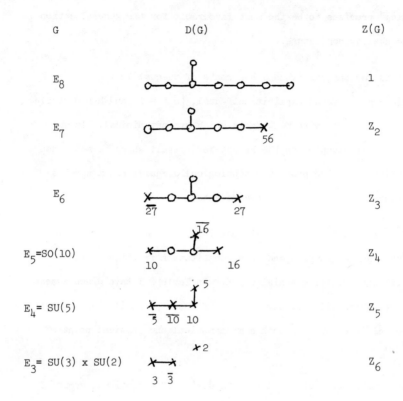

On each Dynkin diagram in Table 2 we have indicated the points corresponding to fundamental weights which are also minimal by a cross rather than a **circle**. These points are labelled by the dimensionality of the representation whose highest weight is the weight in question. These are the minimal representations mentioned in Table 1. We see that in each case except E_8 the amputated point corresponds to a minimal weight so that the symmetry breaking is special in the sense above.

I explained in section 2 the importance of the centre of the group. In the third column of Table 2 appears the centre of the group E_n. Remarkably it is always a cyclic group of order 9-n which increases as n diminishes[32]. It can be checked that 9-n is one more than the number of minimal representations (see Table 1) as it should be.

As was said before the groups E_8 to E_3 (with possible U(1) factors) are candidates for grand unified groups of the fundamental interactions with the fundamental fermions assigned to the minimal representations[31]. The exception is E_8 which has no minimal representation and for which the fermions would lie in the adjoint representation. In these schemes the natural scheme of symmetry breaking is via the special mechanisms we have described (again with the exception of E_8). More discussion of this scheme is given in my Erice 1981 lecture[32] and the intriguing point is the relationship between ideas in monopole theory and this scheme which has some relation to reality.

Mathematically there is nothing very special about the groups SU(3) x SU(2), SU(5) or SO(10) per se, but they do have a special significance as the lowest members of the sequence of E_n groups in Table 2. I therefore wish to say that any 'explanation' of the role of these groups in physics will very likely involve this complete sequence and its interrelationship through the special symmetry breaking discussed above.

It is not my brief to discuss the main activity presented at this meeting, the construction of exact solutions to the self-dual equations but in view of its relevance I shall make two comments.

The first is that the understanding of exact solutions describing several monopoles, particularly for arbitrary groups, should shed much light on the ideas discussed here. It would aid a better understanding of the intermonopole interactions as discussed in section 3 and it should resolve the difficulty mentioned in this section by determining which of the spherically symmetric monopole solutions should be regarded as elementary.

The second point is that there have been many different approaches to the construction of exact self-dual solutions. Nearly all of them have in common[59] an emphasis I made myself here in summer 1980[60], namely a connection with the inverse scattering method which applies to certain field theories in two dimensions. Indeed Atiyah has repeatedly emphasized that the twistor approach to the Atiyah Ward[61] and ADHM[62] constructions should be regarded as a generalization of this approach. In a sense then, the arguments of my talk have moved full circle since they started with reference to the Sine Gordon-massive Thirring model quantum equivalence, models in which the inverse scattering method has been exploited[26,27]. In the future lies the theory of quantum monopoles, presumably N=4 extended supersymmetric.

Finally it seems to me that I am entitled to say that the duality conjectures are still alive and indeed healthy in the sense of providing us with new questions and results.

REFERENCES.

1. P. Goddard, J. Nuyts, and D. Olive, Nucl. Phys. B125, 1 (1977).

2. C. Montonen and D. Olive, Phys. Lett. 72B, 117 (1977).

3. G. 'tHooft, Nucl. Phys. B79, 276 (1974).

4. A.M. Polyakov, JETP. Lett. 20, 194 (1974).

5. P.A.M. Dirac: Lectures on Quantum Field Theory. Academic Press (New York (1966).

6. Yu. S. Tyupkin, V.A. Fateev and A.S. Shvarts JETP Letts. 21, 42 (1975).

7. M.I. Monastyrsky and A.M. Perelomov. JETP Letts. 21, 43 (1975).

8. S. Coleman: in New Phenomena in Subnuclear Physics. (Proc. 1975. Int. School of Physics 'Ettore Majorana') ed. A. Zichichi (New York: Plenum) p.297.

9. E. Lubkin: Ann. Phys. New York 23, 233 (1963).

10. T.T. Wu and C.N. Yang: Phys. Rev. D12, 3845 (1975).

11. P. Goddard and D. Olive, Rep. Prog. Phys. 41 , 1357 (1978).

12. P.A.M. Dirac: Proc. Roy. Soc. A133, 60 (1931).

13. E. Corrigan and D. Olive: Nucl. Phys. B110, 237 (1976).

14. P. Goddard and D. Olive: Nucl. Phys. B191, 511 (1981).

15. G. Racah: "Lectures on Lie groups". CERN yellow report 61-8 and in "Group Theoretical Concepts and methods in Elementary Particle Physics" pl-36, Gordon and Breach, New York (1964).

16. J. Humphreys, Introduction to Lie algebras and representation theory (Springer, Berlin 1972).

17. F. Englert and P. Windey, Phys. Rev. D14,2728 (1976).

18. R. Brandt and F. Neri, Nucl. Phys. B161, 253 (1979).

19. S. Coleman, Seminars given at CERN (1979) unpublished.

20. P. Goddard and D. Olive: Nucl. Phys. B191, 528 (1981).

21. E. Weinberg, Nucl. Phys. B167 (1980) 500.

22. E. Weinberg: private communication.

23. T.H.R. Skyrme: Proc. Roy. Soc. A262, 237 (1961).
R.Streater and I.F. Wilde: Nucl. Phys. B24, 561 (1970).
S. Coleman: Phys. Rev. D11, 2088 (1975).
S. Mandelstam: Phys. Rev. D11, 3026 (1975).

24. D. Olive in Mathematical Problems in Theoretical Physics (Ed. K. Osterwalder, Lecture Notes in Physics 116 (Springer, Berlin, 1980). p.249.

25. D. Zwanziger: Phys. Rev. 137, B647 (1965), and Talk at this meeting.

26. L.A. Takhtadzhyan and L.D. Faddeev. Teor. Mat. Fiz. 21, 160 (1974).

27. M. Ablowitz, D. Kaup, A.C. Newell and H. Segur. Phys. Rev. Lett. 30, 1262 (1973).

28. M.K. Prasad and C.M. Sommerfeld: Phys. Rev. Lett. 35, 760 (1975).

29. S. Mandelstam: Physics Reports 23C, 245 (1976).

30. L. Susskind and J. Kogut; Physics Reports 23C, 348 (1976).

31. E_3: S. Weinberg: Phys. Rev. Lett. 19, 1264 (1967).
A. Salam: Proc. 8th. Nobel Symposium ed. N. Svartholm (Wiley, New York, (1968).
E_4: H. Georgi and S. L. Glashow: Phys. Rev. Lett. 32, 438 (1974).
E_5: H. Fritzsch and P. Minkowski: Ann. Phys. 93, 193 (1975).
E_6: F. Gursey, P. Ramond and P. Sikivie: Phys. Lett. 60B, 177 (1976)
Y. Achiman and B. Stech: Phys. Lett. 77B, 389 (1978).

31.contd. E_7: P. Sikivie and F. Gursey: Phys. Rev. D16, 816 (1977).
E_8: N.S. Baaklini: Phys. Lett. 91B, 376 (1980).
I. Bars and M. Günaydin: Phys. Rev. Lett. 45, 859 (1980).

32. D.I. Olive: 'Relations between Grand Unified and Monopole Theories" ICTP/81/82-6. Invited talk at study conference on "Unification of the fundamental interactions II". Erice, October 1981.

33. E. Corrigan, D. Olive, D.B. Fairlie and J. Nuyts: Nucl. Phys. B106, 475 (1976).

34. P.W. Higgs: Phys. Rev. Lett. 12, 132(1964):
Phys. Rev. Lett. 13, 508 (1964); Phys. Rev. 145, 1156 (1966).
T.W.B. Kibble: Phys. Rev. 155, 1557 (1967)
F. Englert and R. Brout: Phys. Rev. Lett. 13, 321 (1964)

35. D. Olive: Physics Reports 49, 165 (1979).

36. D. Olive: Nucl. Phys. B153, 1 (1979).

37. E.B. Bogomolny: Sov. J. Nucl. Phys. 24, 449 (1976).

38. S. Coleman, S.Parke, A. Neveu and C.M. Sommerfield: Phys. Rev. D15, 554 (1977).

39. M. Lohe: Physics Letters 70B, 325 (1977).

40. B. Julia and A. Zee: Phys. Rev. D11, 2227 (1975).

41. N. Manton: Nucl. Phys. B126, 525 (1977)
S. Magruder: Phys. Rev. 17, 3257 (1978).
W. Nahm: Phys. Lett. 79B, 426 (1978)

42. A. d'Adda, R. Horsley and P. Di Vecchia: Phys. Lett. 76B, 298 (1978).

43. F. Gliozzi, J. Scherk and D. Olive. Nucl. Phys. B122, 253 (1977).

44. L. Brink, J. Schwarz and J. Scherk: Nucl. Phys. B121, 77 (1977)

45. H. Osborn: Talk at this meeting.

46. H. Osborn: 83B, 321 (1979).

47. C.N. Yang: Talk at the meeting.

48. P. West: Nucl. Phys. B106, 219 (1976)

49. E. Witten and D. Olive: Phys. Lett. 78B, 97 (1978).

50. P. Zizzi: Nucl. Phys. B189, 317 (1981).

51. P. Rossi, Phys. Lett. 99B, 229 (1981).

52. F.A. Bais and W. Troost: Nucl. Phys. B178,125 (1981).

53. W. Nahm: Seminar at Imperial College, Spring 1979.

54. S. Ferrara and B. Zumino: Nucl. Phys. B79, 413 (1974).
D.R.T. Jones: Phys. Lett. 72B, 199 (1977).
H.N. Pendleton and E.C. Poggio: Phys. Lett 72B, 200 (1977).
L.V. Avdeev, O.V. Tarasov and A.A. Vladimirov:
Phys. Lett. 96B, 94 (1980).
M.T. Grisaru, M. Roček and W. Siegel:
Phys. Rev. Lett. 45, 1063 (1980).
W.E. Caswell and D. Zanon: Phys. Lett. 100B, 152 (1980).

55. M. Sohnius and P. West. Phys. Lett. 100B, 245 (1981).

56. F.A. Bais: Phys. Rev. D18, 1206 (1978).

57. N. Ganoulis, P. Goddard and D. Olive: "Self dual monopoles and Toda Molecules". Imperial College preprint ICTP/81/82-4.

58. R. Gilmore: "Lie groups, Lie algebras and Some of their applications" Wiley New York (1974) p.314.

59. Talks at this meeting by: M.F. Atiyah, F.A. Bais, E. Corrigan, N. Ganoulis, Z. Horváth, D. Maison and W. Nahm.

60. D. Olive: Imperial College Preptint ICTP/80/81-1, to be published in Proceedings of International Summer Institute on Theoretical Physics organised by Wuppertal University at Bad Honnef - Plenum Press, New York/London.

61. M.F. Atiyah and R.S. Ward. Comm. Math. Phys. 55, 117 (1977).

62. M.F. Atiyah, N. Hitchin, V. Drinfeld and Yu. Manin: Phys. Lett. 65A, 185 (1978).

Semiclassical Methods for Quantising Monopole Field Configurations

H. Osborn

DAMTP, University of Cambridge, Silver Street, Cambridge CB3 9EW, UK.

The semiclassical procedure for quantising stationary solutions is reviewed and its application to monopole solutions in gauge theories is outlined. The discussion is given in terms of the background gauge and angular momentum properties are discussed. It is shown how monopoles can be present in supersymmetric gauge theories when the monopole states form supermultiplets and possess spin. A calculation which would give the electric dipole moment of a monopole with spin is described.

1. Introduction

Given the recent extensive progress[1] in obtaining explicit monopole solutions of the static field equations of gauge theories in the Bogomolny, Prasad, Sommerfield[2] limit it is natural to consider if there are any similar simplifications of the corresponding quantum field theories. There are various possible avenues to explore but the most obvious is to straightforwardly embed the classical solutions in the quantum field theory. In this semiclassical approach it is necessary to consider the problems of quantising systems with extended stable configurations, the dynamical variables being explicit functions of various parameters.

Concerning monopoles in quantum field theory natural questions which arise are

1. What are the quantum properties of monopoles, quantum numbers such as spin, and what are the quantum corrections to classical values of quantities such as mass? What is the spectrum of excited states and how are static quantities such as form factors and, with spin, dipole moments to be calculated? Further how are scattering amplitudes with monopole external states to be obtained?
2. How does the renormalisation procedure work in the presence of monopole background fields?
3. Are there any ambiguities in quantisation in the presence of monopoles or new parameters not present at the classical level?
4. Do monopoles have a significant role in determining the dynamics of particular quantum field theories?

5. Are there any special quantum field theories with monopoles with dramatic properties? A very nice possibility would be the Montonen-Olive[3] conjecture of duality between the elementary particles and the topological monopole states.

The semiclassical approach answers some of these questions, mainly those under (1), and may provide hints for the others. The basic input is to expand the quantum fields about the classical configurations and is valid if the fluctuations remain small. This programme provides a strong stimulus for obtaining further knowledge of the classical theory since, in order to be able to undertake explicit calculations, it is necessary to have Greens functions for differing spin differential operators in the presence of the classical monopole solutions, zero modes for these operators and their normalisation matrix and in some cases also their functional determinants. In the BPS limit, which is the only case in which actual monopole solutions have been obtained, this information should be realisable in terms of the ADHMN construction of these solutions[4].

Concerning the important question of ambiguities raised under (3) above the semiclassical procedure, subject to minor provisos, allows in principle well defined calculations to all orders of perturbation theory for quantum matrix elements in the presence of monopole field configurations. While for differing monopole charges it is necessary to define separate sectors of the quantum field theory, each with its own space of states, a strong constraint is provided by cluster decomposition as when the vacuum sector contains well separated monopole anti-monopole pairs.

In section 2 we outline a version of the semiclassical method in a form which would be applicable to multi-monopole configurations and following this discuss the problems of applying this to gauge theories, considering mainly the simplest SU(2) monopole. Various papers have considered this some time ago[5], their results are recapitulated in part here but with a treatment which has differences in detail. Subsequently in section 5 we consider monopoles in N=4 supersymmetric gauge theories and finally describe a calculation which would give the electric dipole moment of supersymmetric monopoles.

2. Semiclassical Method

In order to discuss the quantisation of a system which contains a stationary configuration depending on parameters which are taken to be time dependent dynamical variables we employ a Hamiltonian approach[6] in which the quantum theory is defined in terms of operators obeying standard canonical commutation relations. To prune the discussion of inessential complications we restrict attention initially to the schematic model constituted by a single multi component real dynamical variable ϕ_i (i may denote continuous variables as for space dependent fields) with Lagrangian

$$L = \tfrac{1}{2}\dot{\phi}^T\dot{\phi} - V(\phi) , \tag{2.1}$$

and corresponding Hamiltonian

$$H = \tfrac{1}{2}\pi\pi^T + V(\phi) , \quad \pi = \dot{\phi}^T . \tag{2.2}$$

This framework is sufficient to encompass the later treatment of monopoles in gauge theories. The quantum theory for the dynamical system described by (2.1) and (2.3) is obtained for

ϕ,π operators obeying the canonical commutation relations at equal times,

$$[\phi,\phi^T] = [\pi^T,\pi] = 0 \ , \qquad\qquad (2.3)$$

$$[\phi,\pi] = i\hbar 1 \ .$$

The representation space for these commutation relations then defines the set of states whose matrix elements give observable physical amplitudes.

We assume that for this model there exists a fixed configuration $\phi_s(\alpha)$, depending on continuous real parameters α_a varying over some parameter space M, which is dynamically stable in that ϕ remains close to ϕ_s under time evolution. The dominant motion of the system is then given by the variation in time of the parameters, $\alpha_a(t)$. Thus

$$\phi = \phi_s + f \ , \qquad\qquad (2.4)$$

with the fluctuations f small and evaluated perturbatively, while the dynamics of the collective variables $\{\alpha_a\}$ are treated without approximations. For the quantum theory, with ϕ an operator, the $\{\alpha_a\}$ are a set of hermitian commuting operators, also commuting with f. However, unless constraints are imposed, the dynamical variables have been artificially increased resulting in a spurious gauge freedom. It is most convenient to eliminate this by taking f to be orthogonal to the variations in ϕ_s induced by infinitesimal variation of the parameters α_a. Thus the constraint

$$z^a = \frac{\partial}{\partial\alpha_a}\phi_s \ , \qquad z^{aT}f = 0 \ , \qquad\qquad (2.5)$$

is imposed on f . In terms of the $\{z^a\}$ the normalisation matrix

$$\eta^{ab} = z^{aT} z^b \tag{2.6}$$

is assumed to be non-singular $\det \eta \neq 0$, corresponding to no redundant parameters, and so $\eta(\alpha)$ is positive definite and symmetric and so provides a natural metric on M , with inverse η_{ab}, $\eta_{ac}\eta^{cb} = \delta_a^b$. To provide an appropriate realisation of the operator algebra (2.3) it is possible to follow the quantisation procedure due to Dirac for constrained dynamical systems which applied in this context[6,7] gives

$$\begin{gathered} \pi = (\bar{P}^a - \bar{\pi} \frac{\partial}{\partial \alpha_a} f)\, O^{-1}_{ab} z^{bT} + \bar{\pi} , \quad \bar{\pi} z^a = 0 , \\ O^{ab} = \eta^{ab} - \frac{\partial}{\partial \alpha_b} z^{aT} f = z^{aT} \frac{\partial \phi}{\partial \alpha_b} , \end{gathered} \tag{2.7}$$

where it is necessary that

$$\frac{\partial}{\partial \alpha_a} (\bar{\pi} f) = 0 . \tag{2.8}$$

In terms of the basic dynamical variables, α_a, $\bar{P}^b$, f, $\bar{\pi}$ the non-zero commutation relations are

$$\begin{gathered} -\frac{i}{\hbar} [\alpha_a, \bar{P}^b] = \delta_a^b , \quad -\frac{i}{\hbar} [f, \bar{\pi}] = 1 - P_Z , \\ -\frac{i}{\hbar} [f, \bar{P}^a] = \frac{\partial}{\partial \alpha_a} f, \quad -\frac{i}{\hbar} [\bar{\pi}, P^a] = \frac{\partial}{\partial \alpha_a} \bar{\pi} , \end{gathered} \tag{2.9}$$

with $P_Z = z^a \eta_{ab} z^{bT}$ the projector onto the space spanned by $\{z^a\}$ and O^{-1}_{ab} in (2.7) defined by expansion about η_{ab} . It is necessary to carefully choose the operator ordering in (2.9) so that π is hermitian and then obeys the commutation relations

(2.3), by virtue of (2.9) the result is guaranteed to $O(\hbar)$ by Dirac's procedure. Such ordering problems are resolvable in simple cases but are neglected here. The dependance of $f, \bar{\pi}$ on $\{\alpha_a\}$ is arbitrary, consistent with $P_Z f = 0$, $\bar{\pi} P_Z = 0$ and (2.8). From (2.7)

$$\bar{P}^a = \pi \frac{\partial \phi}{\partial \alpha_a} . \tag{2.10}$$

The representation of (2.9) determines the quantum theory in the presence of the configuration ϕ_s .

In semiclassical calculation ϕ_s is taken to be a solution of the classical equations

$$\left.\frac{\delta V(\phi)}{\delta \phi}\right|_{\phi = \phi_s} = 0 , \tag{2.11}$$

with energy $E_s = V(\phi_s)$, independent of α . Defining

$$M_{ij} = \left.\frac{\delta^2 V(\phi)}{\delta \phi_i \delta \phi_j}\right|_{\phi = \phi_s} , \tag{2.12}$$

then, from (2.2) and (2.7)

$$H = E_s + \tfrac{1}{2}(\bar{P}^a - \bar{\pi} \frac{\partial}{\partial \alpha_a} f)\, O^{-1}_{ab} \Omega^{bc} O^{-1}_{cd}\, (\bar{P}^d - \bar{\pi} \frac{\partial}{\partial \alpha_d} f) + H_F + H_I ,$$

$$H_F = \tfrac{1}{2}(\bar{\pi}\bar{\pi}^T + f^T M f) , \tag{2.13}$$

and $H_I = O(f^3)$, by virtue of (2.11). In (2.13) operator ordering is again neglected but this is unambiguously determined for H by (2.2) once that of π in (2.7) is fixed. The quadratic piece H_F in (2.13) determines the semiclassical spectrum in the presence of the classical configuration ϕ_s . This may be calculated in terms of the eigenvalues ω_n^2 of M

$$Mv_n = \omega_n^2 v_n , \quad v_n^T v_m = \delta_{nm} , \tag{2.14}$$

for real eigenvectors v_n. By virtue of local stability of ϕ_s $\omega_n^2 \geq 0$ and the zero modes for which $\omega_n^o = 0$ are assumed to lie entirely in the space spanned by the $\{z^a\}$ in (2.5). Consistent with (2.8) the commutation relations of $f, \bar{\pi}$ are realised by taking, at some initial time

$$f = \sum_n \left(\frac{\hbar}{2\omega_n}\right)^{\frac{1}{2}} (a_n + a_n^\dagger) v_n , \quad \bar{\pi} = \sum_n \left(\frac{\omega_n \hbar}{2}\right)^{\frac{1}{2}} \frac{1}{i} (a_n - a_n^\dagger) v_n^T ,$$

$$[a_n, a_m^\dagger] = \delta_{nm} , \tag{2.15}$$

where $\omega_n > 0$, since $f, \bar{\pi}$ are orthogonal to the zero modes. With the expansion in (2.16)

$$H_F = \delta E_s + \sum_n \hbar\omega_n a_n^\dagger a_n , \quad \delta E_s = \tfrac{1}{2}\hbar \sum_n \omega_n , \tag{2.16}$$

and the zero point energy for the harmonic oscillator provides the first quantum correction to the classical energy E_s. In field theories the corresponding vacuum energy shift has to be subtracted and a finite result necessitates careful regularisation and renormalisation.

The semiclassical ground state for H is conveniently given by $|S(\alpha)\rangle$ for $\alpha \in M$, which are simultaneous eigenvectors of the operators $\{\alpha_a\}$, satisfying

$$a_n |S(\alpha)\rangle = 0 ,$$

$$\langle S(\alpha')|S(\alpha)\rangle = \frac{1}{(\det \Omega^{ab}(\alpha))^{\frac{1}{2}}} \prod_a \delta(\alpha'_a - \alpha_a) . \tag{2.17}$$

so that the expectation value of ϕ for these states is the classical configuration $\phi_s(\alpha)$. A basis of excited states is given by the action of the creation operators $\{a_n^\dagger\}$. Neglecting the fluctuations $f,\bar{\pi}$ in H , since for $\omega_n^2 > 0$ these may be supposed to remain small, the dynamics of the variables $\{\alpha_a\}$ is isolated and from (2.13) is governed by the Hamiltonian

$$H_s = \tfrac{1}{2}\bar{P}^a \mathcal{M}_{ab}(\alpha)\bar{P}^b \ . \tag{2.18}$$

This breaks the classical degeneracy in energy of the set of states $\{|S(\alpha)\rangle\}$ for $\alpha \in M$, the shifted energy eigenstates being determined by the Schrödinger equation obtained by taking $\bar{P}^a \to -\hbar i\frac{\partial}{\partial\alpha_a}$ in H_s . Solutions of the ordering problem in other contexts[8] suggests that in this case the correct result is $H_s = -\tfrac{1}{2}\hbar^2\nabla^2$ where ∇^2 is the Laplacian acting on scalar functions on M in terms of the metric $\mathcal{M}(\alpha)$. With $H_o = E_s + H_s + H_F$ now assumed to be treated exactly a well defined propagator for f exists, due to the orthogonality to the zero modes, and a sensible perturbation expansion to all orders can be set up, expanding H about H_o . To $O(\hbar^2)$ careful treatment of ordering is essential. For this calculational scheme to be justified it is necessary that the initial configuration ϕ_s is quantum mechanically as well as classically stable, there are no lower energy configurations which may be reached by tunnelling.

The treatment of the dynamics of the collective variables $\{\alpha_a\}$ may be further simplified if the general configuration $\phi_s(\alpha)$ can be obtained by a group of linear transformations, M is a group manifold,

$$\phi_s(\alpha) = U(\alpha)\phi_s \ , \quad U(\alpha)^T U(\alpha) = 1 \ . \tag{2.19}$$

If for infinitesimal transformations $U(\alpha) \to 1 + \alpha_a T^a$ then $z^a = T^a \phi_s$ is a zero mode and the field operators can be represented by

$$\phi = U(\alpha)(\phi_s + f) , \qquad z^{aT} f = 0 ,$$

$$\pi = (\bar{P}^a - \bar{\pi} T^a f)\, O^{-1}_{ab} z^{bT}\, U(\alpha)^T + \bar{\pi}\, U(\alpha)^T , \quad \bar{\pi} z^a = 0 , \tag{2.20}$$

where the only dependence on α is as displayed. O remains as in (2.7) with

$$\mathcal{N}^{ab} = z^{aT} z^b = -\phi_s^T T^a T^b \phi_s , \tag{2.21}$$

and if $[T^a, T^b] = c^{ab}{}_c T^c$ the fundamental non-zero commutation relations are

$$[U(\alpha), \bar{P}^a] = i\hbar U(\alpha) T^a , \qquad [f, \bar{\pi}] = i\hbar (1 - P_Z) ,$$

$$[\bar{P}^a, \bar{P}^b] = -i\hbar c^{ab}{}_c \bar{P}^c . \tag{2.22}$$

Now in (2.18) $\mathcal{N}_{ab}$ is independent of α so that, for instance,

$$\frac{i}{\hbar} [H_s, \bar{P}^a] = -c^{ab}{}_c \omega_b \bar{P}^c , \quad \omega_b = \mathcal{N}_{bd} \bar{P}^d , \tag{2.23}$$

and corresponding to (2.10)

$$\bar{P}^a = \pi U(\alpha) T^a U(\alpha)^T \phi = R^a{}_b(\alpha) P^b , \quad P^b = \pi T^b \phi . \tag{2.24}$$

It is straightforward to calculate with (2.23) that

$$[H_s, P^a] = 0 , \quad [P^a, P^b] = i\hbar c^{ab}{}_c P^c . \tag{2.25}$$

Hence R represents a transformation from space to body fixed axes and the $\{P^a\}$ are thus conserved charges to leading order in the semiclassical approximation. To higher orders in the perturbation expansion the conserved charges are restricted to those generators of the exact symmetries, for which $V(U\phi) = V(\phi)$. For abelian transformations $U(\alpha)$ there is obviously no distinction between $\bar{P}^a$ and P^a.

When fermions are present as well the semiclassical treatment is easily extended. For a Lagrangian of the generic form

$$L_D = \psi^\dagger \dot{\psi} - \psi^\dagger \mathfrak{M}(\phi)\psi , \tag{2.26}$$

the associated Hamiltonian is

$$H_D = \psi^\dagger \mathfrak{M}(\phi)\psi , \tag{2.27}$$

and for quantization it is necessary to require the standard operator anti-commutation relations

$$\{\psi,\psi^\dagger\} = 1\hbar . \tag{2.28}$$

Semiclassically ϕ is expanded about ϕ_s and the leading term is obtained by taking $\phi \to \phi_s$ in (2.27). It is then appropriate to expand ψ in terms of eigenvectors of $\mathfrak{M}(\phi_s)$,

$$\mathfrak{M}(\phi_s)u_n = \omega_n u_n , \tag{2.29}$$

where, decomposing the eigenvectors as $u_n^\pm$ for $\omega_n^\pm \gtrless 0$ and u_i^o $i = 1 \ldots p$ for $\omega_i = 0$,

$$\hbar^{-\frac{1}{2}}\psi = \sum_i a_i u_i^o + \sum_n (a_n u_n^+ + b_n^\dagger u_n^-) , \tag{2.30}$$

with

$$\{a_n,a_m^\dagger\} = \delta_{nm} , \quad \{b_n,b_m^\dagger\} = \delta_{nm} . \tag{2.31}$$

The presence of zero modes in (2.30) is essential in order to have a complete set. From (2.30) we get

$$\psi^\dagger \mathfrak{M}(\phi_s)\psi = -\delta E_D + \sum_n \hbar(\omega_n^+ a_n^\dagger a_n + |\omega_n^-| b_n^\dagger b_n) \ ,$$

$$\delta E_D = \tfrac{1}{2}\hbar \sum_n (\omega_n^+ + |\omega_n^-|) \tag{2.32}$$

The semiclassical ground state is now also defined by

$$a_n|S\rangle = b_n|S\rangle = 0 \ , \tag{2.33}$$

ensuring no fermionic excited states, and where there is now an extra negative fermionic zero point energy $-\delta E_D$. However, in order to realise the algebra of the zero mode operators $\{a_i, a_j^\dagger\} = \delta_{ij}$, it is crucially necessary to have additionally a degenerate multiplet of states $\{|S_r\rangle\}$, degeneracy 2^p , corresponding to the unexcited classical configuration ϕ_s . If ψ carries quantum numbers, such as fermion number or spin, the operators a_i, $a_j^\dagger$ will define a representation on the states $\{|S_r\rangle\}$ with definite differing values of these quantum numbers. Thus the number and specific properties of the fermion zero modes for (2.29) plays a crucial role in determining the possible allowed quantum numbers for semiclassical states. If the multiplet $\{|S_r\rangle\}$ does not correspond to an irreducible representation of the exact symmetries of the theory the degeneracy may be expected to be removed in higher orders of the perturbation expansion.

3. Background Field Method for Gauge Theories

The semiclassical analysis outlined above applies to classical monopole fields in gauge theories but with some complications due to the requirements of handling the gauge freedom. Suppose for simplicity the gauge field A_μ is coupled to a single scalar field ϕ in the adjoint representation,

$$\mathcal{L} = -\frac{1}{4}(F_{\mu\nu}F^{\mu\nu}) + \tfrac{1}{2}(\mathcal{D}_\mu\phi\mathcal{D}^\mu\phi) - V(\phi) \ , \qquad (3.1)$$

where

$$F_{\mu\nu} = \partial_\mu A_\nu - \partial_\nu A_\mu + e A_\mu \wedge A_\nu \ ,$$
$$\mathcal{D}_\mu\phi = \partial_\mu\phi + e A_\mu \wedge \phi \ , \qquad (3.2)$$

with the notation $A_\mu = A^a_\mu t_a$, $(t_a t_b) = \delta_{ab}$, $t_a \wedge t_b = c_{abc}t_c$. For this theory there is the standard conserved gauge invariant energy momentum tensor

$$\theta^{\mu\nu} = -(F^{\mu\alpha}F^\nu_\alpha) + (\mathcal{D}^\mu\phi\mathcal{D}^\nu\phi) - g^{\mu\nu}\mathcal{L} \qquad (3.3)$$

which defines the energy and momentum density

$$\mathcal{E} = \tfrac{1}{2}(\underset{\sim}{E}.\underset{\sim}{E}) + \tfrac{1}{2}(\underset{\sim}{B}.\underset{\sim}{B}) + \tfrac{1}{2}(\pi\pi) + \tfrac{1}{2}(\underset{\sim}{\mathcal{D}}\phi.\underset{\sim}{\mathcal{D}}\phi) + V(\phi) \ ,$$
$$\underset{\sim}{\mathcal{P}} = (\underset{\sim}{E} \times \underset{\sim}{B}) - (\pi\,\underset{\sim}{\mathcal{D}}\phi) \ . \qquad (3.4)$$

with $E_i = F_{io}$, $B_i = \tfrac{1}{2}\varepsilon_{ijk}F_{jk}$, $\pi = \mathcal{D}_o\phi$. In the presence of monopole fields it is necessary to expand the quantum fields A^q_μ, ϕ^q of the quantum field theory about static solutions $\underset{\sim}{A},\phi$ of the classical equations. It is convenient to adopt the gauge $A^q_o = 0$ (or a non-dynamical external static c number field), which eliminates gauge arbitrariness in the time evolution but still leaves the freedom of t independent gauge transformations

$$\underset{\sim}{A}^q \to g^{-1}\underset{\sim}{A}^q g + \frac{1}{e}\, g^{-1}\underset{\sim}{\partial} g \ , \qquad \phi^q \to g^{-1}\phi^q g \tag{3.5}$$

and then expand about the classical solutions

$$\underset{\sim}{A}^q = \underset{\sim}{A} + \underset{\sim}{a} \ , \qquad \phi^q = \phi + f \ . \tag{3.6}$$

$-\underset{\sim}{E}, \pi$ are the momenta conjugate to the dynamical variables $\underset{\sim}{a}, f$ but they are constrained by the non-dynamical equation (Gauss's law)

$$\chi \equiv \underset{\sim}{\mathcal{D}}^q . \underset{\sim}{E} - e\phi^q \wedge \pi = 0 \ . \tag{3.7}$$

Consistent operator commutation relations with this constraint can be imposed only on the gauge fields modulo gauge transformations. The unconstrained variables are identified by imposition of a gauge condition. The natural gauge choice is to take $\underset{\sim}{a}, f$ orthogonal to infinitesimal gauge transformations, $g \to 1 + \omega$, of the classical solutions $\underset{\sim}{A}, \phi$ or

$$\int d^3x \ \{(\underset{\sim}{a} . \underset{\sim}{\mathcal{D}}\omega) + e(f\phi \wedge \omega) = 0 \tag{3.8}$$

where ω is arbitrary and $\underset{\sim}{\mathcal{D}}$ is the covariant derivative for the classical gauge field $\underset{\sim}{A}$. For $\omega \to 0$ as $r \to \infty$ this implies

$$\underset{\sim}{\mathcal{D}} . \underset{\sim}{a} + e\phi \wedge f = 0 \quad \text{or} \quad \tilde{\mathcal{D}} . V = 0 \ , \qquad V = \begin{pmatrix} \underset{\sim}{a} \\ f \end{pmatrix} . \tag{3.9}$$

The resulting background gauge condition fixes the gauge uniquely, if $g \to 1$ as $r \to \infty$, for fluctuations in the neighbourhood of $\underset{\sim}{A}, \phi$ but preserves manifest gauge covariance with respect to gauge transformations on the background gauge fields $\underset{\sim}{A}, \phi$. In this treatment there is no forced requirement for transforming the classical solution to unnatural gauges. With $\underset{\sim}{a}, f$ restricted by (3.9) the conjugate momenta can be identified by letting

$$\begin{pmatrix} -\underset{\sim}{E} \\ \pi \end{pmatrix} = \Pi + \mathcal{D}\mu, \quad \mathcal{D}\mu = \begin{pmatrix} \underset{\sim}{\mathcal{D}}\mu \\ e\phi \wedge \mu \end{pmatrix}, \quad \tilde{\mathcal{D}}.\Pi = 0 . \tag{3.10}$$

μ can then be determined by solving the constraint (3.7)

$$\Box\mu = \rho , \quad \rho = eV^T \wedge \Pi ,$$

$$\Box\mu = -\tilde{\mathcal{D}}^q.\mathcal{D}\mu = \Delta\mu - e\underset{\sim}{a} \wedge .\underset{\sim}{\mathcal{D}}\mu - e^2 f \wedge (\phi \wedge \mu) , \tag{3.11}$$

$$\Delta\mu = -\tilde{\mathcal{D}}.\mathcal{D}\mu = -\underset{\sim}{\mathcal{D}}^2\mu - e^2\phi \wedge (\phi \wedge \mu) .$$

Δ is positive definite so there is no problem with inversion, $\Box^{-1}$ is defined by expansion about Δ^{-1}. The basic canonical equal time commutation relations to be imposed on the dynamical variables is then

$$[V^a(x), \Pi^b(y)^T] = i\hbar G^{ab}(x,y) \tag{3.12}$$

where $G^{ab}(x,y)$ is the kernel for the projector onto the space orthogonal to the gauge modes

$$1 + \mathcal{D}\Delta^{-1}\tilde{\mathcal{D}} . \tag{3.13}$$

The Hamiltonian operator is then given by the energy density in (3.4)

$$H = \int d^3x\, \mathcal{E}^q . \tag{3.14}$$

With (3.6), (3.10) and (3.11) it can be decomposed (for $\underset{\sim}{A}, \phi$ zero as in Gribov[9])

$$H = E_S + H_C + H_Q + H_I ,$$

$$H_C = \tfrac{1}{2}\int d^3x((\mathcal{D}\mu)^T\mathcal{D}\mu) = \tfrac{1}{2}\int d^3x\,(\rho\,\Box^{-1}\Delta\,\Box^{-1}\rho) , \qquad (3.15)$$

$$H_Q = \tfrac{1}{2}\int d^3x\,\{(\Pi^T\Pi) + (V^T M V)\} .$$

E_S is the energy of the classical fields $\underset{\sim}{A},\phi$, as given by the energy density (3.4) with $\underset{\sim}{E},\underset{\sim}{\pi}$ zero, H_I is the interaction containing terms of cubic or higher order in $\underset{\sim}{a},f$ and in the quadratic piece H_Q M is defined by

$$M\begin{pmatrix}\underset{\sim}{a}\\ f\end{pmatrix} = \begin{pmatrix}\underset{\sim}{\mathcal{D}}\times(\underset{\sim}{\mathcal{D}}\times\underset{\sim}{a}) - e^2\phi\wedge(\phi\wedge\underset{\sim}{a}) + e\underset{\sim}{B}\times\wedge\underset{\sim}{a} + e\phi\wedge\underset{\sim}{\mathcal{D}}f - e\underset{\sim}{\mathcal{D}}\phi\wedge f\\ -\underset{\sim}{\mathcal{D}}^2 f + \phi\wedge\underset{\sim}{\mathcal{D}}.\underset{\sim}{a} + 2e\underset{\sim}{\mathcal{D}}\phi\wedge.\underset{\sim}{a} + V''(\phi)f\end{pmatrix}. \qquad (3.16)$$

V,Π may then be expanded over a complete set of eigenvectors of M in the background gauge. The semiclassical asymptotic elementary particle states may be identified with the eigenstates of H_Q. To be precise the operator ordering in H has to be carefully specified but this has been resolved elsewhere[8].

Given such a quantisation of the gauge fields the momentum and angular moment operators, which serve to define the spin of any state, of the theory are

$$\underset{\sim}{P} = \int d^3x\,\underset{\sim}{\mathcal{P}}^q , \quad \underset{\sim}{J} = \int d^3x\,\underset{\sim}{x}\times\underset{\sim}{\mathcal{P}}^q \qquad (3.17)$$

with $\underset{\sim}{\mathcal{P}}^q$ constructed out of the operator fields $\underset{\sim}{A}^q,\phi^q$ and their conjugates as in (3.4). These results can be rewritten in canonical form using

$$\underset{\sim}{\mathcal{P}}^q = \underset{\sim}{\mathcal{P}}^q_c - \partial_i(E_i\underset{\sim}{A}^q) + (\chi\underset{\sim}{A}^q) ,$$
$$\underset{\sim}{\mathcal{P}}^q_c = -(-\underset{\sim}{E}\;\pi)\,\underset{\sim}{\partial}\begin{pmatrix}\underset{\sim}{A}^q\\ \phi^q\end{pmatrix}. \qquad (3.18)$$

Thus, with the constraint (3.7),

$$\underset{\sim}{P} = \int d^3x \, \underset{\sim c}{\mathcal{P}^q} \, , \tag{3.19}$$

so that $\underset{\sim}{P}$ is manifestly the generator of translations on the basic fields, up to an infinitesimal gauge transformation. For the angular momentum the canonical result using (3.18) is

$$\underset{\sim}{J}_c = \int d^3x \, (-\underset{\sim}{E} \, \pi) \, \underset{\sim}{\mathcal{J}} \begin{pmatrix} \underset{\sim}{A}^q \\ \phi^q \end{pmatrix} , \tag{3.20}$$

$$\mathcal{J}_i A_j = -(\underset{\sim}{x} \times \underset{\sim}{\partial})_i A_j - \varepsilon_{ijk} A_k , \; \underset{\sim}{\mathcal{J}} \phi = -(\underset{\sim}{x} \times \underset{\sim}{\partial}) \phi \; , \; [\mathcal{J}_i , \mathcal{J}_j] = \varepsilon_{ijk} \mathcal{J}_k \; .$$

However in this case, as shown explicitly later for monopoles, there may be a boundary surface contribution for long range gauge fields,

$$\underset{\sim}{J}_B = - \int_{r \to \infty} d^2\underset{\sim}{S} \, . \, (\underset{\sim}{E} \; \underset{\sim}{x} \times \underset{\sim}{A}^q) \; . \tag{3.21}$$

If present this modifies the usual straightforward discussion from (3.20) of the angular momentum for differing field configurations in the theory.

4. Application to Monopoles

If the potential $V(\phi)$ in (3.1) is such that its minimum value is achieved for

$$|\phi|^2 \equiv (\phi^2) \to F^2 \, , \tag{4.1}$$

then, with ϕ in the adjoint representation, the gauge group is broken to $U(1) \times K$ where , when (4.1) is realised, the generator of U(1) is ϕ . The corresponding U(1) abelian gauge field is naturally regarded as the electromagnetic field so that asymptotically, in the limit (4.1),

$$A_\mu^{e.m.} \sim \frac{1}{F}(\phi A_\mu) \ . \tag{4.2}$$

There then exist finite energy classical solutions representing monopoles. For $V = 0$, while retaining the boundary condition (4.1), these can be given explicitly in terms of solutions of the Bogomolny equations

$$\underset{\sim}{B} = \pm \underset{\sim}{\mathcal{D}} \phi . \tag{4.3}$$

The elementary monopole solutions are those which are spherically symmetric up to gauge transformations and can be written in a gauge so that[10]

$$\mathcal{J}_i'\begin{pmatrix} A_j \\ \phi \end{pmatrix} \equiv \mathcal{J}_i\begin{pmatrix} A_j \\ \phi \end{pmatrix} + t_i \wedge \begin{pmatrix} A_j \\ \phi \end{pmatrix} = 0 \tag{4.4}$$

with $\mathcal{J}_i$ defined by (3.20) and t_i generators of a SU(2) subgroup $t_i \wedge t_j = \varepsilon_{ijk} t_k$. For gauge group SU(2) then, with t_i the generators, spherical symmetry implies

$$\phi(\underset{\sim}{x}) = \frac{\underset{\sim}{x}.\underset{\sim}{t}}{er^2} H(r) \ , \quad \underset{\sim}{A}(\underset{\sim}{x}) = \frac{\underset{\sim}{x} \times \underset{\sim}{t}}{er^2}(K(r) - 1) \ , \quad r = |\underset{\sim}{x}| \tag{4.5}$$

with asymptotically $H(r) \to \mp Fer$, $K(r) \to 0$ for magnetic charges $g = \pm 4\pi/e$. In the BPS limit, with $\phi, \underset{\sim}{A}$ satisfying (4.3), H,K can be easily found as functions of the single dimensionless variable Fer and the classical energy or mass $M = 4\pi F/e$[2] . For multi-monopoles with magnetic charge g in this case $M = F|g|$.

Such classical solutions can be embedded into the quantum field theory associated to (3.1) within the framework described in the previous two sections. The development of the semiclassical method in the context of the commutation relations (2.3) for the schematic model there is extended for the gauge theory to a corresponding treatment of (3.12). Thus the parameters of the general monopole solution, for given magnetic charge, are treated as dynamical variables with the fluctuation fields orthogonal to the zero modes of M in the background gauge. Global stability of the solution is guaranteed by conservation of magnetic charge.

It is straightforward to identify at least some of the zero modes. Due to translational symmetry there are three arising from infinitesimal translation of the classical solution

$$Z_i^t = -\partial_i \begin{pmatrix} \underset{\sim}{A}_j \\ \phi \end{pmatrix} + \mathcal{D} A_i , \quad \underset{\sim}{n} . \underset{\sim}{Z}^t = - \begin{pmatrix} \underset{\sim}{B} \times \underset{\sim}{n} \\ \underset{\sim}{n} . \mathcal{D} \phi \end{pmatrix} , \tag{4.6}$$

so that Z^t satisfies (3.9). The normalisation matrix is readily computed,

$$\int d^3x (Z_i^{tT} Z_j^t) = \delta_{ij} M , \tag{4.7}$$

since

$$S_{ij} = (Z_i^{tT} Z_j^t) = \theta_{ij} + \delta_{ij} \mathcal{E} \tag{4.8}$$

with θ_{ij}, given by (3.3), (3.4) (with $\pi, \underset{\sim}{E}$ zero) and for static solutions $\partial_i \theta_{ij} = 0$ implying $\int d^3x \theta_{ij} = 0$. Following the treatment in (2.19) and (2.22) a collective variable $\underset{\sim}{X}$ for translations and conjugate generator $\underset{\sim}{P}$ for the translation group may be introduced by

$$U^t(\underset{\sim}{X})\begin{pmatrix} \underset{\sim}{A}^q(\underset{\sim}{x}) \\ \phi^q(\underset{\sim}{x}) \end{pmatrix} = \begin{pmatrix} \underset{\sim}{A}^q(\underset{\sim}{x}-\underset{\sim}{X}) \\ \phi^q(\underset{\sim}{x}-\underset{\sim}{X}) \end{pmatrix} \tag{4.9}$$

$$[U^t(\underset{\sim}{X}),\underset{\sim}{P}] = -i\hbar U^t(\underset{\sim}{X})\underset{\sim}{\partial}$$

where $U^t(\underset{\sim}{X}) = e^{-\underset{\sim}{X}.\underset{\sim}{\partial}}$, $[X_i,P_j] = i\hbar\delta_{ij}$. From (2.18) and (4.7) the contribution to the Hamiltonian H_s acting on the monopole states is

$$\frac{\underset{\sim}{P}^2}{2M} \tag{4.10}$$

which with $\underset{\sim}{P}$ the momentum corresponds to the usual non-relativistic expansion of the relativistic energy $(M^2 + \underset{\sim}{P}^2)^{\frac{1}{2}}$. Acting on the monopole state then by (2.20) the field theory canonical momenta, neglecting the fluctuating fields, have contributions

$$\underset{\sim}{E}^t(\underset{\sim}{x}) = -\frac{\underset{\sim}{P}}{M}\times\underset{\sim}{B}(\underset{\sim}{x}-\underset{\sim}{X}) \ , \quad \pi^t(\underset{\sim}{x}) = -\frac{\underset{\sim}{P}}{M}\cdot\underset{\sim}{\mathcal{D}}\phi(\underset{\sim}{x}-\underset{\sim}{X}) \tag{4.11}$$

so that the associated momentum density from (3.4) becomes

$$\mathcal{P}_j^t(\underset{\sim}{x}) = \frac{P_i}{M}S_{ij}(\underset{\sim}{x}-\underset{\sim}{X}) \ , \tag{4.12}$$

with S_{ij} given by (4.8). Thus $\underset{\sim}{P} = \int d^3x\,\underset{\sim}{\mathcal{P}}^t$, reproducing the exact result (3.17). For spherically symmetric monopoles centred at $\underset{\sim}{x} = \underset{\sim}{0}$, and obeying (4.4), it is easy to see that

$$\int d^3x\, x_k S_{ij}(\underset{\sim}{x}) = 0$$

and hence the contribution from (4.12) to the piece of the angular momentum operator that acts on the semiclassical monopole state becomes

$$\underset{\sim}{J}_S = \int d^3x \; \underset{\sim}{x} \times \underset{\sim}{\mathcal{P}}^t = \underset{\sim}{X} \times \underset{\sim}{P} \quad . \tag{4.13}$$

This is of course the standard expression for orbital angular momentum. In the absence of any additional terms in $\underset{\sim}{J}$ acting on the monopole state alone the spin of such spherically symmetric monopoles is zero.

Besides the translational modes (4.6) there is a further general normalisable zero mode which has the form of a gauge mode obeying also the background gauge condition,

$$z^g = -\frac{1}{F}\begin{pmatrix} \underset{\sim}{\mathcal{D}}\phi \\ 0 \end{pmatrix} + \mathcal{D}\omega = \begin{pmatrix} \underset{\sim}{a}^g \\ f^g \end{pmatrix} \tag{4.14}$$

with $\omega = 0(1/r)$ as $r \to \infty$ and determined uniquely by

$$\Delta\omega = \frac{1}{F} V'(\phi) \quad . \tag{4.15}$$

z^g is orthogonal to the translational modes z_i^t and has a normalisation

$$N = \int d^3x (z^{gT} z^g) = -\frac{1}{F}\int_{r\to\infty} d^2\underset{\sim}{S} . (\underset{\sim}{a}^g \phi) \quad . \tag{4.16}$$

In the BPS limit for $V = 0$ then $\omega = 0$ in (4.14) and in this case (4.16) gives $N = M/F^2$. Since this mode is necessary for completeness in the expansion of the fluctuation fields $\underset{\sim}{a}, f$ in the background gauge it is necessary to identify the appropriate parameter in the classical solution that serves as the associated dynamical variable. This is achieved by considering the unbroken global $U(1)$ subgroup, parameterised in terms of a phase θ , of the overall gauge group. The associated transformation on the quantum fields is then

$$U^g(\theta)\begin{pmatrix} \underset{\sim}{A}^q \\ \phi^q \end{pmatrix} = \begin{pmatrix} g_\theta^{-1}\underset{\sim}{A}^q g_\theta + \frac{1}{e}\, g_\theta^{-1}\underset{\sim}{\partial} g_\theta \\ g_\theta^{-1}\phi^q g_\theta \end{pmatrix} \tag{4.17}$$

with asymptotically for $r \to \infty$

$$g_\theta(\underset{\sim}{x}) = \exp\left(-\frac{e\theta}{F}\,\phi(\underset{\sim}{x})\right)\ , \tag{4.18}$$

so that $e\theta$ is an angle, defined modulo 2π. Apart from the limit shown in (4.18) g_θ can be chosen arbitrarily so long as it forms a U(1) subgroup of the gauge group. A possible choice is to take $F \to |\phi(\underset{\sim}{x})|$ for all $\underset{\sim}{x}$ in (4.18). The conjugate operator to θ , which generates these U(1) transformations, can be defined, in the limit when (4.18) is valid, by

$$[g_\theta(\underset{\sim}{x}), Q] = -i\hbar e\,\frac{\phi(\underset{\sim}{x})}{F}\, g_\theta(\underset{\sim}{x})\ . \tag{4.19}$$

Hence Q has eigenvalues $ne\hbar$ and, identifying the unbroken U(1) gauge symmetry with that for electromagnetism, Q is the quantised electric charge operator. This interpretation is confirmed by considering, analogously to (4.11), the associated contributions containing Q to the canonical momenta acting on monopole states,

$$\underset{\sim}{E}^g = -\frac{Q}{N}\,\underset{\sim}{a}_\theta^g\ , \quad \pi^g = \frac{Q}{N}\, f_\theta^g \tag{4.20}$$

where $X_\theta = g_\theta^{-1} X g_\theta$. Thus we obtain

$$Q = \int_{r\to\infty} d\underset{\sim}{S}.\underset{\sim}{E}^{e.m.} \tag{4.21}$$

using, by (4.2), in this limit $\underset{\sim}{E}^{e.m.} = (\underset{\sim}{E}^g\phi_\theta)/F$ and also (4.16). Just as in (4.10) there is a piece

$$\frac{Q^2}{2N} \tag{4.22}$$

in the Hamiltonian H_s acting on monopole states. The treatment of θ as a dynamical variable thus ensures that the classical monopole corresponds in the quantum theory to a tower of states of differing electric charges, as well as momentum. This is related to classical dyon solutions[11]. In the BPS limit the result (4.22), with $N = |g|/F$, is exactly what would be expected by expansion of the classical dyon energy $F(g^2 + Q^2)^{\frac{1}{2}}$ where the charge Q is unquantised. The analogous result when the potential V is non-zero has been shown by Affleck and Manton[5].

For a single monopole and gauge group SU(2) $\underset{\sim}{Z}^t, Z^g$ exhaust the normalisable zero modes so the monopole state is uniquely specified by the quantum numbers $\underset{\sim}{P}, Q$[12]. For multimonopoles in SU(2) there are $4|n| - 1$ parameters, for $g = 4\pi n/e$, which can be interpreted as the positions and relative U(1) phases of the individual monopoles. Including the gauge mode discussed above there are then $4|n|$ zero modes[12] requiring consideration in a semiclassical treatment. The Hamiltonian H_s constructed according to (2.18) from the zero modes in the background gauge describes, as discussed by Manton[13], the classical motion of slowly moving multi-monopoles along geodesics in the parameter space with Ω as the metric. The dependence of H_s on the relative phase is associated with fluctuations in charge of the individual monopoles, the overall charge remaining conserved. For gauge groups larger than SU(2), even for spherically symmetric monopoles, there are zero modes associated with the unbroken gauge groups K[14].

The angular momentum properties of the excited states in the presence of a monopole is also of interest and can be discussed in this framework. Supposing that for a spherically symmetric classical monopole solution the asymptotic behaviour is

$$\phi(\underset{\sim}{x}) \sim \mp F\hat{\underset{\sim}{x}}.\underset{\sim}{t} , \quad \underset{\sim}{A}(\underset{\sim}{x}) \sim - \frac{\underset{\sim}{x} \times \underset{\sim}{t}}{er^2} , \tag{4.23}$$

as from (4.5) for SU(2), then assuming this to be the leading behaviour of the quantum fields the surface term (3.21) gives

$$\underset{\sim}{J}_B = \frac{1}{e} \int_{r\to\infty} d^2\underset{\sim}{S} . (\underset{\sim}{E}\underset{\sim}{t}) \pm \frac{1}{e} \int_{r\to\infty} d^2\underset{\sim}{S} . \underset{\sim}{E}^{e.m.}\hat{\underset{\sim}{x}} . \tag{4.24}$$

Given that the equations of motion should determine $\underset{\sim}{E}^{e.m.} \sim Q\underset{\sim}{x}/4\pi r^3$ then the second term in (4.14) should be zero. Using the constraint (3.7) the first term can be written as a volume integral and added to $\underset{\sim}{J}_C$ in (3.20) gives a result of the same form but with $\underset{\sim}{\mathcal{J}} \to \underset{\sim}{\mathcal{J}}'$, defined by (4.4). Thus by (4.4), and inserting (3.6), the classical fields $\underset{\sim}{A}, \phi$ are not present in the total angular momentum but acting on the fluctuation fields $\underset{\sim}{a}, f$ there is an extra contribution containing the generators $\underset{\sim}{t}$ of the gauge group. If additional fields are present in different representations of the gauge group there may then be spin ½ states despite the theory containing only bosonic fields.

These features may be clearly seen by considering, following Hasenfratz and 't Hooft[5], a non-relativistic spinless particle coupled to the gauge field and neglecting the fluctuations $\underset{\sim}{a}, f$. Such a particle obeys the equations of motion[15]

$$m\ddot{\underset{\sim}{r}} = e\{(\underset{\sim}{E}(\underset{\sim}{r})I) + \dot{\underset{\sim}{r}} \times (\underset{\sim}{B}(\underset{\sim}{r})I)\} ,$$

$$\dot{I} + e\dot{\underset{\sim}{r}}.\underset{\sim}{A}(\underset{\sim}{r}) \wedge I + eA_0(\underset{\sim}{r}) \wedge I = 0 \tag{4.25}$$

where $I = I^a t_a$ denotes the dynamical gauge group charge vector. There may be additional interactions, which may allow bound states with the monopole, but they are not relevant here. The gauge fields obey the obvious non-abelian generalisations of Maxwell's equations with, apart from contributions due to the ϕ field, a charge, current density $eI\delta(\underset{\sim}{x}-\underset{\sim}{r})$, $eI\dot{\underset{\sim}{r}}\delta(\underset{\sim}{x}-\underset{\sim}{r})$. Then the angular momentum is defined to be

$$\underset{\sim}{J} = m\underset{\sim}{r} \times \dot{\underset{\sim}{r}} + \int d^3x\underset{\sim}{x} \times \underset{\sim}{\mathcal{P}} . \tag{4.26}$$

which is manifestly gauge invariant and conserved with the above equations of motion. Applying the same analysis as in eqs. (3.17)-(3.21), but now with the constraint $\chi(\underset{\sim}{x}) = eI\delta(\underset{\sim}{x}-\underset{\sim}{r})$

$$\underset{\sim}{J} = \underset{\sim}{r} \times \underset{\sim}{p} + \underset{\sim}{J}_C + \underset{\sim}{J}_B \tag{4.27}$$

where $\underset{\sim}{J}_C$, $\underset{\sim}{J}_B$ are given by (3.20), (3.21) and $p = m\dot{\underset{\sim}{r}} + e(\underset{\sim}{A}(\underset{\sim}{r})I)$. Then following the discussion of (4.23), (4.24), neglecting $\underset{\sim}{a}$, f and contributions arising from the motion of the monopole, using the constraint equation once more, the angular momentum of the non-relativistic particle in the presence of a spherically symmetric monopole becomes

$$\underset{\sim}{J} = \underset{\sim}{r} \times \underset{\sim}{p} + (I\underset{\sim}{t}) . \tag{4.28}$$

This result displays the additional terms arising from the gauge field interaction explicitly. Under quantisation $\underset{\sim}{r}$, p are conjugate operators and I becomes an operator satisfying $[I^a, I^b] = i\hbar c_{abc} I^c$, so that $(I\underset{\sim}{t})$ are then generators of SU(2). In appropriate representations for I this may obviously give spin ½ states for the particle in the presence of the monopole.

5. Supersymmetric Monopoles

In supersymmetric gauge theories monopoles enjoy an especial theoretical interest since the quantum corrections may be controlled and calculable. Also it is for these theories that the intriguing hints of a duality between the elementary and topological sectors have the greatest chance of being realised. The maximally symmetric such theory is that corresponding to N=4 supersymmetry. Since theories for lower N are contained as special cases we restrict attention to the N=4 case here. The basic field content is a vector A_μ, 6 scalars S_i and 4 Majorana fields $\psi, \bar{\psi} = \psi^T C$ (C being the charge conjugation matrix $C^T = -C$, $(C\gamma^\mu)^T = C\gamma^\mu$), all in the adjoint representation of an arbitrary gauge group. The Lagrangian, which may easily be obtained by dimensional reduction from simple supersymmetry in 10 dimensions, is, in terms of these component fields

$$\mathcal{L} = -\frac{1}{4}(F_{\mu\nu}F^{\mu\nu}) + \tfrac{1}{2}(\mathcal{D}_\mu S_i \mathcal{D}^\mu S_i) + \tfrac{1}{2}\, i\, (\bar{\psi}\gamma^\mu \mathcal{D}_\mu \psi)$$

$$-\frac{1}{2}\, i\, e(\bar{\psi}\lambda_i S_i \wedge \psi) - V(S) \, , \tag{5.1}$$

$$V(S) = \frac{1}{4}\, e^2 (S_i \wedge S_j S_i \wedge S_j) \, .$$

$\lambda_i = \gamma^0 \lambda_i^\dagger \gamma^0$ are a set of matrices acting on ψ, linear in γ_5, and, with $\lambda_i \gamma^\alpha = \gamma^\alpha \tilde{\lambda}_i$, $\tilde{\lambda}_i \lambda_j = \delta_{ij} + \Sigma_{ij}$, $\Sigma_{ij} = -\Sigma_{ji}$ being generators of SO(6). This is a global symmetry of $\mathcal{L}$ in (5.1), with ψ belonging to a 4-dimensional spinor representation. The conserved supercurrent corresponding to N=4 supersymmetry is

$$\mathcal{J}^\mu = \frac{i}{4}\, [\gamma^\alpha, \gamma^\beta]\, \gamma^\mu (F_{\alpha\beta}\psi) + i\gamma^\alpha\, \gamma^\mu \lambda_i\, (\mathcal{D}_\alpha S_i \psi)$$

$$-\frac{i}{2}\, \gamma^\mu \Sigma_{ij} (S_i \wedge S_j \psi) \, .$$

The quantum field theory corresponding to (5.1) may well be finite[16]. If so, in the standard realisation when all the fields are massless, it would be conformally invariant and there may be infra red problems and difficulties of interpretation. However an external mass scale may be introduced by hand by requiring that the scalar fields have a vacuum expectation value $\mathring{S}_i$ with

$$(\mathring{S}_i\mathring{S}_i) = F^2 , \quad \mathring{S}_i \wedge \mathring{S}_j = 0 \tag{5.3}$$

so that $V(\mathring{S}) = 0$ but otherwise $\mathring{S}_i$ is unconstrained. With supersymmetry $\mathring{S}_i$ is also undetermined in the quantum field theory. For F non-zero spontaneous symmetry breakdown occurs and, with the Higgs mechanism, there is a massive supermultiplet of elementary particles consisting of one spin 1 , four spin ½ and 5 spinless states, together with their antiparticles, all with $m^2 = e^2F^2\hbar^2$. This result for the mass is associated with the presence of central charges in supersymmetry algebra[17,18]. Defining

$$Q = \int d^3x\, \mathcal{J}^0 , \quad \bar{Q} = Q^T C , \tag{5.4}$$

this is

$$\frac{1}{\hbar}\{Q,\bar{Q}\} = 2\gamma^\mu P_\mu - 2\lambda_i(Q_i + i\gamma_5 T_i) ,$$

$$Q_i = \int_{r\to\infty} d\underset{\sim}{S}.(\underset{\sim}{E}S_i) , \quad T_i = \int_{r\to\infty} d\underset{\sim}{S}.(\underset{\sim}{B}S_i) . \tag{5.5}$$

With $S_i \to \mathring{S}_i$ for $r \to \infty$ the charges Q_i, T_i need not be zero. For massive states with spins ≤ 1 there is a unique representation of the algebra (5.5), described above for the elementary particles, of dimensions 2^4 (× 2 by CPT), which requires that the mass is determined by the central charges

$$M^2 = Q_iQ_i + T_iT_i \quad . \tag{5.6}$$

If $\mathring{S}_i = n_i \mathring{S}$ for n_i a fixed 6-dimensional unit vector, so that the global SO(6) symmetry is broken to SO(5), then monopole solutions of the Bogomolny equations (4.3) can be embedded as the static classical solutions for this theory by taking $S_i = n_i\phi$. In this case in (5.5) $Q_i = FQn_i, T_i = FGn_i$, with Q, G the electric, magnetic charges as discussed earlier, and (5.6) gives the standard dyon mass formula. The treatment of monopoles in the supersymmetric quantum field theory proceeds semiclassically in the same fashion as in section 4 but in this case there are additional effects due to the Fermi fields. In the presence of the classical monopole solutions the classical Dirac Hamiltonian, which gives the semiclassical energy spectrum of excited states, for this theory is defined by

$$\begin{aligned} \mathcal{M}_s\psi &= -i\gamma^o\underset{\sim}{\gamma}\cdot\underset{\sim}{\mathcal{D}}\,\psi + e\gamma_5 P\phi \wedge \psi \, , \\ P &= i\gamma_5\gamma^o\lambda_i n_i = P^\dagger \, , \ P^2 = 1 \, . \end{aligned} \tag{5.7}$$

It is straightforward[19] to show that the eigenvalues and eigenvectors of $\mathcal{M}_s$ in (5.7) are simply related to those of M in (3.16), with $V = 0$, using supersymmetry and suitable boundary conditions, so that the leading semiclassical corrections to the classical monopole mass exactly cancel. The zero modes, satisfying $\mathcal{M}_s\psi_o = 0$, determine the multiplet structure of the semiclassical monopole state, in accord with the discussion in section 2. In the BPS limit there are zero modes obtainable by supersymmetry transformations of the classical solution

$$\psi_o = \frac{1}{M^{\frac{1}{2}}} \underset{\sim}{\sigma}.\underset{\sim}{B}u \ , \quad Pu = \pm u \ ,$$

$$\int d^3x(\psi_o^\dagger\psi_o) = u^\dagger u \tag{5.8}$$

for M the classical monopole mass and $\underset{\sim}{\gamma} = \gamma^o\gamma_5\underset{\sim}{\sigma}$, $[\sigma_i,\sigma_j] = 2i\varepsilon_{ijk}\sigma_k$. If the quantum fermion field is expanded as $\psi = \psi_o + \psi_f$, with ψ_f orthogonal to ψ_o, then it is necessary, in order to realise the standard anticommutation relations of the form (2.28), that

$$\{u,u^\dagger\} = \frac{\hbar}{2}(1 \pm P) \ , \quad u^\dagger\gamma^o = u^T C \ . \tag{5.9}$$

The part of $\mathcal{J}^o$ in (5.2) relevant for computing the supersymmetry charges acting on the monopole states is

$$\gamma^o\underset{\sim}{\sigma}.\{(\underset{\sim}{B}\psi) + P(\underset{\sim}{\mathcal{D}}\phi\psi)\} + \gamma^o\gamma_5\{i(\underset{\sim}{\sigma}.\underset{\sim}{E}\psi) + P(\pi\psi)\} \tag{5.10}$$

and then neglecting quantum fluctuations using (5.8), (4.11) and (4.20)

$$Q_s = \{\gamma^o - \tfrac{1}{2}\underset{\sim}{\gamma}.\frac{\underset{\sim}{P}}{M} - \tfrac{1}{2}\lambda_i n_i \frac{FQ}{M}\}2M^{\frac{1}{2}}u \ . \tag{5.11}$$

Thus (5.9) ensures that the supersymmetry algebra (5.5) is realised on the monopole states, satisfying the mass condition (5.6), to within the limitations of the semiclassical method. The full supersymmetry charge contains additionally terms of quadratic and higher order in the fluctuation fields.

The spin content of the monopole may be obtained by recognizing that the generators of supersymmetry (5.4) have spin ½

$$[\underset{\sim}{J},Q] = -\tfrac{1}{2}\hbar\,\underset{\sim}{\sigma}\,Q \ . \tag{5.12}$$

Acting on the static monopole state the effective spin operator which arises from these fermion zero modes is then

$$\underset{\sim}{S} = \frac{1}{4} u^\dagger \underset{\sim}{\sigma} u , \quad [\underset{\sim}{S},u] = -\tfrac{1}{2}\hbar \underset{\sim}{\sigma} u \tag{5.13}$$

as this realises (5.12) when acting on static monopole states for which $Q \to 2M^{\frac{1}{2}}u$. To display the spin content explicitly u may be expanded, while satisfying the Majorana condition, in terms of conventional fermion creation annihilation operators by using a complete orthonormal set of spin ½ eigenvectors of the form $u_{s,n}$, $(C\gamma^o)^\dagger u^*_{s,n}$, $s = \pm\tfrac{1}{2}$, $n = 1,\ldots,$

$$u = \hbar^{\frac{1}{2}} \sum_{sin} (u_{s,n}\, a_{s,n} + (C\gamma^o)^\dagger u^*_{s,n}\, a^\dagger_{sin}) ,$$
$$\{a_{s,n}, a^\dagger_{s'n'}\} = \delta_{ss'}\delta_{nn'} . \tag{5.14}$$

This automatically ensures (5.9) and

$$\underset{\sim}{S} = \frac{\hbar}{2} \sum_{n,s,s'} a^\dagger_{s'n}(\underset{\sim}{\sigma})_{s's}a_{s,n} , \quad u^\dagger_{s',n'}\underset{\sim}{\sigma}\, u_{s,n} = \delta_{n'n}(\underset{\sim}{\sigma})_{s's} , \tag{5.15}$$

so that the Fermi vacuum has spin zero and $\{a^\dagger_{s,n}\}$ are creation operators for spin ½ .

For $N=4$ supersymmetry and $\tfrac{1}{2}(1 \mp P)u = 0$ u has 8 independent components and hence $n=1,2$ in (5.14). Thus there are 2^4 degenerate monopole states comprising the spin spectrum described before. In the case of $N=2$ a similar analysis can be undertaken but $n=1$ only so the monopole has just a spin ½ and two spin 0 states.

For a single monopole and gauge group SU(2) (5.8) gives all the normalisable Dirac zero modes in the adjoint representation and so there are no more degenerate monopole states than required for the elementary massive representation of the supersymmetry algebra. With large gauge groups further zero modes are present which may be associated with monopole multiplets characterised by quantum numbers of the unbroken gauge group K.

6. Electric Dipole Moment

Just as electrically charged particles have static magnetic dipole moments proportional to their spin we may expect a magnetic monopole with spin to possess an electric dipole moment. This can be calculated by considering the response of the magnetic monopole to a weak external static electric field. At the classical level the static equations from (3.1), with $\underset{\sim}{E} = \underset{\sim}{\mathcal{D}} A_o$

$$\underset{\sim}{\mathcal{D}} \times \underset{\sim}{B} = -e\ \phi \wedge \underset{\sim}{\mathcal{D}}\phi + eA_o \wedge \underset{\sim}{\mathcal{D}} A_o\ , \quad \underset{\sim}{\mathcal{D}} . \underset{\sim}{E} = e^2 \phi \wedge (A_o \wedge \phi)\ ,$$
$$-\underset{\sim}{\mathcal{D}}^2 \phi + V'(\phi) = -e^2 A_o \wedge (A_o \wedge \phi) \tag{6.1}$$

are such that the solutions for A_o small can be taken as $\underset{\sim}{A}, \phi + O(A_o^2)$ with $\underset{\sim}{A}, \phi$ a classical monopole solution and A_o satisfying the homogeneous linear equation

$$\Delta A_o = 0 \tag{6.2}$$

where Δ is defined by (3.11). There are no normalisable solutions of (6.2) but following (4.2), if there is an external applied electric field $\underset{\sim}{\mathcal{E}} = \underset{\sim}{\nabla} \varphi$, we may allow A_o to have asymptotically for $r \to \infty$ the behaviour

$$A_o \sim \frac{1}{F} \phi \varphi\ , \quad \underset{\sim}{E} \sim \frac{1}{F} \phi\ \underset{\sim}{\mathcal{E}} . \tag{6.3}$$

For any $\varphi, \nabla^2 \varphi = 0$, the asymptotic behaviour (6.3) should determine A_o uniquely (if $\varphi = 1$, $A_o = \phi/F - \omega$ from (4.15)). Computing the energy of this configuration with energy density (3.4) and $\pi = eA_o \wedge \phi$ then, since this is stationary at the classical solution and using $(\underset{\sim}{E}.\underset{\sim}{E}) + (\pi^2) = \underset{\sim}{\partial}.(\underset{\sim}{E}A_o)$, we get

$$E = M + E^{e.m.} + O(A_o^3) \ , \quad E^{e.m.} = \tfrac{1}{2} \int_{r\to\infty} d\underset{\sim}{S}.(\varphi \underset{\sim}{\mathcal{E}}) \ . \tag{6.4}$$

$E^{e.m.}$ is just the volume energy of the external electric field. This result implies that classically there is no electric polarisability of the monopole. For a constant external electric field and a spherically symmetric SU(2) monopole satisfying (4.5) it is sufficient to take in (6.2) and (6.3)

$$\varphi(\underset{\sim}{x}) = \underset{\sim}{x} . \underset{\sim}{\mathcal{E}} \ ,$$
$$A_o(\underset{\sim}{x}) = \underset{\sim}{x}.\underset{\sim}{t} \ \underset{\sim}{x}.\underset{\sim}{\mathcal{E}} \ f(r) + \underset{\sim}{t}.\underset{\sim}{\mathcal{E}} \ g(r) \ , \tag{6.5}$$

where asymptotically $f(r) \to \mp 1/r$, $g(r) \to 0$ to satisfy (6.3) and f,g satisfying coupled linear ordinary differential equations.

When an external electric field is present there is however a shift to first order in $\underset{\sim}{\mathcal{E}}$ in the eigenvalues of the Dirac Hamiltonian. In the adjoint representation this becomes

$$\mathcal{M}_s - i e A_o \wedge \tag{6.6}$$

which now contains an extra term linear in A_o , determined by (6.2) and (6.3), in addition to the unperturbed result $\mathcal{M}_s$ expressed in terms of the monopole solutions $\underset{\sim}{A}, \phi$. In the BPS limit $\mathcal{M}_s$ may be taken as in (5.7), but without any necessary specific form for P . The degeneracy of the zero modes $\{\psi_{oi}\}$, where this is an orthonormal set, of $\mathcal{M}_s$ is now broken. To first order, using standard perturbation theory, the eigenvalues ω_{oi} , which go to zero as $\underset{\sim}{\mathcal{E}} \to \underset{\sim}{0}$, are given by diagonalising

$$- i e \int d^3x (\psi_{oi}^{\dagger} A_o \wedge \psi_{oj}) \ . \tag{6.7}$$

The degenerate monopole multiplet, which is defined semiclassically as the representation space for the fermion creation and annihilation operators of the zero modes $\{\psi_{oi}\}$, is now split by the electric field into states with energy shifts $\hbar\omega_{oi}$. For a constant applied field $\underset{\sim}{\mathcal{E}}$ and in the BPS limit using (5.8) for ψ_{oi}, with $u_i^\dagger u_j = \delta_{ij}$, and the explicit form of the gauge fields (4.5) together with (6.5) then (6.7) gives the perturbed energy in terms of eigenvalues of the matrix

$$\delta E_{ij} = -\frac{\lambda\hbar}{Me} u_i^\dagger \underset{\sim}{\sigma} u_j . \underset{\sim}{\mathcal{E}} ,$$

$$\lambda = \frac{2}{3}\int d^3x \, \{K'^2(f + \frac{1}{r^2} g) + \frac{2}{r^3} K(K-1)g^2\} . \tag{6.8}$$

Since $\underset{\sim}{S}_{ij} = \frac{\hbar}{2} u_i^\dagger \underset{\sim}{\sigma} u_j$ is effectively the matrix element of the spin operator $\underset{\sim}{S}$ in this basis (6.8) is equivalent to the monopole possessing an electric dipole moment

$$\frac{2\lambda}{Me} \underset{\sim}{S} . \tag{6.9}$$

A similar result should still apply outside the BPS limit. In this simplified case however dimensional considerations show that λ is dimensionless and independent of any parameters so the electric dipole moment is $O(gh/M)$ with g the magnetic charge. Explicit calculation requires only determination of f,g in (6.5).

References

(1) See these Proceedings.

(2) E.B. Bogomolny, Sov. J. Nucl. Phys. 24, 449 (1976).
M.K. Prasad and C.M. Sommerfield, Phys. Rev. Lett. 35, 760 (1975).

(3) C. Montonen and D. Olive, Phys. Lett. 72B, 117 (1977).
F.A. Bais, Phys. Rev. D18, 1206 (1978).

(4) M.F. Atiyah, N.J. Hitchin, V.G. Drinfeld and Yu.I. Manin, Phys. Lett. 65A, 185 (1978).
W. Nahm, Phys. Lett. 90B, 413 (1980); 93B, 42 (1980); CERN preprint TH-3172; ICTP preprint IC/81/238 and in these proceedings.

(5) R. Jackiw, Rev. Mod. Phys. 49, 681 (1977).
E. Tomboulis and G. Woo, Nucl. Phys. B107, 221 (1976).
R. Jackiw and C. Rebbi, Phys. Rev. D13, 3398 (1976); Phys. Rev. Lett. 36, 1116 (1976).
P. Hasenfratz and G. 't Hooft, Phys. Rev. Lett. 36, 1119 (1976).
P. Hasenfratz and D.A. Ross, Nucl. Phys. B108, 462 (1976); Phys. Lett. 64B, 78 (1976).
S.B. Libby, Nucl. Phys. B113, 50 (1976).
N.H. Christ, A.H. Guth and E.J. Weinberg, Nucl. Phys. B114, 174 (1976).
A. Sinha, Phys. Rev. D16, 1828 (1979).
M.M. Ansourian, Phys. Rev. D14, 2732 (1976).
S.R. Wadia, Phys. Rev. D15, 3615 (1977).
K. Huang and D.R. Stump, Phys. Rev. Lett. 37, 545 (1976); Phys. Rev. D15, 3660 (1977).
O. Steinmann, Particles and Fields 6, 139 (1980).
I.K. Affleck and N.S. Manton, Nucl. Phys. B194, 38 (1982).

(6) E. Tomboulis, Phys. Rev. D12, 1678 (1975).

(7) E. Tomboulis and G. Woo, Ann. Phys. 98, 1 (1976).
A. Hosoya and K. Kikkawa, B101, 271 (1975).
J.L. Gervais, A. Jevicki and B. Sakita, Phys. Rev. 23, 281 (1978).

(8) N.H. Christ and T.D. Lee, Phys. Rev. D22, 939 (1980).

(9) V.N. Gribov, Nucl. Phys. B139, 1 (1978).

(10) D.I. Olive, Nucl. Phys. B113, 413 (1976).
D. Wilkinson and A.S. Goldhaber, Phys. Rev. D16, 1221 (1977).

(11) B. Julia and A. Zee, Phys. Rev. D11, 2227 (1975).

(12) E.J. Weinberg, Phys. Rev. D20, 936 (1979).

(13) N.S. Manton, Santa Barbara preprint NSF-ITP-81-116 .

(14) E.J. Weinberg, Nucl. Phys. B167, 500 (1980); and in preparation.

(15) S.K. Wong, Nuovo Cimento 65A, 689 (1970).

(16) K.S. Stelle preprint LPTENS 81/24, in Proceedings of the 1981 Nuffield Quantum Gravity Workshop.

(17) E. Witten and D.I. Olive, Phys. Lett. 78B, 99 (1978).
D.I. Olive, Nucl. Phys. B153, 1 (1979).

(18) H. Osborn, Phys. Lett. 83B, 321 (1979).

(19) A. D'Adda and P. di Vecchia, Phys. Lett. 73B, 162 (1978).

DUALITIES AND FERMIONS

B. Julia

Laboratoire de Physique Théorique de l'Ecole Normale Supérieure,
24 rue Lhomond, 75231 Paris, Cedex 05, France.

Dirac's introduction of magnetic monopoles in Maxwell's theory can be described as :

1) the observation of a symmetry (in four space-time dimensions) of some equations of motion ; U(1) is an invariance of the system :

$$\mathcal{F} \equiv \begin{pmatrix} F_{\mu\nu} \\ G_{\mu\nu} \end{pmatrix} = d\mathcal{A} \equiv \begin{pmatrix} A_\mu \\ B_\mu \end{pmatrix} , \quad \mathcal{F} = \Omega \tilde{\mathcal{F}} \equiv \begin{pmatrix} 0 & -1 \\ 1 & 0 \end{pmatrix} \begin{pmatrix} \frac{1}{2} \varepsilon_{\mu\nu}{}^{\rho\sigma} F_{\rho\sigma} \\ \frac{1}{2} \varepsilon_{\mu\nu}{}^{\rho\sigma} G_{\rho\sigma} \end{pmatrix}$$

2) the extension of this symmetry to matter fields coupled minimally. In non-abelian situations even step 1) has not been generalized, the new results (in particular the progresses that motivated this meeting) use a generalization of (anti) self-duality and lead to solutions of Yang-Mills equations or Einstein vacuum equations. My goal will be to review the essentially abelian generalizations of 1) that arise in supergravity theories (on their bosonic restriction). These are by now well-known : for example in "N = 8 supergravity" U(1) is replaced by E_7 global x SU(8) $_{\text{Local}}$ where some of the global transformations are combinations of duality transformations on the vector field strengths, chiral rotations on the spinor fields (coupled through Pauli interactions) and complex transformations of the spin 0 fields. What is not so widely recognized is the analogy between this mysterious E_7 and the equally mysterious SL(2,R) duality group of Ehlers and Papapetrou. The latter is an invariance of the set of stationary (or x^3 independent) solutions of Einstein's equations, and with some technical restriction an invariance of the equations of stationary axially symmetric space-times (or gravitational plane-waves). It is relevant for the construction of multimonopole solutions (L. Witten, Eötvös group), it is crucial for the generation of infinite

dimensional (Kac-Moody) "Lie" groups of symmetries of two-dimensional models (Einstein, monopoles, σ -models, integrablë systems and their non-local charges). A better understanding of this SL(2,R) is potentially useful for a geometrical interpretation of colour SU(3) (as a subgroup of the above SU(8)) if the conjectures of Cremmer-Julia and Zumino et al. apply to the real world, but also it might be useful for the discovery of a criterion of integrability of Hamiltonian systems, in particular the realistic system of equations describing gravitational plane waves : I am thinking of some signature of gravitational pulses that might help detection.

It seems rather discouraging not to have found any single monopole in fifty years, and the BPS monopoles are even more theoretical than Dirac's monopole. Yet, duality is most important as it transforms a strong doupling theory into a weakly coupled one that might be amenable to a perturbative treatment and it fits nicely with supersymmetry (D. Olive and H. Osborn in these proceedings).

We may now recall the two main motivations for supergravity :

1) Unification of fermions and bosons : why are the bosonic and fermionic Regge slopes equal ?

2) Hope for quantum finiteness. We recall Pauli's speculation of 1951 in his lectures on quantization of fields that the zero point energy of the vacuum could be made to vanish by compensating the boson contribution against the fermionic one. This dream was realized in supersymmetric theories (Zumino 1974). This feature might be required in renormalizable theories (Dirac : Lectures on Quantum Field Theory 1966), in the case of (super) gravitation however it is mandatory for predictivity of the theory and not only for aesthetic reasons. One recalls that the two theories that are candidates for finiteness are self-dual (CPT self-conjugate) : N = 4 super Yang-Mills and N = 8 supergravity.

I would like to conclude with a personal reason for confidence in the supersymmetry program : it led to a better understanding of the Kac-Moody symmetries of Einstein vacua and Einstein-Maxwell electrovacs (namely the bosonic sectors of N = 1 and N = 2 supergravities). These two theories are physically realistic and we can build on them to stay in contact with experiment. In particular, it is a tantalizing conjecture that one can define a duality "à la Dirac" for gravitation in four dimensions (the translation group being abelian) using the fermionic extension as a guide.

SOLITON SUPERMULTIPLETS

P.A. Zizzi
Department of Mathematics, King's College, Strand,
London WC2, England.

In this paper we will consider supersymmetric Yang-Mills theories which have solitons:

1) The N = 2 supersymmetric SU(2) Yang-Mills theory in 3+1D [1].
2) The N = 4 supersymmetric SU(2) Yang-Mills theory in 3+1D [2].
3) The USp(2) supersymmetric SU(2) Yang-Mills theory in 4+1D [3].

The theories 1) and 2) have solitons, the 't Hooft-Polyakov monopoles in the phase of broken gauge symmetry. In this phase, both the electric charge E of the unbroken Abelian subgroup U(1) and the magnetic charge G of the monopoles, appear in the supersymmetry algebra as central charges (electric-magnetic duality). The main result of Witten and Olive [1] was that the mass formula: $M = \langle \phi \rangle \sqrt{E^2 + G^2}$ (where M is the mass given by the Higgs mechanism to elementary particles and solitons) survives quantization because of the presence of E and G in the supersymmetry algebra as central charges. Osborn [2] extended the results of Witten and Olive to N = 4 and constructed the soliton supermultiplet. He found that the soliton supermultiplet in 3+1D has not only the same number of states as the massive supermultiplet of the elementary particles (because of the Higgs mechanism) but also an identical spin structure (because of the electric-magnetic duality), i.e. one spin 1 state, four spin $\frac{1}{2}$ states and five spin 0 states.

The theory 3) has solitons (instantons in the four-dimensional Euclidean space) in the phase of unbroken gauge symmetry. In this phase the supersymmetry algebra has a central charge that is the topological charge of the solitons [3]:

$$Z_T = -\frac{1}{4}\int d^4\vec{x}\, F^a_{\ell m}\tilde{F}^{\ell m a} = \frac{8\pi^2}{g^2}\, q \qquad (1)$$

(with $\ell, m = 1,2,3,4$, $a = 1,2,3$) where g is the coupling constant that in 4+1D has the dimension of a mass, and q is the topological number of the instantons. Then, the one-soliton ($q = 1$) has a mass: $M = Z_T$ (even in the absence of the Higgs mechanism, because of the self-duality of the instanton solution). In the phase of broken gauge symmetry, instead, the supersymmetry algebra has a central charge that is the electric charge E of the unbroken Abelian subgroup U(1). In this phase the elementary particles acquire a mass: $M = \langle \phi \rangle E$, given by the Higgs mechanism. Thus the supersymmetry algebra for the 4+1-dimensional theory does not involve both the electric charge and the topological charge in the same phase (no electric-magnetic duality).

In a recent paper [4] we investigated the structure of the soliton supermultiplet in 4+1D We considered the $SO(4) = SU(2) \otimes USp(2)$ Clifford algebra (obtained by the USp(2) supersymmetry algebra [3] in the phase of unbroken gauge symmetry):

$$\{a_\alpha , a^*_\beta\} = \delta_{\alpha\beta} \qquad \alpha,\beta = 1,2$$

$$\{a_\alpha , a_\beta\} = \{a^*_\alpha , a^*_\beta\} = 0 \ . \tag{2}$$

From the Clifford vacuum $|\Omega\rangle$ defined as: $a_\alpha|\Omega\rangle = 0$ we can construct the four states:

$$|\Omega\rangle \ , \quad a^*_1|\Omega\rangle \ , \quad a^*_2|\Omega\rangle \ , \quad a^*_1 a^*_2|\Omega\rangle \ . \tag{3}$$

The Clifford vacuum is a singlet of the SU(2) (Clifford) spin, but it belongs to a doublet under the USp(2) spin. Moreover, choose the Clifford vacuum to be the soliton state, that is a singlet of the little group O(4), then the lowest massive supermultiplet (the soliton supermultiplet) will contain two scalars and one fermion (the number of states must be doubled by CPT). In conclusion, the soliton supermultiplet in 4+1D has the same number of states (even in the absence of the Higgs mechanism) as the massive supermultiplet of the elementary particles, but a different spin structure (because of the absence of the electric-magnetic duality).

REFERENCES

[1] E. Witten and D. Olive, Phys. Letters B78, 97 (1978).

[2] H. Osborn, Phys. Letters B83, 321 (1979).

[3] P.A. Zizzi, Nucl. Phys. B189, 317 (1981).

[4] T. Marinucci and P.A. Zizzi, to appear in Nucl. Phys. B.

III: POINT MONOPOLES AND QUANTUM FIELD THEORY

BOUND STATES FOR THE e-g SYSTEM

Chen Ning Yang
Institute for Theoretical Physics
State University of New York
Stony Brook, New York 11794

THE PROBLEM

The motion of a Dirac electron in the field of a magnetic monopole has been studied in the papers listed in reference 1. A very interesting feature found in these papers was the existence of bound states. All the bound states were found for angular momenta

$$j = |q| - \frac{1}{2}, \quad (q = eg),$$

but not for higher angular momenta. The purpose of this report is to complete this analyses and discuss all bound states for angular momenta

$$j > |q| - \frac{1}{2} \tag{1}$$

The Hamiltonian studied is

$$H = \alpha\cdot(p - ZeA) + \beta M - \kappa q \beta(\sigma\cdot r)(2Mr^3)^{-1},$$

where we have added an extra magnetic moment $Ze\kappa(2M)^{-1}$ in order to eliminate the Lipkin-Weisberger-Peshkin difficulty discussed in reference 1.

Using the idea of sections[2] and using the following representations of the Dirac matrices

$$\alpha = \begin{pmatrix} 0 & \sigma \\ \sigma & 0 \end{pmatrix}, \qquad \beta = \begin{pmatrix} 1 & 0 \\ 0 & -1 \end{pmatrix}$$

it is convenient to write the wave function in the following form

$$\psi_{jm} = r^{-1} \begin{pmatrix} h_1 \xi^{(1)}{}_{jm} + h_2 \xi^{(2)}{}_{jm} \\ -i\,[h_3 \xi^{(1)}{}_{jm} + h_4 \xi^{(2)}{}_{jm}]\ \kappa q/|\kappa q| \end{pmatrix}. \tag{2}$$

As was shown in reference 1, the radial wave functions $h_i(r)$ satisfy the following equation

$$\Omega^{(o)}\, h^{(o)} = 0, \tag{3}$$

where
$$h^{(o)} = \begin{bmatrix} h_1 \\ h_2 \\ h_3 \\ h_4 \end{bmatrix}$$

$$\Omega^{(o)} = \partial_\rho - \mu a_3 \rho^{-1} + b_1 \rho^{-2} + A_o a_1 b_1 + iB_o a_1 b_2, \tag{4}$$

$$r = |\kappa q| \rho (2M)^{-1},$$

$$\mu = [(j + \tfrac{1}{2})^2 - q^2]^{1/2} > 0\ ,$$

$$A_o = \kappa q/2 \neq 0,\ B_o = \kappa q E (2M)^{-1}$$

and a_1, a_2, a_3 and b_1, b_2, b_3 are two sets of Pauli matrices. The boundary conditions are

$$\lim_{\rho\to o} h^{(o)}(\rho) = \lim_{\rho\to\infty} h^{(o)}(\rho) = 0. \tag{5}$$

THE METHOD

Eq. (3) is an eigenvalue equation for four ordinary differential equations of the first order in four unknowns. It is thus a generalization of the Sturm-Liouville problem. For the Sturm-Liouville problem the fundamental trick which allows for an elegant analysis of the eigenvalues is the definition of a phase angle (related to the logarithmic derivative of the wave function) which is monotonic with respect to the energy. It turns out that the eigenvalue problem (3) allows for a similar analysis through the definition of two phase angles.

Multiplying (3) on the left by ia_1b_2 we obtain

$$[(ia_1b_2)\partial_\rho - \frac{\mu}{\rho} a_2b_2 + \frac{a_1b_3}{\rho^2} + A_ob_3 - B_o]\, h^{(o)} = 0. \tag{6}$$

Choosing a representation where a_1, a_3, b_1, b_3 are real symmetrical and a_2, b_2 are equal to i times real antisymmetrical matrices, we find (6) to be of the form

$$[\omega_o\partial_\rho + V_o]h^{(o)} = B_oh^{(o)} \tag{7}$$

where $$\omega_o = \left(\begin{array}{cc|cc} & & & 1 \\ & & 1 & \\ \hline & -1 & & \\ -1 & & & \end{array}\right)$$

and V_o is real symmetrical.

It is convenient to make a further similarity transformation with

$$T = \begin{pmatrix} 1 & 0 & & \\ 0 & 1 & & \\ & & 0 & 1 \\ & & 1 & 0 \end{pmatrix}:$$

$$h^{(1)} = Th^{(o)}, \quad \omega_1 = T\omega_oT, \quad V_1 = TV_oT.$$

Then $$[\omega_1\partial_\rho + V_1]h^{(1)} = B_oh^{(1)} \tag{8}$$

where $$\omega_1 = \left(\begin{array}{cc|cc} & & 1 & \\ & & & 1 \\ \hline -1 & & & \\ & -1 & & \end{array}\right)$$

and V_1 is real symmetrical.

Equation (8) is now ready for an analysis which is similar to the Sturm-Liouville theory. We consider a finite internal $\rho = a$ to $\rho = b$ and first take the boundary conditions to be

$$\lambda = K_a \mu \text{ at } \rho = a, \tag{9a}$$

$$\lambda = K_b \mu \text{ at } \rho = b, \tag{9b}$$

where λ and μ are the upper and lower 2 components of $h^{(1)}$:

$$h^{(1)} = \begin{pmatrix} \lambda \\ \mu \end{pmatrix} . \tag{10}$$

We further assume K_a and K_b to be real symmetrical 2 x 2 matrices. That such conditions are relevant for the boundary condition (5) can be demonstrated after a simple analysis.

The key concepts in the generalized Sturm-Liouville analysis of (8) is contained in the following:

If Ψ is a real 4 x 2 matrix consisting of two columns each of which satisfies (8) and (9a), and are linearly independent, then

$$\tilde{\Psi} \omega_1 \Psi = 0 \text{ for all } \rho .$$

Further writing $\Psi = \begin{pmatrix} \xi \\ \eta \end{pmatrix}$,

where ξ and η are 2 x 2 matrices, then for any ρ $\xi\eta^{-1}$ is real symmetrical and $\frac{\partial}{\partial B_0}\left(\xi\eta^{-1}\right)$ is negative definite or semi-definite, if the determinant of η is not zero. [For the case when the determinant of η is zero at a point $\rho = \rho_0$, the analysis can be easily extended.] We can now define two phase angles from the eigenvalues of $\xi\eta^{-1}$. Both phase angles would then be monotonic in E.

THE RESULT

Through such an analysis we obtain the following bound state diagram for angular momenta j satisfying (1).

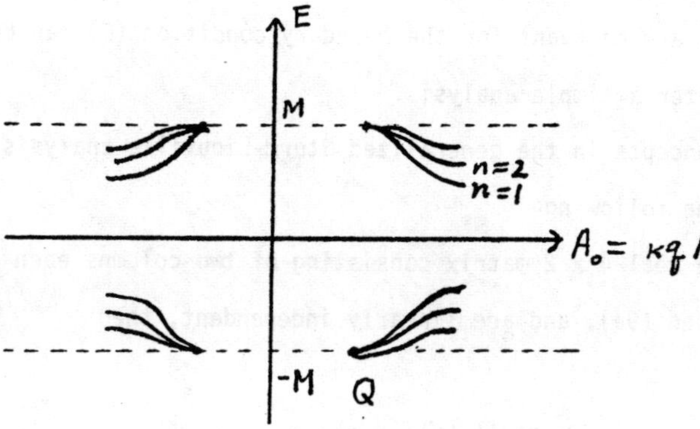

Bound-state energy for $j > |q| - \frac{1}{2}$. Solid lines show bound-state energies E_n as functions of A_o. The dashed lines are thresholds outside of which there are only scattering states. E = o is a bound state for all $A_o \neq 0$. The curves are symmetrical with respect to a reflection in either the A_o or the E axis. Each integral value of n > o gives rise to four curves. Only those for n = 1, 2 and 3 are shown. At the end point Q, the value of A_o is $\frac{1}{2}(\mu^2 - \frac{1}{4})$ where μ is given in the text.

REFERENCES

[1]Y. Kazama, Chen Ning Yang and Alfred S. Goldhaber, Phys. Rev. D15, 2287 (1977) and Y. Kazama and Chen Ning Yang, Phys. Rev. D15, 2300 (1977).

[2]Tai Tsun Wu and Chen Ning Yang Nuclear Physics B107, 365 (1976).

Point Monopoles in Quantum Field Theory

Daniel Zwanziger

Department of Physics
New York University
4 Washington Place
New York, New York 10003
U.S.A.

Abstract

Aspects of the quantum field theory of point charges and poles are reviewed. String independence of gauge invariant Euclidean Green's functions is exhibited by expressing them as functional integrals over classical charged particle loops. Electric and magnetic interactions are expressed in terms of the mutual inductance and mutual solid angle of pairs of loops in Euclidean 4-space.

Introduction

Fifty years ago, in an article worth reading today, Dirac[1] introduced the magnetic monopole into quantum mechanics. As the monopole has not yet been observed in nature, the present lively research on the subject is a tribute to the sheer beauty of his idea. Its progress may be gauged by comparing the present Proceedings with the proceedings of the Meeting on Monopoles held November 18-19, 1977 at New York University.[2]

Most of the recent effort is devoted to monopoles in non-Abelian gauge theories. These are the everywhere regular, classical soliton solutions first discovered by 't Hooft and Polyakov. I will not discuss these but, rather, the older magnetic monopole which is a point singularity in the Abelian theory. The two, however, are related because in the singular gauge where only, say, the third internal component of the 't Hooft-Polyakov solution is non-zero, it agrees,[3] at large distances, with the static vector potential introduced by Dirac, namely, in spherical coordinates,

$$\underset{\sim}{A}(\underset{\sim}{x}) = \frac{-g}{4\pi}\,\frac{1}{r}\,\frac{\sin\theta}{1-\cos\theta}\,\hat{\phi}\ . \tag{1}$$

This potential has a singularity along the positive z-axis called a "string" and may alternately be written

$$\underset{\sim}{A}(\underset{\sim}{x}) = \frac{g}{4\pi}\int_0^\infty ds\ \hat{n}\times\underset{\sim}{\nabla}\ \frac{1}{|\underset{\sim}{x}-\hat{n}s|}\ , \tag{2}$$

where $\hat{n}$ is a fixed unit vector. I will briefly review some aspects of monopole theory that I have worked on. The only new result will be an improvement in the argument which shows that the quantum field theory of charges and poles is independent of the path of the string. Whereas this property of the gauge-invariant Minkowskian Green's functions was exhibited previously,[4] here it will be done for the Euclidean Green's functions which are better behaved.

2. Two Fundamental Constants

Schwinger[5] has introduced the name "dyon" for particles with both electric and magnetic charge. The theory of dyons admits not one but two fundamental constants.[6] Let particle of type i have electric charge e_i and magnetic charge g_i which may be joined in the real two-vector

$$\vec{q}_i = (e_i g_i) \ . \tag{3}$$

As will be seen in Sec. 5, Dirac's quantization condition, generalized to dyons, reads, in suitable units,

$$\vec{q}_i \times \vec{q}_j \equiv (e_i g_j - g_i e_j) = Z_{ij} \tag{4}$$

where the Z_{ij} are integers. Let $\vec{q}_1$ and $\vec{q}_2$ be two charge vectors with minimal non-zero cross product, $\vec{q}_1 \times \vec{q}_2 = 1$. It follows that with

$\vec{q}_1$ and $\vec{q}_2$ as basis vectors the generic charge vector is given by

$$\vec{q}_i = Z_{i2}\vec{q}_1 + Z_{1i}\vec{q}_2 \ . \tag{5}$$

The Z_{i2} and Z_{1i} are arbitrary integers and this is a sufficient condition for all the Z_{ij} to be integers, so Eq. (4) provides no further constraints. Physical quantities are independent of direction in the 2-plane, as will be seen in Sec. 5. The only invariants which may be from the two vectors $\vec{q}_1$ and $\vec{q}_2$ are their lengths and the included angle. Call these three parameters e, g, and θ. As $\vec{q}_1 \times \vec{q}_2 = 1$ they are related by

$$eg \sin \theta = 1 \ . \tag{6}$$

Any independent pair of them may be chosen to characterize the theory of dyons.

3. Angular Distributions in Charge-Pole Processes

The S-matrix for the scattering of charged particle is invariant under local gauge transformations that vanish sufficiently rapidly at infinity. However the magnetic monopole has a string which extends out to infinity, and the S-matrix is not invariant, in general, under the type of gauge transformation extending to infinity that is necessary to change the direction $\hat{n}$ along which the string recedes to to infinity. In consequence, the S-matrix picks up a gauge transformation under rotations, so it does not transform like a scalar, and in general depends upon $\hat{n}$. This dependence is quite simple however and may be found explicitly for both non-relativistic and relativistic kinematics.[7] Although observables are of course n-independent, the fact that the S-matrix is not a scalar leads to characteristic angular distributions. For example, consider the decay at rest of a dyon with spin-one and polarization vector $\underset{\sim}{\varepsilon}$ into two spin-zero daughter particles with momentum directions $\hat{k}$ and $-\hat{k}$ respectively. Let the charge vectors $\vec{q}_i = (e_i, g_i)$ of each particle be $\vec{q}_1$, $\vec{q}_2$ and $\vec{q}_3$ respectively, so $\vec{q}_1 = \vec{q}_2 + \vec{q}_3$ because electric and magnetic charge is conserved. It follows that $\vec{q}_2 \times \vec{q}_3 = \vec{q}_2 \times \vec{q}_1 = \vec{q}_1 \times \vec{q}_3$. If these cross products vanish, the S-matrix $S = S(\hat{k})$ is a scalar, so $S(\hat{k}) = c\underset{\sim}{\varepsilon}\cdot\hat{k}$, and the angular distribution is $|S(\hat{k})|^2 = |c|^2 \underset{\sim}{\varepsilon}\cdot\hat{k}\,\hat{k}\cdot\underset{\sim}{\varepsilon}^*$, where c is a constant. However if the cross product is unity, corresponding to unit angular momentum in the crossed asymptotic electric and magnetic fields of the daughter particles, one has instead

$$S(\hat{k}) = c\underset{\sim}{\varepsilon}\cdot\underset{\sim}{a}_n(\hat{k}) \quad , \tag{7}$$

where

$$\underset{\sim}{a}_n(\hat{k}) = |\hat{k}\times\hat{n}|^{-1}\,[\hat{k}\times\hat{n} + i\,\hat{k}\times(\hat{k}\times\hat{n})] \; . \tag{8}$$

The angular distribution works out to be

$$|S(\hat{k})|^2 = |c|^2 \, [\underset{\sim}{\varepsilon}\cdot\underset{\sim}{\varepsilon}^* - \underset{\sim}{\varepsilon}\cdot\hat{k}\;\hat{k}\cdot\underset{\sim}{\varepsilon}^* - i(\underset{\sim}{\varepsilon}\times\underset{\sim}{\varepsilon}^*)\cdot\hat{k}] \; . \tag{9}$$

It is string independent of course, but violates parity and time reversal invariance. If the initial particle has spin up, $\vec{\varepsilon} = (1,i,0)/\sqrt{2}$, the decay angular distribution displays a characteristic up-down asymmetry

$$|S(\hat{k})|^2 = \frac{1}{2}\,|c|^2 \;(1-\cos\theta)^2 \tag{10}$$

instead of $\frac{1}{2}\,|c|^2\;(1-\cos^2\theta)$.

4. Local Lagrangian Quantum Field Theory of Charges and Poles

In 1947, Dirac[8] constructed a relativistic quantum field theory of charges and poles with dynamical variables to describe the motion of the strings. In 1966 Schwinger[9] developed a formulation free of string variables in terms of a non-local Hamiltonian density. It is also possible to formulate the quantum field theory of charges and poles in terms of a Lagrangian density which is a local function of the pair of electric and magnetic potentials A_μ and B_μ, and of the matter fields ψ_i, that also depends on a fixed direction n, $L = L(A,B,\psi_i,n)$. The electromagnetic field tensor $F_{\mu\nu} = -F_{\nu\mu}$ is expressed locally in terms of A and B according to

$$F = (1/n^2)(\{n\wedge[n\cdot(\partial\wedge A)]\} - {}^*\{n\wedge[n\cdot(\partial\wedge B)]\} , \tag{11}$$

where $(a\wedge b)$ is the anti-symmetric tensor $(a\wedge b)_{\mu\nu} = a_\mu b_\nu - a_\nu b_\mu$; $[n\cdot(a\wedge b)]_\mu = n\cdot a b_\mu - n\cdot b a_\mu$ and ${}^*F$ is the dual tensor ${}^*F_{\mu\nu} = \frac{1}{2}\varepsilon_{\mu\nu\kappa\lambda}F^{\kappa\lambda}$. The local lagrangian field equations are

$$(1/n^2)n\cdot\partial(n\cdot\partial A^\mu - \partial^\mu n\cdot A - n^\mu\partial\cdot A - \varepsilon^\mu{}_{\nu\kappa\lambda}n^\nu\partial^\kappa B^\lambda) = j^\mu \tag{12a}$$

$$(1/n^2)n\cdot\partial(n\cdot\partial B^\mu - \partial^\mu n\cdot B - n^\mu\partial\cdot B + \varepsilon^\mu{}_{\nu\kappa\lambda}n^\nu\partial^\kappa A^\lambda) = k^\mu . \tag{12b}$$

Here j^μ and k^μ are the conserved electric and magnetic currents

$$(j_\mu, k_\mu) = \sum_i q_i^a \bar{\psi}_i \gamma_n \psi_i , \tag{13}$$

where $q_i^a = (e_i, g_i)$. Provided n has vanishing time component, $n^\mu = (\underset{\sim}{n}, 0)$, Eqs. (12) are of first order in $\partial/\partial t$ instead of second order as in the one-potential formalism where the field equation, in the absence of magnetic source, is $\partial^2 A_\mu = j_\mu$. Thus the number of dynamical variables is not doubled in the two-potential formalism.

Because of current conservation, $\partial\cdot j = \partial\cdot k = 0$, Eqs. (12) imply

$$\partial^2 n\cdot A = \partial^2 n\cdot B = 0 , \tag{14}$$

so $n\cdot A$ and $n\cdot B$ are free fields. They are the generators respectively of restricted local electric and magnetic gauge transformations. Moreover, Eqs. (12) and (14) together imply Maxwell's equations in operator form, generalized to include magnetic sources

$$\partial^{\mu}F_{\mu\nu} = j_{\nu} \qquad \partial^{\mu\,*}F_{\mu\nu} = k_{\nu} \ , \tag{15}$$

with F given in Eq. (11). In contrast, quantization of the one-potential formalism by the Gupta-Bleuler method yields only one free field, $\partial\cdot A$, with $\partial^2\partial\cdot A = 0$ (as follows from $\partial^2 A_{\mu} = j_{\mu}$ and $\partial\cdot j = 0$) which is the generator of restricted local electric gauge transformations, and Maxwell's equation is satisfied only in expectation value, $<\phi|\partial^{\mu}F_{\mu\nu}|\phi> = 0$, on the gauge invariant states defined by the subsidiary condition $\partial\cdot A^{(-)}(x)|\phi> = 0$. In the two-potential method gauge invariant states are defined by the pair of subsidiary conditions

$$n\cdot A^{(-)}(x)|\phi> = n\cdot B^{(-)}(x)|\phi> = 0 \tag{16}$$

which assure positivity of the metric.

5. String Independence in the Quantum Field Theory of Charges and Poles

Only gauge invariant objects can be expected to be string independent. We will exhibit this property for the generating functional

$W[v]$ of electric and magnetic currents $J^a_\mu(x) = \sum_i q^a_i \bar{\psi}_i \gamma_\mu \psi_i$, where ψ_i is a Dirac field carrying charge $q^a_i = (e_i, g_i)$. It is defined by

$$<T\,[J^a_\mu(x) J^b_\nu(y) \cdots]> = \frac{1}{i}\frac{\delta}{\delta v^a_\mu(x)} \frac{1}{i}\frac{\delta}{\delta v^b_\nu(y)} \cdots W[v] \tag{17}$$

and may be calculated from the Gell-Mann-Low formula

$$W[v] = N<T\,\exp[\,i\int \sum_i q^a_i \bar{\psi}_i (V^a_\mu + v^a_\mu)\psi_i d^4x]>_o \quad , \tag{18}$$

where $V^a_\mu = (A_\mu, B_\mu)$ represents the pair of potentials discussed previously. Our method will be to systematically eliminate gauge dependent fields. The propagator of the free field V ,

$$D_{F\,\mu\nu}^{\;ab}(x-y) \equiv <T\,[V^a_\mu(x) V^b_\nu(y)]> \quad , \tag{19}$$

is found to be

$$\begin{aligned} D_{F\,\mu\nu}^{\;ab}(x) = \{[-g_{\mu\nu} + (\partial_\mu n_\nu + n_\mu \partial_\nu)(n\cdot\partial)^{-1}]\delta^{ab} \\ - \varepsilon_{\mu\nu\sigma\tau} n^\sigma \partial^\tau (n\cdot\partial)^{-1}\ \varepsilon^{ab}\}\ \Delta_F(x) \end{aligned} \tag{20}$$

where $\varepsilon^{ab} = -\varepsilon^{ba}$, with $\varepsilon^{12} = 1$, and $\Delta_F(x)$ is the propagator of a massless scalar field. Here $(n\cdot\partial)^{-1}$ is the integral operator with kernel

$$(n\cdot\partial)^{-1}(x) = \frac{1}{2}\int_0^\infty ds\,[\delta^4(x-ns)-\delta^4(x+ns)] \quad , \tag{21}$$

and n has vanishing time component. The Euclidean propagator is obtained by continuing the Minkowski propagator to $x^o = -ix^4$, x^4 real:

$$\begin{aligned} D_{MN}^{ab}(x) = \{[\delta_{MN}-(\partial_M n_N + n_M\partial_N)(n\cdot\partial)^{-1}]\delta^{ab} \\ - i\varepsilon_{MNKL}n_K\partial_L(n\cdot\partial)^{-1}\,\varepsilon^{ab}\}\,\Delta(x) \;. \end{aligned} \tag{22}$$

Here the indices K,L,M,N run from 1 to 4, $x = x^M$, the metric is Euclidean throughout, and $\Delta(x) = (2\pi)^{-2}(x^2)^{-1}$ satisfies $-\partial_M\delta_{MN}\partial_N\Delta(x) = \delta^4(x)$. Note that the Euclidean propagator has a real and imaginary part which are respectively symmetric and anti-symmetric in the charge indices.

The generating functional of Euclidean Green's functions may be expressed as a functional integral over anti-commuting fields ψ_i and $\bar{\psi}_i$,

$$\begin{aligned} W_E[v] = N\int\prod_i d\psi_i d\bar{\psi}_i\, \exp\{-\int\sum_i\bar{\psi}_i(\gamma^M\nabla_{i_M}-m_i)\psi_i d^4x\} \\ \times \exp(\frac{1}{2}\,JDJ)\Big|_{J\,\equiv\,0} \end{aligned} \tag{23}$$

where $\nabla_{i_M} = \partial_M + iq_i^a(V_M^a + v_M^a)$, and $\{\gamma_M,\gamma_N\} = 2\delta_{MN}$, with γ_M hermitian, $\gamma_M^\dagger = \gamma_M$. Here V_M^a is the operator $V_M^a(x) = \delta/\delta J_M^a(x)$ that acts on the last factor

$$\exp(\frac{1}{2}\, JDJ) \equiv \exp\, [\frac{1}{2} \int J_M^a(x) D_{MN}{}^{ab}(x-y) J_N{}^b(y) d^4x d^4y] \ .$$

This gives

$$W_E[v] = N\Pi_i \det\, [\gamma^M \nabla_{iM} - m_i]\, \exp(\frac{1}{2}\, JDJ) \ . \tag{24}$$

We next manipulate this formal expression to write it as a functional integral over classical particle trajectories. Note that

$$\det\, [\gamma^M \nabla_M - m] = \det\, [\gamma^5 \gamma^M \nabla_M - \gamma^5 m] = \det^{\frac{1}{2}}[(\gamma^5 \gamma^M \nabla_M - \gamma^5 m)^2] \ , \quad \text{so}$$

$$\det(\gamma^M \nabla_M - m) = \det^{\frac{1}{2}}[-\nabla^2 + \sigma \cdot F + m^2] \quad ,$$

where $\sigma \cdot F \equiv \frac{1}{2}\, \sigma_{MN} F_{MN}$, $[\gamma_M, \gamma_N] = 2i\sigma_{MN}$, and $F_{MN} = q^a[\partial_M(V_N^a + v_N^a) - \partial_N(V_M^a + v_M^a)]$. Note that the argument of the last determinent is a positive operator since it is the square of a hermitian operator. This gives

$$W_E[v] = N \exp\, [\frac{1}{2} \sum_i \mathrm{tr}\, \ell n(-\nabla_i^2 + \sigma \cdot F_i + m_i^2)] \times \exp(\frac{1}{2}\, JDJ) \Big|_{J=0} \ , \tag{25}$$

where $N = \exp\, [-\frac{1}{2} \sum_i \mathrm{tr}\, \ell n(-\partial^2 + m_i^2)]$. From the operator identity

$$\ell nA - \ell nB = -\int_0^\infty d\tau \; [\exp(-\frac{1}{2} A\tau) - \exp(-\frac{1}{2} B\tau)]/\tau$$

we obtain

$$W_E[v] = N' \exp[-\frac{1}{2} \sum_i \int_0^\infty d\tau \; \tau^{-1} \mathrm{tr}(\exp e^{-H_i \tau})] \times \exp(\frac{1}{2} JDJ) \Big|_{J=0} \tag{26}$$

where $H_i = \frac{1}{2}(-\nabla_i^2 + \sigma \cdot F_i + m_i^2)$. Henceforth we shall drop the $\sigma \cdot F$ term because it complicates the argument which we outline here, without altering the conclusion. (See Ref. 4 for more details.) Recalling that $\nabla_i = \partial + iq_i^a(V^a + v^a)$ we may write tr $\exp(-H\tau)$ as the functional integral over closed trajectories of classical particles,

$$\begin{aligned} \mathrm{tr}\, e^{-H\tau} &= \int Dx(\cdot) \exp[-\frac{1}{2} \int_0^\tau (\dot{x}^2 + m^2) d\tau'] \\ &\quad \exp i\oint q^a (V_M^a + v_M^a) dx_M \,. \end{aligned} \tag{27}$$

The potentials appear only in the gauge invariant line integral around a closed loop.

Expand the first exponential which appears in Eq. (26) in a power series. The n'th term represents the sum of all Feynman diagrams containing exactly n charged particle loops. When this is done the operator $V_\mu^a(x) = \delta/\delta J_\mu^a(x)$ appears only in a product of exponentials of the type $\exp(i\oint q^a V_M^a dx_M)$ which merely induces a shift to J in $\exp(\frac{1}{2}JDJ)$. The result, after setting J = 0, is that all string dependence is contained in factors of the type

$$\exp[-\oint_i dx_M \oint_j dy_N q_i^a q_j^b D_{MN}^{ab}(x-y)] \quad , \tag{28}$$

where the integrations extend around loops i and j. With reference to Eq. (22) write

$$D_{MN}^{ab} = \delta^{ab} D_{MN}^{e\ell} + i\epsilon^{ab} D_{MN}^{mag} \; . \tag{29}$$

The electric propagator contributes to (28) the positive factor

$$\exp(-\vec{q}_i \cdot \vec{q}_j \; I_{ij}) \quad , \tag{30}$$

where

$$I_{ij} = \oint_i dx_M \oint_j dy_M \; \Delta(x-y) \tag{31}$$

and $\Delta(x) = (2\pi)^{-2}(x^2)^{-1}$. This represents the mutual inductance of a pair of current loops in Euclidean 4-space. The n-dependence of the electric part of the propagator (22) has disappeared because $\oint dx_M \partial_M f(x) = 0$ for all f.

The magnetic propagator contributes a pure phase factor to (28) that may be written

$$\exp\,[i\vec{q}_i \times \vec{q}_j\, \Omega_{ij}] \tag{32}$$

where

$$\Omega_{ij} = \oint_i dx_M \oint_j dy_N\, \varepsilon_{MNKL}\, n_K \partial_L (n\cdot\partial)^{-1}\, \Delta(x-y)\ . \tag{33}$$

Stoke's theorem and identities of the ε-symbol give

$$\Omega_{ij} = \int_i {}^*dS_{LM} \oint_j dy_L\, [n_M (n\cdot\partial)^{-1}(x-y) + \partial_M \Delta(x-y)]\quad , \tag{34}$$

where $-\partial^2\Delta(x) = \delta^4(x)$ has been used. All of the string dependence is in the first term of Eq. (34). By Eq. (21), this term is found to have the value Z(n)/2, where Z(n) is a topological integer that counts the number of times the cylinder along n erected on loop j intersects the surface bounded by loop i. Thus, provided the charge quantization condition, $\vec{q}_i \times \vec{q}_j = 4\pi \times$ (integer) holds,[10] the phase factor (32) is independent of n and Eq. (34) may be replaced by

$$\Omega_{ij} = \int_i {}^*dS_{LM} \oint dy_L \partial_M \Delta(x-y)\ . \tag{35}$$

String independence is now established. We note moreover from this expression, that by Stoke's theorem, $\Omega_{ij} = -\Omega_{ji}$ + (integer) and that, by Gauss's theorem, as the surface bounded by loop i is varied Ω_{ij} changes in integral steps. Thus Ω_{ij} in Eq. (35) is really defined modulo unity and may be interpreted as a kind of mutual solid angle of two loops in 4-space. The weaker condition,[10] $\vec{q}_i \times \vec{q}_j = 2\pi \times$ (integer) is sufficient to make the phase factor (32) independent of the surface in Eq. (35). To sum up, the Euclidean Green's function is expressed as a sum of functional integrals over classical charged particle loops. Interaction is accounted for by the factor $\exp(\frac{1}{2}\vec{q}_i^{\,2} I_{ii})$ for each loop and the factor

$$\exp(-\vec{q}_i \cdot \vec{q}_j I_{ij} + i\vec{q}_i \times \vec{q}_j \Omega_{ij}) \tag{36}$$

for each pair of loops, where I_{ij} and Ω_{ij} are respectively the mutual inductance and mutual solid angle of each pair of loops.

References

1. P.A.M. Dirac, Proc. Roy Soc. (London), A133 (1931) 60.
2. The proceedings of the New York University Conference on Monopoles, including a historical introduction by P.A.M. Dirac, were published in the International Journal of Theoretical Physics 17 (1978) 235-308.
3. W. Marciano, International Journal of Theoretical Physics 17, (1978) 275, Eqs. (2.3), (2.10) and (4.5a).
4. R. Brandt, F. Neri and D. Zwanziger, Phys. Rev. D19, (1979) 1153.
5. J. Schwinger, Science 165, (1969) 757; 166, (1969) 690.
6. D. Zwanziger, Phys. Rev. 176, (1968) 1489.
7. D. Zwanziger, Phys. Rev. 176, (1968) 1480; Phys. Rev Phys. Rev. D6, (1972) 458.
8. P.A.M. Dirac, Phys. Rev. 74, (1948) 817.
9. J. Schwinger, Phys. Rev. 144, (1966) 1087; Phys. Rev. 151, (1966) 1048; Phys. Rev. 151, (1966) 1055; Phys. Rev. 173, (1968) 1536.
10. The condition, $(\vec{q}_i \times \vec{q}_j)/(4\pi)$ = integer, reflects the ½ which appears in the symmetrical string, Eq. (21). It is possible that a consistent theory satisfying the weaker condition, $(\vec{q}_i \times \vec{q}_j)/(2\pi)$ = integer, could be constructed. (If, in the present formalism, the asymmetrical string $\int_0^\infty ds n \delta^4(x-ns)$ had been used instead of Eq. (21), the condition $e_i g_j/(2\pi)$ = integer would have been obtained instead of $(\vec{q}_i \times \vec{q}_j)/(2\pi)$ = integer.)

EQUATIONS OF MOTION FOR NON-ABELIAN MONOPOLES

Chan Hong-Mo
Rutherford Laboratory, Chilton, Didcot, OXON., UK.

and

Tsou Sheung Tsun
Mathematical Institute, Oxford University, UK.

This is a brief preliminary report of our attempt to derive the equations of motion for non-Abelian monopoles with some recent contributions from P Scharbach. A fuller account of it will be presented elsewhere.

Since the subject has not as yet been discussed in this meeting, we shall start with a couple of definitions to clarify what questions are involved.

1. Definition of the non-Abelian monopole charge

For every closed loop C in space-time $\mathcal{M}$, passing through a fixed point ξ_μ^o introduce the phase-factor:

$$\psi_C : P \exp ie \oint_C a_\mu dx_\mu \tag{1}$$

If we parametrise C as:

$$C: \begin{cases} \xi_\mu(s), & s = o \to 2\pi \\ \xi_\mu(o) = \xi_\mu(2\pi) = \xi_\mu^o \end{cases} \tag{2}$$

then, ψ_C can be written as

$$\psi_C = P_s \exp ie \int_0^{2\pi} a_\mu(\xi) \frac{d\xi_\mu}{ds} ds \tag{3}$$

where P_s denotes an ordering with respect to s, ψ_C gives an element of the gauge group G for each closed loop C in space-time $\mathcal{M}$.

Consider now a one-parameter family of closed curves C^t, t = $0 \to 2\pi$, which we may parametrise as:

$$C^t : \begin{cases} \xi_\mu^t(s) & s = o \to 2\pi, \quad t = o \to 2\pi \\ \xi_\mu^t(o) = \xi_\mu^t(2\pi) = \xi_\mu^o \\ \xi_\mu^o(s) = \xi_\mu^{2\pi}(s) = \xi_\mu^o \end{cases} \tag{4}$$

As t varies from o to 2π, C^t loops over a closed surface $\sum$ in space-time $\mathcal{M}$, as illustrated in Fig 1.

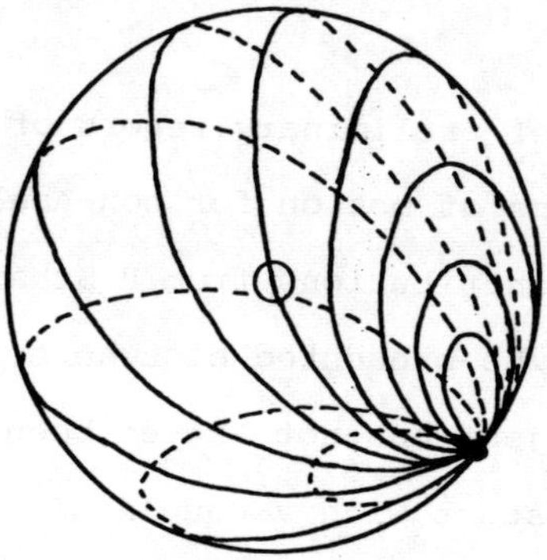

Fig 1

The expression (1) gives one group element $\psi_{C^t} \varepsilon G$ for each value of t. For t = o or 2π, C^t shrinks to a point, so that

$$\psi_{C^o} = \psi_{C^{2\pi}} = \text{identity} \tag{5}$$

Thus as t varies from o to 2π, ψ_{C^t} describes a closed curve Γ in group space.

We <u>define</u> now the monopole charge enclosed inside the surface $\sum$ as the homotopy class of the closed curve Γ in G. This is independent of the manner how C^t loops over $\sum$ or how the loops C^t are parametrised, and is equivalent to definitions given earlier by Lubkin[1], Wu-Yang[2] and 'tHooft[3].

We claim that this definition of the non-Abelian monopole charge is a natural extension of Dirac's magnetic charge for the following reasons:

(a) For electromagnetism (Abelian U(1) theory) it reduces to the Dirac charge, since homotopy classes there of closed curves Γ in U(1) can be labelled by the the winding number n, which equals $(2\pi)^{-1}$ times the phase change as C^t loops over $\sum$ and the latter in turn can be equated via the Stoke's theorem,

$$\oint_{C^t} a_\mu dx_\mu = \oiint f_{\mu\nu} d\sigma_{\mu\nu} \text{ over surface bounded by } C^t. \qquad (6)$$

to the total magnetic flux emerging from the surface.

(b) The beauty of the Dirac charge lies in the fact that it is an intrinsic geometrical attribute of the gauge theory, and that it is automatically quantised and conserved. These fundamental features are shared by the extension above to non-Abelian theories.

We note, however, that depending on the gauge group G the non-Abelian monopole charges so-defined can sometimes have rather unfamiliar properties. For example,

(i) for $G = SO(3) = SU(2)/Z_2$ (eg pure Yang-Mills theory for isospin), which is doubly-connected, the monopole charge can take only two values and operate multiplicatively, like a sign, $\zeta = \pm$.

(ii) for $G = SU(3)/Z_3$ (eg pure Yang-Mills theory for colour) which is triply-connected, the monopole charge can take only three values and operate multiplicatively like the cube-roots of unity: $\zeta = \exp i2\pi n/3$ or triality.

Further we note that these charges are to be distinguished from:

(i) the soliton-monopoles originated by 'tHooft[4] and Polyakov[5], which occupied the first one and a half days of this conference. These though sometimes also called non-Abelian monopoles, are actually only U(1)-monopoles embedded in non-Abelian theories.

(ii) the "magnetic weights" conjectured to exist by some authors[6] and discussed earlier in the meeting by Olive. These latter quantities are not in general conserved[2) 7)], and even if they turn out to be so in some particular theories yet to be specified, they will not be as intrinsic an attribute of gauge theories as the Dirac magnetic charge and its extensions above , and are therefore not interesting for our purpose.

2. "Derivation" of equations of motion for monopoles

An ordinary magnetic monopole (in an Abelian U(1) theory) satisfies the following equations:

(Mg) the Maxwell equation

$$\bar{f}_{\mu\nu,\nu}(\xi) = -4\pi g \int d\tau \frac{dx^\mu}{d\tau} \delta^4(\xi - X(\tau)) \tag{7}$$

(Lg) the Lorentz equation

$$M \frac{d^2 X_\nu}{d\tau^2} = -g \bar{f}_{\mu\nu} \frac{dx^\mu}{d\tau} , \tag{8}$$

where

$$\bar{f}_{\mu\nu} = \frac{1}{2}\xi_{\mu\nu\rho\sigma} \quad f^{\rho\sigma} \tag{9}$$

and $x_\mu(\tau)$ represents the monopole world-path. The equations (8) and (9) can be obtained by a dual transformation $f_{\mu\nu} \to \bar{f}_{\mu\nu}$ from the usual Maxzell equation (Me) and Lorentz equation (Le) for electric charges, or equivalently by extremising the dual action

$$\bar{\mathcal{A}} = -\frac{1}{16\pi}\int d^4\xi \bar{f}_{\mu\nu}\bar{f}^{\mu\nu} - M\int d\tau + g\int \bar{a}_\mu \frac{dx^\mu}{d\tau} d\tau \tag{10}$$

with respect to $\bar{a}_\mu$ and x_μ, with

$$\bar{f}_{\mu\nu} = \partial_\nu \bar{a}_\mu - \partial_\mu \bar{a}_\nu \tag{11}$$

Notice the introduction of an interaction term

$$\bar{\mathcal{A}}_1 = g\int \bar{a}_\mu \frac{dx^\mu}{d\tau} d\tau \tag{12}$$

in the action (10).

There is however an alternative method of arriving at the same equations when g is considered as a monopole charge defined topologically in the manner outlined above, as follows:

(a) The Maxwell equation (Mg) is just a statement that the total magnetic flux emerging from the monopole is equal to $4\pi g$, which according to the discussion in 1 is identical to the topological definition of the monopole charge

b) The Lorentz equation (Lg) can be obtained by extremising the 'free' action:

$$\bar{\mathcal{A}}_o = -\frac{1}{16\pi}\int d^4\xi\, \bar{f}_{\mu\nu}\,\bar{f}^{\mu\nu} - M\int d\tau \tag{13}$$

with respect to X_μ under the constraint (Mg)[8].

A virtue of this latter approach is that it by-passes the explicit introduction of an interaction term in the action $\bar{\mathcal{A}}$. The equations of motion appear then as a consequence of the very definition of monopole charges whose existence, in turn, is a consequence of topology, so that all quantities and equations now have a geometrical origin. For us, in particular, this approach is attractive in that it permits a generalisation to non-Abelian theories, where we have no duality on which the first approach relies.

3. Generalised Maxwell equation

To be specific we consider here only theories with gauge group SO(3), although our method is immediately seen to be generalisable to other theories. For $G = SO(3)$, then the monopole charges $\zeta = \pm$, where $\zeta = +$ corresponds to the vacuum and $\zeta = -$ denotes a monopole.

As noted above, the Maxwell equation is equivalent to the topological definition of the monopole charge given in terms of the phase factor ψ_{C^t}. For an Abelian theory, because of the Stoke's theorem (6), this definition can be reduced to the equation (7) which is local in space-time $\mathcal{M}$.

For non-Abelian theories however, there is no direct generalisation of the Stoke's theorem so that a Maxwell equation local in $\mathcal{M}$ is unlikely to exist. What we mean by the Maxwell equation then should just be an explicit expression in terms of the variables a_μ and X_μ, of the definition of the monopole charge given only abstractly so far as a homotopy class of closed curves in G.

For the Abelian theory, we were able to obtain such an expression by writing the winding number n which labels the homotopy classes in terms of the change in phase of ψ_{C^t} as C^t loops over a surface enclosing the monopole. To do the same for the SO(3) theory, therefore, we seek an equivalent of the phase change. Now the phase of ψ_C in U(1) may be interpreted as the copy of ψ_C in the covering space R. So an equivalent of the 'phase of ψ_C' in SO(3) would be the copy of ψ_C in the covering group SU(2).

Write therefore;

$$\Psi_C = P_s \exp ie \int_0^{2\pi} A_\mu(\xi) \frac{d\xi^\mu}{ds} ds \tag{14}$$

where

$$A_\mu(\xi) = a_\mu^i(\sigma_i/2), \tag{15}$$

$\sigma_i (i = 1,2,3)$ being 2x2 Pauli matrices. (14) gives an element of SU(2) for each loop C in space-time $\mathcal{M}$, or for each point in loop space $\Omega^1\mathcal{M}$. Introduce

$$F_\mu(s,C) = \frac{\delta}{\delta\xi_\mu(s)} \Psi_C \cdot \Psi_C^{-1} \tag{16}$$

being the 'logarithmic derivative' of Ψ_C when C is changed at the point corresponding to the value s in the direction μ, as illustrated in Fig 2. (For a proper definition of $F_\mu(s,C)$, see the contribution of Tsou to this meeting).

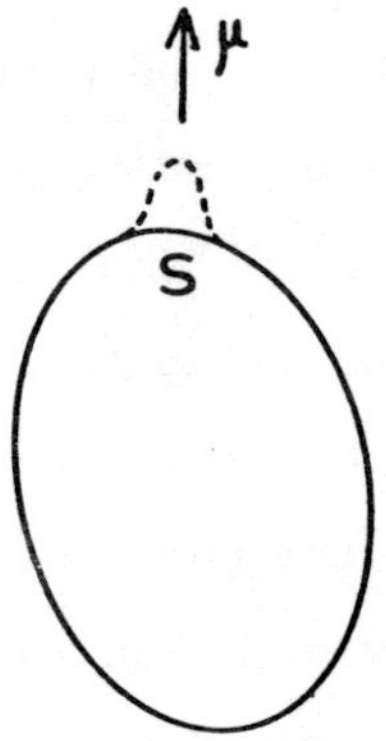

Fig 2

$F_\mu(s,C)$ can be interpreted as the "change in phase" of ψ_C from one point in loop space $\Omega^1\mathcal{M}$ to a nieghbouring point, or in other words a connection in $\Omega^1\mathcal{M}$.

The total change in Ψ_{C^t} as C^t loops over surface Σ is :

$$\Theta_\Sigma = P_t \exp \int_0^{2\pi} F_\mu(s,C^t) \frac{\partial \xi_\mu^t(s)}{\partial t} \, ds dt \tag{20}$$

where P_t denotes ordering in t. The expression (4) may be compared to (3) which gives the phase factor ψ_C in terms of the connection A_μ in ordinary space-time $\mathcal{M}$. The expressions are similar apart from the fact that the connection $F_\mu(s,C)$

in $\Omega^1\mathcal{M}$ has one more index s which has also to be summed (integrated) over. Θ_Σ is therefore just the phase factor (holonomy element, Wilson operator) defined for the closed loop Σ in loop space $\Omega^1\mathcal{M}$.

From the definition of the monopole charge then, we have that

$$\Theta_\Sigma = -I$$

if Σ encloses a monopole, and

$$\Theta_\Sigma = +I$$

otherwise, or equivalently

$$\Theta_\Sigma = Z_\Sigma \equiv \exp i\pi \int_{\text{inside } \Sigma} \frac{dX^\mu}{d\tau} \delta^4(\xi - X(\tau))\, dV_\mu d\tau \tag{21}$$

to hold for all Σ in $\Omega^2\mathcal{M}$, this being the space of all loops in loop space $\Omega^1\mathcal{M}$. Since the condition (21) contains now all the information in the definition of the monopole charge, it should, according to our logic, be considered at the Maxwell equation (Mg') generalised for the gauge group SO(3).

The same arguments if applied to the Abelian U(1) theory would lead to the statement that the total phase change of ψ_{C^t} as C^t loops over Σ is equal to the magnetic charge enclosed namely

$$\int_0^{2\pi} F_\mu(s,C^t)\, \frac{\partial \xi_\mu^{\ t}(s)}{\partial t}\, ds dt = 2n\pi \int_{\text{inside } \Sigma} \frac{dX^\mu}{d\tau} \delta^4(\xi - X(\tau))\, dV_\mu d\tau \tag{22}$$

where the ordering of quantities is now immaterial.

Using the Stoke's and Gauss Theorems and the expression (17) for $F_\mu(s,C)$, the left-hand side of (22) can be transformed into an integral of $\bar{f}_{\mu\nu,\nu}$ over the volume inside Σ, from which the validity of (22) for all Σ then implies the usual Maxwell equation (Mg) of (7), local in space-time , For non-Abelain theories, on the other hand, phase factors (holonomy elements) overfinite loops such as Θ in (20) cannot in general be expressed in terms of field tensor $f_{\mu\nu}$ (curvature) local in $\mathcal{M}$. Hence we are stuck with a Maxwell equation which is local only in $\Omega^2\mathcal{M}$.

4. Generalised Lorentz Equation

According to 2, this is to be obtained by extremising the 'free' action

$$\mathcal{O}\!\mathcal{L}_o(A,X) = -\frac{1}{16\pi}\int \mathrm{Tr}\{F_{\mu\nu}F^{\mu\nu}\}\, d^4x - M\int d\tau \qquad (23)$$

under the Maxwell equation (Mg') of (21) as constraint. There is one constraint for each Σ in $\Omega^2\mathcal{M}$.

We note first some peculiarities of this variational problem:

a) the constraint is 'non-holonomic' in the sense that it involves derivatives of the variables of variation $A_\mu(X)$ and $X_\mu(\tau)$, but is of the 'isoperimetric' type, namely of the form

$$\int dx\, F_j(q,\dot{q}) = \text{const},$$

for which it appears that the Lagrange method of undetermined multipliers apply, with constant multipliers. We introduce therefore one such multiplier Λ_Σ for each constraint labelled by Σ, where Λ_Σ is an element of the gauge Lie algebra.

b) The constraint is given in group space (ie global) and not as usual in the Lie algebra (local), so that the Euler-Lagrange equations are slightly modified:

$$\delta \mathcal{A}_0 + \int \delta\Sigma \; \mathrm{Tr}\left\{\Lambda_\Sigma \delta\left[\Theta_\Sigma Z_\Sigma^{-1}\right]\left[\Theta_\Sigma \; Z_\Sigma^{-1}\right]^{-1}\right\} = 0 \tag{24}$$

where the integral $\delta\Sigma$ is to be taken over all Σ in $\Omega^2\mathcal{M}$.

The derivation of (24) for variations with respect to $A_\mu(x)$ is fairly straightforward and gives:

$$\begin{aligned}
&\frac{1}{4\pi} D^\nu F_{\mu\nu}(x) \\
&= -ie\int \delta\Sigma \int_0^{2\pi} dt \int_0^{2\pi} ds \; \frac{\partial(\xi_\mu^t(s), \xi_\nu^t(s))}{\partial(s,t)} \; \frac{\Delta}{\Delta\xi_\nu^t(s)} \Lambda_\Sigma^T(s,t) \\
&\qquad \times \; \delta^4(\xi^t(s) - x) \\
&- (ie)^2 \int \delta\Sigma \int_0^{2\pi} dt \int_0^{2\pi} ds \int_0^s ds' \; \frac{\partial\xi_\rho^t(s)}{\partial t} \; \frac{\partial\xi_\sigma^t(s)}{\partial s} \; \frac{\partial\xi_\mu^t(s')}{\partial s'} \\
&\qquad \left[F_{\rho\sigma}^T(s') . \Lambda_\Sigma^T(s't)\right] \; \delta^4(\xi^t(s') - x)
\end{aligned} \tag{25}$$

where $\Sigma = C^t$ denotes elements of $\Omega^2\mathcal{M}$ parametrised as in (4), and the covariant loop derivative is defined as:

$$\frac{\Delta}{\Delta\xi^t_\mu(s)} = \frac{\delta}{\delta\xi_\mu(s)} - ie\left[A_\mu(\xi^t(s)), \cdots\right] \tag{26}$$

Superscripts T denote parallelly transported quantities

$$F^T_{\mu\nu}(s') = \Psi_{C^t}(s',s)\; F_{\mu\nu}(\xi^t(s))\; \Psi^{-1}_{C^t}(s',s) \tag{27}$$

$$\Lambda^T_\Sigma(s,t) = \Psi^{-1}_{C^t}(o,s)\; \Theta^{-1}_\Sigma(o,t)\,\Lambda_\Sigma\; \Theta_\Sigma(o,t)\; \Psi_{C^t}(o,s) \tag{28}$$

along the curves C^t, via the transport operators (18) and

$$\Theta_\Sigma(t_1,t_2) = P_t \exp \int_{t_1}^{t_2} dt \int_o^{2\pi} ds\; F_\mu(s,C^t)\; \frac{\partial\xi^t(s)}{\partial t} \tag{29}$$

The derivation of the stationary condition (24) for variations with respect to $X_\mu(\tau)$ is more delicate since Z_Σ makes a finite jump from $-I$ to I as X passes outside Σ. Our conclusion is that the condition then reads

$$M\frac{d^2X_\sigma}{d\tau^2} = \int \delta\Sigma \int_o^{2\pi} dt \int_o^{2\pi} ds\; \mathrm{Tr}\left[\kappa\Lambda_\Sigma\right]$$

$$\xi_{\mu\nu\rho\gamma}\; \frac{\partial\xi^t_\mu(s)}{\partial t}\;\frac{\partial\xi^t_\nu(s)}{\partial s}\;\frac{dX_\rho}{d\tau}\;\; \delta^4(\xi^t(s) - X(\tau))$$

(30)

for all κ satisfying:

$$\exp \kappa = -I \qquad (31)$$

Equations (25) and (30) together then represent the Lorentz equations for our SO(3) monopole in parametric form in terms of the undetermined multipliers Λ_{Σ}. They are recent results, and, being of a rather unfamiliar form, are not guaranteed to be correct, and certainly not fully understood. Similar equations written for an Abelian U(1) theory do appear however to reduce to the usual Lorentz equation (8) as expected.

5. Remarks

Equations (21), (25), (30) together represent the equations of motion for a non-Abelian SO(3) monopole. At first sight, the equations look complicated and it is not clear whether in this form they will ever be useful for studying monopole dynamics. The exercise, however, is we believe, interesting in that.

(I) the equations are a consequence of the topological definition of the monopole charge, so that not merely the existence of monopoles but also their dynamics are now implied by geometry.

(II) the equations appear to reduce on the one hand to the usual Maxwell and Lorentz equations for Abelian U(1) theories, and on the other to be generalisable readily to more complicated gauge groups.

REFERENCES

1) Elihu Lubkin, Ann. Phys. (N.Y.) 23 (1963) 233.

2) Tai Tsun Wu and Chen Ning Yang, Phys. Rev. D12 (1975) 3845.

3) G. 'tHooft, Nucl. Phys. B138 (1978) 1.

4) G. 'tHooft, Nucl. Phys. B79 (1974) 276.

5) A Polyakov, Zh. Ekop. Teor. Fiz. Pis'ma 20 (1974) 430; (JETP Lett. 20 (1974) 194).

6) P. Goddard, J Nuyts and D. Olive, Nucl. Phys. B125 (1977) 1.

7) Chan Hong-Mo amd Tsou Sheung Tsun, Phys.Letters 95B (1980) 395.

8) eg Gu Chao-Hao and Yang Chen-Ning, Scientia Sinica, 20 (1977) 177.

FIELD THEORY OF DIRAC MONOPOLES

W. Deans *

Faculté des Sciences, Université Libre de Bruxelles **

A generalization of Dirac's theory of point monopoles is presented and used to derive the renormalization of the charge quantization condition.

I start with the 1948 theory [1], with action

$$S = S_0 - e\int dx\, j^\mu A_\mu - \frac{g}{2}\int dx\, K^\alpha(x)\int_{-\infty}^{0} d\eta\, h^{\mu\nu}_{\alpha}(x,\eta)\, {}^*F_{\mu\nu}(\xi(x,\eta))$$

$$-\frac{1}{4}\int dx\, {}^*F^{2}_{\mu\nu} + \int dx\, A_\mu \partial_\nu F^{\mu\nu} \tag{1}$$

Here j^μ, K^α are the electric and magnetic currents, So is the free particle action, and ${}^*F_{\mu\nu}$ is the dual field strength tensor.

$$h^{\mu\nu}_{\alpha} = \frac{\partial\xi^\mu}{\partial x^\alpha}\frac{\partial\xi^\nu}{\partial\eta} - \frac{\partial\xi^\nu}{\partial x^\alpha}\frac{\partial\xi^\mu}{\partial\eta} \tag{2}$$

describes the surface swept out by the string $\xi^\mu(x,\eta)$. $\xi^\mu(x,0) = x^\mu$ is the end point of the string, where the monopole sits.

The form (1) is equivalent to that of Dirac if $K^\alpha(x) = \int d\tau \frac{dz^\alpha}{d\tau}\delta(z-x)$

I now generalize to the case $K^\alpha = \bar{\chi}\gamma^\alpha\chi$, $j^\mu = \bar{\Psi}\gamma^\mu\Psi$

where χ, Ψ are fermionic fields and S_0 is modified accordingly.

The action is invariant under the transformation

$$\Delta\chi(x) = -ig\int_{-\infty}^{0} d\eta\, a^\mu(x,\eta)\, {}^*F_{\mu\nu}(\xi(x,\eta))\frac{\partial\xi^\nu}{\partial\eta}(x,\eta)\,\chi(x)$$

$$\Delta\bar{\chi}(x) = ig\,\bar{\chi}(x)\int_{-\infty}^{0} d\eta\, a^\mu(x,\eta)\, {}^*F_{\mu\nu}(\xi(x,\eta))\frac{\partial\xi^\nu}{\partial\eta}(x,\eta)$$

$$\Delta A_\mu(x) = \frac{g}{2}\int dy\, K^\alpha(y)\int_{-\infty}^{0} d\eta\, \epsilon_{\mu\nu\lambda\rho}\, h^{\nu\lambda}_{\alpha}(y,\eta)\, a^\rho(y,\eta)\,\delta(\xi(y,\eta)-x)$$

$$\Delta\xi_\mu(x,\eta) = a_\mu(x,\eta) \tag{3}$$

---

* Supported by the Belgian State under the contract A.R.C. 79/83-12.

** Postal address : Pool de Physique, Campus Plaine, C.P. 225, Bld du Triomphe, 1050 Bruxelles, Belgium

except for the "veto term"
This corresponds to a motion of the string $\xi^{\mu} \to \xi^{\mu} + a^{\mu}$.

We wish the string to be unobservable, which means using the veto so that (3) is a genuine gauge transformation. This can be done if the commutator of two gauge transformations is also a gauge transformation, which can only happen if

$$eg = 2\pi n \tag{4}$$

This is the "classical" theory ; in order to derive the analogous condition in the complete theory, I use the general form of the action after loop corrections have been included [(2)]. The condition (4) becomes

$$e_R g_R = 2\pi n \tag{5}$$

where e_R and g_R are the physical renormalized charges. The coupling constants e, g are found to obey the same renormalization group equations

$$\frac{1}{e} \mu \frac{\partial e}{\partial \mu} = \frac{1}{g} \mu \frac{\partial g}{\partial \mu} \tag{6}$$

so that both e and g vanish in the infrared limit.

Reference : (1) P.A.M. Dirac, Phys. Rev. 74 (1948) 817.
(2) W. Deans, Nucl. Phys. B, to be published

BOHM-AHARONOV EFFECT IN SU(N) GAUGE THEORY

P.A. Horváthy *
Istituto di Fisica Matematica, Università di Torino, Italy,
and
Centre de Physique Theorique, CNRS Marseille, France,

and

J. Kollár
Istituto di Fisica Matematica, Università di Torino, Italy,
and
Maths. Dept., Brandeis University, Waltham, Mass., USA.

In [1] a generalized Bohm-Aharonov (BA) experiment was proposed for testing the existence of Yang-Mills (Y-M) fields. Here we study it for gauge group G = SU(N).

Let A_μ denote a Y-M field with vanishing field strength tensor outside of a "non-Abelian solenoid" located at the origin. The field is characterized by the "non-integrable phase factor" [1] $\Phi := (\exp - \oint_\gamma A_\mu dx^\mu)$ where γ circles once around 0. A suitable gauge transformation brings Φ to diagonal form : $\underset{\sim}{\Phi} = \mathrm{diag}(\exp i\lambda_j)$, $\lambda_j \in R$, $j = 1,\ldots N$. The λ_j's are unique (mod 2π) if we require $\sum \lambda_j = 0$, $\lambda_1 \geq \ldots \geq \lambda_{N-1}$. [2, 5].

Let φ stand for a spin zero field representing a test particle moving in the external Y-M vacuum A_μ. It satisfies the Klein-Gordon equation

(1) $$(D_\mu D^\mu + m^2)\varphi = 0 \qquad (g = \hbar = 1)$$

where, as usually, $D_\mu = \partial_\mu - iA^k_\mu X_k$. The X_k's generate a unitary irreducible representation (irrep) of SU(N).

Let us cut the plane along the positive x axis and work over $[0, 2\pi] \times R^+$. $F_{\mu\nu} = 0$ implies that

(2) $$g(x) := \mathcal{P}(\exp - \int_{x_0}^{x} A_\mu dx^\mu) \quad , \quad x_0 \text{ fixed}$$

is independent of the path from x_0 to x. It satisfies $\partial_\mu g = -A_\mu g$, $g(2\pi, r) = \Phi$, $\forall\ r \geq 0$. Let us apply the gauge transformation

(3) $$A_\mu \rightarrow g^{-1}A_\mu g + g^{-1}\partial_\mu g = 0$$
$$\varphi \rightarrow U(g)\varphi =: \Psi$$

$U(g)$ is here the unitary operator representing $g \in SU(N)$ in our irrep.

(3) carries (1) to a free equation but with boundary conditions:

(4) $$(\Box + m^2)\psi = 0 \qquad \psi(2\pi) = U(\phi)\,\psi(0).$$

The solutions of (1) in different vacua A_μ and A'_μ are thus identical if the boundary conditions are the same. This question is studied the best in terms of the Kirillov-Kostant version of the Borel-Weil theorem on irreps. [3, 4]. Any irrep of SU(N) is associated in fact to a prequantizable coadjoint orbit $\mathcal{O}_q = Ad^*_G\, q$, where q is

* A.v. Humboldt Fellow at the Fak. fur Physik, Univ. Bielefeld, W. Germany.

a fixed element in $\mathcal{G}^*$,the dual of the Lie algebra $\mathcal{G}$ of SU(N). $\mathcal{O}_q \simeq G/G_q$, where G_q is the isotropy subgroup of q in the coadjoint action Ad^* of G on $\mathcal{G}^*$. $q \in \mathcal{G}^*$ can be chosen in the form $q = tr\,(diag\ iq_j ,..)$ with $q_1 \geq .. \geq q_{N-1} > 0$, $\sum^{N} q_i = 0$ .Write $m_j := q_1 + \dots 2q_j + \dots + q_{N-1}$. $\mathcal{O}_q$ is prequantizable iff $m_j \in \mathbb{N}\ \forall j$.

Define $\chi(h) := \prod^{N-1} (h_{jj})^{m_j}$. $\widetilde{G}_q := \{ h \in G_q : \chi(h) = 1 \}$ is then a subgroup of G_q. $Y := G/\widetilde{G}_q$ is a principal U(1) bundle with connection;it prequantizes in fact $\mathcal{O}_q$. [4]
Let us consider $\mathcal{H} = \{ \psi : Y \to \mathbb{C} \quad \psi(yz) = z\psi(y) \quad y \in Y, z \in U(1) \}$
G acts on $\mathcal{H}$ according to $(U(g)\psi)(y) := \psi(F_{g^{-1}}\, y)$ where F_g denotes the action of $g \in G$ on Y.As the representation space of the irrep is a subspace of $\mathcal{H}$ [3,4] ,two elements $g_1, g_2 \in G$ are represented by the same operator iff $F_{g_1} = F_{g_2}$. This happens iff

(5) $\quad Ad^*_{g_1}\big|\mathcal{O}_q = Ad^*_{g_2}\big|\mathcal{O}_q \qquad$ and $\qquad \chi(g_1^{-1} g_2) = 1$

Applied to our problem we conclude

<u>Theorem</u> For two Y-M fields A_μ and A_μ' $U(\phi) = U(\phi')$ and thus (1) admits identical solutions iff

(6) $\quad \exp i\lambda_j = \varepsilon \exp i\lambda_j' \quad , j = 1,\dots N \qquad$ where $\varepsilon^N = 1$

and

(7) $\quad \prod_j^{N-1} (\exp i\lambda_j)^{m_j} = \prod_j^{N-1} (\exp i\lambda_j')^{m_j} \quad$ or $\varepsilon^{(\sum^{N-1} m_j)} = 1$

Eqn (3) can be solved actually.In representation $\{k\}$ it stands for k uncoupled free equations with boundary conditions,representing k independent electromagnetic BA effects with enclosed fluxes corresponding to the diagonalmatrix $U(\phi)$ .The most interesting cases are contained in the Table below.

Note that the above results can be interpreted in terms of classical- and quantized motions in an external Yang-Mills field,cf. [5] .

gauge group	m_j		orbit type	represen-tation	physical interpretation	number of fields with the same solutions
SU(2)	1		$S^2 \simeq P^1(\mathbb{C})$ dim 2	{2}	nucleon	1
	2			{3}	meson*	2
SU(3)	1	0	minimal orbit $P^2(\mathbb{C})$ dim 4	{3}	quark	1
	1	1		$\{\bar{3}\}$	antiquark	
	3	0		{10}	decimet*	3
	3	3		$\{\overline{10}\}$	antidecimet*	
	2	1	maximal orbit $PT(P^2\mathbb{C})$ dim 6	{8}	octet *	3
	4	2		{27}	*	3

[1] T.T.Wu,C.N.Yang,Phys.Rev.<u>D12</u>, 3845 (1975).
[2] M.Asorey,J.Math.Phys. <u>22</u>, 179 (1981).
[3] B.Kostant,in Springer Lecture Notes in Math.<u>170</u> , (1970).
[4] C.Duval,Ann.Inst.H.Poincaré ,<u>36</u>,95 (1981).
[5] P.A.Horváthy,J.Kollár,to appear.

INFINITY SUBTRACTION IN A QUANTUM FIELD THEORY OF CHARGES AND MONOPOLES

Costas Panagiotakopoulos
International Centre for Theoretical Physics, Trieste, Italy.

A discussion of the subtraction of infinities in Zwanziger's [1] local quantum field theory of charges and monopoles is given. The theory is by power counting renormalizable since it has the same Feynman graph structure as QED. Nevertheless, the renormalization programme cannot be implemented order by order in perturbation theory due to the presence in the Lagrangian of an arbitrary but fixed four vector n_μ, the remnant of the old Dirac string [2]. Therefore the non-renormalizability of this theory appears to be intimately related to the lack of Lorentz invariance. It is known that, for observable quantities, the string dependence disappears in the full quantum field theory [3], provided Dirac's charge quantization condition is satisfied. The hope thus exists that the restoration of Lorentz invariance will be accompanied by the possibility of absorbing the infinites with an appropriate redefinition of the field normalizations and coupling constant values.

This idea is shown to be realizable. The case of an electric charge and a monopole interacting with the electromagnetic field is examined. Using the n-independence of infinite series of graphs the analogue of the Z_3 of QED is defined. It is an n-independent but cut-off dependent constant. Renormalized charges are defined by $\sqrt{Z_3}\, e = e_R$ and $\sqrt{Z_3}\, g = g_R$. with e and g the bare charges. This means that both charges are renormalized in the same way due to vacuum polarization. This is in accordance with previous work of Schwinger [4]. Under the reasonable assumption that the n-cancellation procedure respects the Ward identities, the renormalization due to vacuum polarization is the only one for both charges. Dirac's quantization condition can still be satisfied if a discrete cut-off is used in the case that the bare charges go to infinity with the cut-off going to infinity. Otherwise Z_3 must be a ratio of appropriate integers. The wave function renormalization of the matter fields is found to be n-dependent. The position of the poles of the matter field propagators can be arranged to be in the right place with the use of a mass counterterm. New vertices, not present in the tree graphs, which are infinite by power counting, are in fact finite because of the gauge invariance and present no problem. The renormalized theory is finite and Lorentz invariant provided the summations of infinite series involved do not give rise to pathological situations. The existence of a gauge invariant regularization that keeps the proof of n-independence of the full theory, necessary for this discussion, is shown [5]. Use has been made throughout of the standard LSZ reduction formula. This means that the correct behaviour of the electromagnetic field at large distances is not taken properly into account and this might give rise to infra-red problems [6] to which we did not address ourselves in this work.

REFERENCES

1) D. Zwanziger, Phys. Rev. D3, 880 (1971).

2) P.A.M. Dirac, Phys. Rev. 74, 817 (1948).

3) R. Brandt, F. Neri and D. Zwanziger, Phys. Rev. D19, 1153 (1979).

4) J. Schwinger, Phys. Rev. 151, 1048 (1966).

5) C. Panagiotakopoulos, to appear in Nucl. Phys. B.

6) M. Blagojevic and P. Senjanovic, Phys. Letters 101B, 277 (1981).

EMBEDDING DIRAC MONOPOLES INTO SU(2) n-MONOPOLES

M. Quirós.

Inst. Estructura de la Materia. Serrano 119. Madrid-6. Spain.

Electromagnetism is described by principal fiber bundles over S^2 with structural group U(1) [1]. After the Milnor-Steenrod theorem [2], there are $\Pi_1(S^1)=Z$ inequivalent fiber bundles (S^2, U(1)) and, thus, Z "different" electromagnetisms. These inequivalent bundles are $B_n=S^3/Z_n$ with transition functions $\Psi_{21}(\phi,\theta)=n\phi$, (ϕ,θ) being the angular coordinates on S^2 and U_i (i=1,2) an open covering of S^2. Elements of B_n are classes $V^{(n)}(\phi,\theta,\Psi)=\{U(\phi,\theta,\Psi+4\Pi m/n)\in SU(2)$; $m=0,\ldots,n-1\}$, where (ϕ,θ,Ψ) are Euler angles parametrizing SU(2), and canonical projections $\Pi_n\, V^{(n)}(\phi,\theta,\Psi)=(\phi,\theta)$. In particular, S^3 is the Hopf bundle and Π_1 the old Hopf map, and B_2 isomorphic to the rotations group SO(3). The Wu-Yang connection form on S^2, $\omega_1^{(n)}= i\, n(1+\cos\theta)/2\, d\phi$, $\omega_2^{(n)}=-i\, n(1-\cos\theta)/2\, d\phi$, can be lifted to a global connection in B_n as, [3], $\omega_D^{(n)}= i\, n(d\Psi+\cos\theta\, d\phi)/2$. The integer n (winding number) is proved to be equal to the first Chern class, and equal to $2eg/\hbar c$ (Dirac quantization condition).

On the other hand, SU(2) monopoles are described by regular connections in the trivial bundle S^2xSU(2). The Hopf bundle S^3 is a U(1)-subbundle of S^2xSU(2) since there is a bundle morphism from S^3 into S^2xSU(2), defined by $f_1(\phi,\theta,\Psi)=(\phi,\theta\ ;\ \phi,\theta,\Psi)$, inducing the identity map between the base spaces $\Phi_1(\phi,\theta)=(\phi,\theta)$. Thus, the 't Hooft-Polyakov monopole [4], which behaves asymptotically as $\omega_{TP}= i\,\vec{\sigma}(\vec{x}\wedge d\vec{x})/2r^2$, can be lifted to a connection ω' in S^3 by $\omega' = ad\,(U^{-1})\omega_{TP} + U^{-1}\, dU$, where $U=U(\phi,\theta,\Psi)$, and the answer is $\omega' = \omega_D^{(1)}\sigma_3$, the Dirac-Wu-Yang connection with unit topological charge. In general, it is not possible to go downstairs (from connection in S^3 to one-form on S^2), but the above formula is reversible, $\omega_{TP} = ad\,(U)\omega' - dU\, U^{-1}$.

There is a bundle morphism from B_n into S^2xSU(2), $f_n(\phi,\theta,\Psi) = (n\phi,\theta\ ;\ ;n\phi,\theta,n\Psi)$, inducing between bases the mapping $\Phi_n(\phi,\theta) = (n\phi,\theta)$ which belongs to the n-th homotopy class of $\Pi_2(S^2) = Z$. Starting from the global connection

$\omega_D^{(n)}$ in B_n we can go downstairs and compute the asymptotic connection on $S^2 \times SU(2)$ with magnetic charge n. The answer is $\omega^{(n)} = i\{d\theta[-\sigma_1 \cos n\phi + \sigma_2 \sin n\phi] + n \sin\theta \, d\phi[\sigma_1 \cos\theta \sin n\phi + \sigma_2 \cos\theta \cos n\phi - \sigma_3 \sin\theta]\}/2$. Some comments: a) In terms of vector potentials, it satisfies the condition [5] $e\vec{A}_\mu = \vec{\Phi} \wedge \partial_\mu \vec{\Phi} + \vec{\Phi}\, \Lambda_\mu$, with $\Lambda_\mu \equiv 0$ for all n; b) It is axisymmetric for $n \geqslant 2$ in agreements with no-go theorems [6], i.e. $\partial_\phi \vec{\Phi} = \vec{\lambda} \wedge \vec{\Phi}$ [7] and $\partial_\phi \vec{A}_i - \epsilon_{ij3} \vec{A}_j = \vec{\lambda} \wedge \vec{A}_i$, for $\vec{\lambda}$ constant; c) It corresponds to a solution in the superimposed limit, as expected by its axisymmetry [8]; d) We hope that an axisymmetric ansatz $A_i^a f(r,\theta)$, with appropriate boundary conditions, has a solution, at least in the BPS limit [9].

References:

1. T.T. Wu, C.N. Yang, Phys. Rev. D12, 3845 (1975).
2. N. Steenrod, The Topology of Fibre Bundles, Princeton Univ. Press (1951).
3. N. Minami, Prog. Theor. Phys. 62, 1128 (1979); L.H. Ryder, J. Phys. A, 13, 437 (1980); M. Quirós, J. Ramirez, E. Rodríguez, J. Math. Phys. (to appear).
4. G. 't Hooft, Nucl. Phys. B79, 276 (1974); A.M. Polyakov, JETP Lett. 20, 194 (1974).
5. E. Corrigan, D. Olive, D. Fairlie, J. Nuyts, Nucl. Phys. B106, 475 (1976).
6. E. Cremmer, F. Schaposnik, J. Scherk, Phys. Lett. 65B, 78 (1976).
7. P. Houston, L.O'Raiteartaigh, Phys. Lett. 93B, 151 (1980).
8. L. O'Raiteartaigh, S. Rhouani, Phys. Lett. 105B, 177 (1981).
9. M.K. Prasad, Commun. Math. Phys. 80, 137 (1981).

STUDY OF MONOPOLES USING LOOP SPACE (Summary)

TSOU Sheung Tsun

Mathematical Institute, 24-29 St. Giles', Oxford, UK.

One may describe gauge theory in terms of the phase factor defined as:

$$\Psi(C) = P \exp i \int_C A_\mu(\xi) \frac{d\xi^\mu}{ds} ds, \tag{1}$$

where $\xi^\mu(s)$ parametrizes the closed loop C and P means path-ordering. In mathematical language this is the holonomy corresponding to the loop C. and can be regarded as a finite version of the curvature $f_{\mu\nu}$. In general one can define its derivative with respect to any vector field along C, but in order to retain spacetime dependence we define the logarithmic derivative of $\Psi(C)$ in terms of only variations of $\Psi(C)$ consisting of "blips" at a given point s along C:

$$F_\mu(s,C) = \frac{\delta \Psi(C)}{\delta \xi_\mu(s)} \cdot \Psi^{-1}(C) = i \int_0^{2\pi} \Psi_{s'} f_{\mu\nu}(\xi(s')) \Psi_{s'}^{-1} \frac{d\xi^\nu}{ds'} \delta(s-s') ds', \tag{2}$$

where $\Psi_{s'} = P \exp i \int_0^{s'} A_\mu(\xi) \frac{d\xi^\mu}{ds} ds$ and $\Psi_{2\pi} = \Psi$. In the absence of electric charges, Yang-Mills equation becomes:

$$\frac{\delta}{\delta \xi_\mu(s)} F_\mu(s,C) = 0. \tag{3}$$

Notice that this would not be the divergence in loop space but for our restriction on variations of C.

$F_\mu(s,C)$ is a connection in loop space. In the absence of magnetic charges one can easily show that

$$G_{\mu\nu}(s,s') = \frac{\delta}{\delta \xi_\nu(s')} F_\mu(s) - \frac{\delta}{\delta \xi_\mu(s)} F_\nu(s') + [F_\mu(s), F_\nu(s')] = 0.$$

In the presence of monopoles, however, one has to define the double derivative carefully so as to obtain the correct expression for $G_{\mu\nu}$. In particular this double derivative has to be symmetric in its two variables, a property that is lacking in some lattice formulation of the theory. Thus:

$$\frac{\delta}{\delta \xi_\nu(s_2)} \left(\frac{\delta \Psi(C)}{\delta \xi_\mu(s_1)} \Psi^{-1}(C) \right) = i \int_0^{2\pi} \Psi_s \cdot \frac{\delta}{\delta \xi_\nu(s_2)} \left\{ f_{\rho\mu}(\xi(s)) \frac{d\xi^\rho(s)}{ds} \right\} \Psi_s^{-1} \frac{d\xi^\mu}{ds} \delta(s-s_1) ds - \int_0^{2\pi} \left[\Psi_s f_{\rho\mu}(\xi) \Psi_s^{-1}, \frac{\delta}{\delta \xi_\nu(s_2)} \Psi_s \cdot \Psi_s^{-1} \right] \frac{d\xi^\rho(s)}{ds} \delta(s-s_1) ds \tag{4}$$

Using the following two formulae:

$$\frac{\delta}{\delta \xi_\nu(s_2)} \Psi_s \cdot \Psi_s^{-1} = i \int_0^s \Psi_{s'} f_{\lambda\nu}(\xi(s')) \Psi_{s'}^{-1} \frac{d\xi^\lambda}{ds'} \delta(s'-s_2) ds' + \Psi_s A_\nu(\xi(s_2)) \Psi_s, \tag{5}$$

and $$\frac{d}{ds}\{\Psi_s f_{\mu\nu}(\xi(s'))\Psi_s^{-1}\} = \Psi_s \{D_\lambda f_{\mu\nu}(\xi(s'))\} \frac{d\xi^\lambda}{ds} \Psi_s^{-1}, \qquad (6)$$

where D denotes covariant differentiation in spacetime, we get after some algebra:

$$G_{\mu\nu}(s,s') = \Psi_s \{D_\lambda f_{\mu\nu} + D_\mu f_{\nu\lambda} + D_\nu f_{\lambda\mu}\} \Psi_s^{-1} \frac{d\xi^\lambda}{ds} \delta(s-s'). \qquad (7)$$

The expression in braces is the L.H.S. of Bianchi's identity, which indeed vanishes in the absence of monopoles. In other words, we have shown that whereas electric charges are sources of divergence in loop space, magnetic charges are sources of curvature in loop space.

In order to make this last statement more explicit, we can rewrite the ordinary Maxwell equation for monopoles:

$$\tilde{f}_{\mu\nu,\nu}(\xi) = -4\pi g \int \frac{dX^\rho}{d\tau} \delta^4(\xi - X(\tau))\, d\tau, \qquad (8)$$

where $X(\tau)$ is the monopoles worldline, in terms of loop variables as:

$$G_{\mu\nu}(s,s') = -4\pi g\, \delta(s-s')\, \epsilon_{\mu\nu\rho\sigma} \int \frac{dX^\rho}{d\tau} \frac{d\xi^\sigma}{ds} \delta^4(\xi(s) - X(\tau))\, d\tau. \qquad (9)$$

(8) and (9) can be shown to be equivalent for abelian theory. For non-abelian theory, however, it is unlikely that a Maxwell's equation for monopoles would be a straightforward generalization of (8), for two reasons. Firstly $f_{\mu\nu}$ is no longer adequate, as shown by Yang and others. Secondly, the existence of a non-abelian monopole depends directly on the global character of the group, and an element of the Lie algebra (which $f_{\mu\nu}$ is) alone cannot characterize it. For the second reason $G_{\mu\nu}$ is also inadequate. It seems to us that the non-abelian analogue of (8) or (9) should be non-local even in loop space. A proposal using $\Omega^2 M$ is reported elsewhere in this Conference.

This work is done in collaboration with Chan Hong-Mo and Peter Scharbach. Details and list of reference will be found in a forth-coming paper by us. I thank Graeme Segal for numerous conversations concerning analysis in loop space.

ON THE QUANTUM FIELD THEORY OF CHARGES AND DIRAC MONOPOLES

M.T. Vallon

Scuola Internazionale Superiore di Studi Avanzati, Trieste, Italy,

and

INFN, Sezione di Trieste, Italy.

In this brief report I will present the formalism introduced by G. Calucci, R. Jengo and myself to study the second quantization of point-like magnetic monopoles [1].

In our treatment the interactions between charges and monopoles are described in terms of functional integration over closed paths, as in the Zwanziger approach to the problem.[2] By using the Stokes theorem we can however write the interaction term in this way:

$$I_{m,n} = -\frac{ig}{2}\int_{T_m} d\tau^{(m)}_{\mu\nu}\,\tilde{F}^{(n)}_{\mu\nu} ,$$

where g is the magnetic charge, T_m is a surface having as a boundary the loop of the mth monopole and $\tilde{F}^{(n)}_{\mu\nu}$ is the dual of the field originated by the nth charged particle loop. We can also show that we can rewrite the interaction term in the following equivalent way:

$$I_{m,n} = \frac{ie}{2}\int_{\Sigma n} d\sigma^{(n)}_{\mu\nu}\,\tilde{M}^{(m)}_{\mu\nu} ,$$

where e is the electric charge, $\Sigma^{(n)}$ a surface having the nth charged particle loop as a boundary and $\tilde{M}^{(m)}_{\mu\nu}$ the dual of the field originated by the mth monopole loop.

For this theory to be consistent we must require the invariance of this term under those displacements of the surface that leave the boundary unchanged. We find out that the terms originated by such displacements are of the form $i(ge)n$ with $n = 0,1,2,\ldots$ If we require the Dirac condition to hold ($ge = 2\pi n$) we are left with a $2\pi ni$ term. Since the $I_{n,m}$ interactions appear at the exponent in the path integral formalism, we can conclude that our theory is well defined. We have also been able to avoid expressing the monopole potential, having introduced only the field, and therefore to maintain the Lorentz invariance at each step of the theory.[3]

Our formalism seems to be essentially non-perturbative. In fact when we try a perturbation expansion some ambiguities can appear, and they are due to those $2\pi ni$ terms which we have analyzed above. We have however considered some cases in which the perturbation expansion is meaningful (due also to the large mass of the monopole). We have evaluated the contributions due to monopole loops to the vacuum polarization and to the light-light scattering. The contributions are very small for the conceivable values of the monopole mass and in the case of the vacuum polarization they are of opposite sign if compared with those due to charged particle loops.

We can make a comparison between our theory and the Zwanziger approach. The non-perturbative descriptions are completely equivalent, however the perturbative expansions differ by an infinite series. Terms equivalent to our $2\pi ni$ pieces appear also in the Zwanziger formalism when one moves the vector defining the direction of the string in the monopole potential. They are probably unavoidable in every theory of this type because they reflect the compactness of the U(1) gauge group of electromagnetism with monopoles.

We are now thinking of applying our formalism to the study of condensed states and confinement. But this is just work in progress.

REFERENCES

1) G. Calucci, R. Jengo and M.T. Vallon, Nucl. Phys. B197, 93 (1982) and references therein.

2) R. Brandt, F. Neri and D. Zwanziger, Phys. Rev. D19, 1153 (1979).

3) An alternative way of maintaining explicit Lorentz invariance has been found by Yang by introducing a potential defined over patches of the space time, C.N. Yang, CERN preprint TH-2886 (June 1980).

IV: THE ROLE OF MONOPOLES IN PHYSICAL THEORIES

THE POSSIBLE ROLE OF MONOPOLES IN THE CONFINEMENT MECHANISM

Stanley Mandelstam
Department of Physics
University of California
Berkeley, California 94720, U.S.A.

My object in this talk is to outline how the confinement phase may be understood as a coherent plasma of monopoles. Let me say at the outset that no-one has yet been able to make use of this characterization to perform quantitative calculations in continuum gauge theories. At the end of my talk I shall refer to a recent proposal in this direction by 't Hooft.

I believe that the concept of the monopole plasma enables us to reach an <u>intuitive</u> understanding of the confinement phase, and to relate it to well-understood phenomena in superconductivity. By ennumeration we know that the existence of monopoles is a necessary and sufficient condition for a system to be able to confine. Continuum Q.E.D. without the explicit introduction of monopoles cannot confine. Lattice Q.E.D., and lattice or continuum non-Abelian theories, possess monopoles and can confine.

The monopole plasma is probably a way of parametrizing the ground state of a confinement phase; it is not necessarily the most efficient way for practical calculations.

I shall introduce the subject by discussing the superconducting, or Higgs phase, in a simple model. I ask the patience of the physicists in the audience, who are thoroughly familiar with this material, but I should like to attempt to present the subject as a coherent whole for the benefit of the mathematicians as well. I shall then discuss magnetic vortices and confinement of monopoles, which will lead us to the electric-magnetic dual of the superconducing phase, namely the confinement phase. From this viewpoint, confinement appears as an ordered rather than a disordered phase.

After this I should like to say something about monopoles in lattice Q.E.D., and to outline work which shows how confinement may be understood in terms of them. I shall also discuss transitions which are observed in lattice Yang-Mills theories, and the understanding of such transitions in terms of monopoles. Finally, I shall mention the recent work of 't Hooft referred to above.

Ordinary Superconductor as the Electric-Magnetic Dual of a Confinement Phase

An ordinary superconductor is a coherent plasma of charged objects, i.e., it is a system where the ground-state expectation value of a charge-annihilation operator is non-zero. In metallic superconductors, the charged objects are Cooper pairs. The

Higgs vacuum, very popular in particle physics, gives a non-zero vacuum expectation value to an elementary charged field.

The simplest Higgs system is the Abelian Higgs model, with the Lagrangian

$$\mathcal{L} = -\frac{1}{4} F_{\mu\nu}F^{\mu\nu} - \left\{(\partial_\mu + ieA_\mu)\Phi^+\right\} \left\{(\partial^\mu - ieA^\mu)\Phi\right\} + \mu^2\Phi^+\Phi - \lambda\left\{\Phi^+\Phi\right\}^2 \quad . \tag{1}$$

It is similar to the Lagrangian for scalar Q.E.D., but the mass term has the "wrong sign". As a consequence the vacuum, i.e., the state of lowest energy, has a non-zero expectation value for Φ.

$$\langle 0|\Phi|0 \rangle \neq 0 \quad . \tag{2}$$

In the semi-classical system which we shall consider

$$|\Phi| \neq 0 \quad . \tag{3}$$

A fundamental property of a superconductor is the <u>Meissner effect</u>. Magnetic flux cannot get into a superconductor unless it is squeezed into vortices of strength $2\pi/e$, the Dirac value, or integral multiples thereof. Such vortices were originally proposed by Landau and Ginsburg, and were treated in the B.C.S. theory by Abrikosov. They were first discussed in particle physics by Nielsen and Olesen.[1]

Let us consider a system where Φ, instead of being constant at infinity, has the behavior

$$\Phi \sim |\Phi| e^{i\phi}, \quad r = \quad (x^2 + y^2) \rightarrow \infty \ , \tag{4a}$$

ϕ is azimuthal angle.

The solution is independent of z. By gauge invariance, this will be equivalent to a system where Φ is constant provided

$$A_\phi \sim er^{-1} \ , \quad r \rightarrow \infty \ . \tag{4b}$$

Near the origin, the formulas (4) would make both Φ and A_ϕ singular. For Φ to be analytic, it must have a zero at some point, which we take to be the origin. If we assume that A_ϕ is finite near the origin, we have the situation depicted in Fig. 1. The magnetic field is non-zero in the central region where (4b) fails to hold, and we have a vortex of magnetic

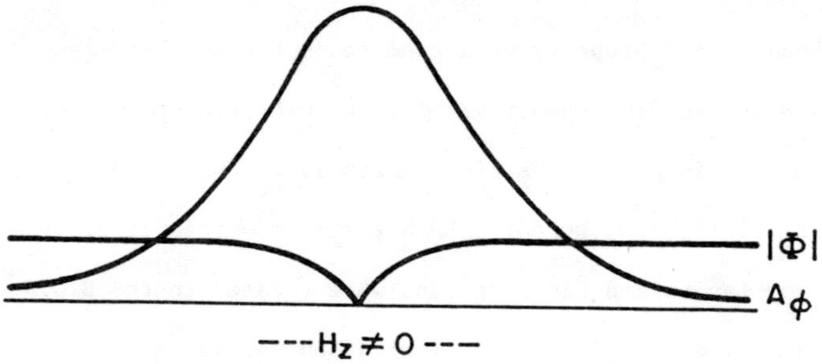

Fig. I

Figure 1. Behavior of $|\Phi|$ and A_ϕ in a magnetic vortex.

flux in the z-direction. The minimization of the energy, subject to the boundary conditions (4), gives us a solution to the classical equations of motion.

We can easily calculate the total flux:

$$\begin{aligned} \text{Total flux} &= 2\pi \int r F_{xy}\, dr = 2\pi \int_0^\infty r \frac{1}{r} \frac{\partial}{\partial r} (r A_\phi) \\ &= 2\pi r A_\phi |_\infty = \frac{2\pi}{e} \text{ , from (4b).} \end{aligned} \tag{5}$$

The vortex thus contains one Dirac unit of flux. A vortex with a non-integral number of units would not be possible, since the boundary condition would then be

$$\Phi \sim |\Phi| e^{i\alpha\phi} ,$$

and Φ would not be single-valued. Thus magnetic flux in a superconductor is thus squeezed into vortices with an integral number of Dirac units.

The Nielsen-Olesen vortices are endless; they are either closed or of infinite length. We could have finite, open vortices if we had Dirac monopoles at their ends to absorb the flux. We then have a typical confinement situation, as was first stated explicitly by Nambu.[2] A single monopole in an infinite superconductor has an infinitely long magnetic vortex attached to it, and its energy is therefore infinite. A monopole anti-monopole pair has a vortex stretched between the monopole and the anti-monopole (Fig. 2). The potential energy grows proportionally to the distance, and the two particles never separate by large distances.

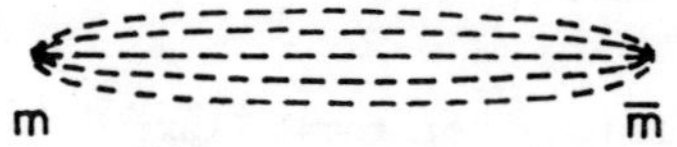

Fig. 2

Figure 2. Magnetic vortex stretched between monopole anti-monopole pair.

In the confinement phases color charges rather than monopoles are confined. We thus require the electric-magnetic dual of the system just described. Since the superconducting phase is a coherent superposition of charges, the confinement phase would be a coherent superposition of monopoles. Q.E.D. does not possess monopoles unless they are introduced explicitly, and continuum Q.E.D. without monopoles cannot confine. In non-Abelian gauge theories, on the other hand, monopole-like field distributions can occur without the explicit introduction of singularities. Thus, even in pure gauge theories, the confinement phase may be realized as a superposition of monopoles as was suggested by Mandelstam[3] and 't Hooft.[4]

We emphasize that by a monopole we mean a field distribution where, in some gauge

$$F^{\alpha}_{ij} \sim \varepsilon_{ijk} r^k r^{-3} \quad r \to \infty \,, \tag{6}$$

F^{α} being one colour component. We do not demand that we have a solution of the classical field equation with this boundary condition. Such a field distribution was proposed by Wu and Yang,[5] though they used a gauge where the monopole character was not evident until the work of 't Hooft and Polyakov. A Wu-Yang monopole may be described as a 't Hooft-Polyakov monopole without the Higgs field. It is not a topological object, and it is therefore not a monopole in the sense used by many people at this conference. There exist gauges, such as those used by Wu and Yang, 't Hooft and Polyakov, where there is no Dirac string. For an SU(N) gauge theory, we can have Wu-Yang monopoles of strength $2\pi Nn/g$. Monopoles of strength $2\pi n/g$ $(1 \leqslant n \leqslant N-1)$ are Dirac monopoles which require a string in all gauges.

To specify the colour directions of the monopoles in the plasma, we use the "abelian" gauge. Within the monopole all three color components of A are non-zero. Outside the monopole, A^3, the only non-zero component, has the usual monopole form, including the string. (One can of course remove the string by changing the gauge.) In the Abelian gauge, all monopoles have the same colour direction; if we attempted to superpose monopoles with different colour directions the Dirac singularity would carry energy. Having specified the state in the Abelian gauge, one can go to any other gauge, where all colour components are present in the monopole field.

The proof that the monopole plasma confines colour is precisely the dual of the proof that the charge plasma confines monopoles. Again we use the Abelian gauge. The effective Lagrangian for the (composite) monopole field has the usual Dirac term associated with the string. By Lorentz invariance it follows that, in writing down the effective Lagrangian, we must make the replacement

$$\frac{\partial \Phi}{\partial x^i} \rightarrow \frac{\partial \Phi}{\partial x^i} - i\left(\frac{2N\pi}{e}\right)B_i\Phi \quad , \tag{7a}$$

where

$$B_i = -\varepsilon_{ij3}\int_{-\infty}^{x_3} dx_3' \, E^j(x_3') \quad , \tag{7b}$$

Φ being the monopole field. It is assumed that the Dirac strings are in the z-direction. We can now repeat the Nielsen-Olesen proof to show that the phase of $\langle \Phi \rangle$ will change by 2π as we go once round an electric vortex of strength $\frac{e}{N}$. Electric flux can only occur in vortices whose strength is a multiple of $\frac{e}{N}$, otherwise we cannot have a single-valued Φ without changing the energy density at infinity. From the fact that electric flux is confined in units of $\frac{1}{N}e$, it follows that colour charges are confined.

One might also try to prove confinement by calculating the Wilson loop integral. If we calculate it straightforwardly from a simple monopole plasma, we find that ln W is proportional to $-L^{3/2}$, where L is the linear dimension of the loop. This is an unreasonable result; an unmodified plasma is a bad choice for a trial vacuum. A better choice would appear to be to weight each spatial configuration by $e^{-\text{const.}\, E_c}$, where E_c is the Coulomb energy of the monopoles. A calculation performed by Polyakov[7] then shows that $W \propto e^{-\text{const.}\, A}$, where A is the area of the loop.

Finally, we must ask the crucial question whether the Yang-Mills vacuum is a coherent superposition of Wu-Yang monopoles. The answer to this question is not known from purely theoretical analysis; if it were we should have a proof of confinement. A plausibility argument is that the energy of a sufficiently spread-out monopole appears to be negative. This is a consequence of the magnetic instability of the vacuum, noticed by Saviddy[8] and Wilczek;[9] a small vacuum-expectation value given to H, the colour magnetic field, decreases the energy by an amount proportional to $H^2 \ln(H\Lambda^{-2})$, where Λ is the scale of the theory. Though their result was obtained in a one-loop calculation, it appears to be a general feature of an asymptotically free theory. For a monopole spread out to a radius R; it follows that

$$\text{Magnetic energy } (r > R) \propto -R^{-1} \ln(\Lambda R), \tag{8a}$$

$$\text{Monopole energy } (r < R) \propto R^{-1}. \tag{8b}$$

If R is sufficiently large, (8a) > (8b).

As you will hear in the following talk, the Copenhagen group propose taking a trial vacuum where $H_c \neq 0$. They find that such a vacuum itself possesses instabilities and, after making further modifications, they eventually restore Lorentz invariance and obtain confinement. The two methods of defining the vacuum are probably not fundamentally different. In both cases, the low-frequency vacuum fluctuations of the magnetic field are much greater than in the bare vacuum. I believe that this feature is common to all models of confinement.

We may note that the immediate solution of the magnetic instability problem, which is to take $\langle H \rangle \neq 0$, corresponds to a ferromagnetic vacuum, whereas the confinement vacuum is a perfect paramagnet ($\mu = \infty$, $t = 0$, the dual of the superconducting case where $\mu = 0$, $t = \infty$). (See Pagels and Tomboulis[10]).

From our "magnetic" viewpoint the confinement vacuum appears as an ordered phase, whereas the conventional "electric" viewpoint treats it as a disordered phase. Usually, the Wilson-loop operator is regarded as the order parameter, while the 't Hooft-loop operator,[11] which creates a closed magnetic vortex, is regarded as the disorder parameter. There is, however, no fundamental distinction between an order and a disorder parameter; it depends one one's viewpoint. The fact that confinement is thought to disappear at high temperatures is in intuitive accord with the viewpoint of confinement as an ordered phase.

We may close this portion of the talk by mentioning the possible phases in SU(2) non-Abelian gauge theories (without quarks). We assume C.P. invariance; we do not examine phases with a non-zero value of θ, the instanton angle.

(i) The "perturbation" phase, with real non-Abelian gluons. Such a phase would probably be difficult to interpret physically.

(ii) The Georgi-Glashow phase, where SU(2) is broken to U(1). Such a phase has "photons", charged particles, and monopoles.

(iii) The superconducting phase, with complete Higgs symmetry breaking. Such a phase supports magnetic vortices, and monopoles are confined.

(iv) The confinement phase, which supports electric vortices and confines (colour) charges.

Phase (i) and (ii) are self-dual under electric-magnetic intercharge, whereas (iii) and (iv) transform into one another.

Monopoles in Lattice Q.E.D.

Unlike continuum Q.E.D., lattice Q.E.D. has monopoles built into it, and it does exhibit a confinement phase. In this part of the talk I should like to consider how these two features may be related.

Associated with each lattice line, we have a U(1) rotation U. The correspondence with the continuum potential is

$$U_\ell \sim \exp\{ieA.d\ell\} \quad , \tag{9}$$

$d\ell$ being the length of a lattice line. The action is a follows:

$$\text{Action} = -\frac{1}{2e^2} \sum_P \left\{ \prod_\ell U_\ell + \prod_\ell U^+_\ell \right\} \quad . \tag{10}$$

The symbol P represents a plaquette, the subscript ℓ goes over the four edges of the plaquette, all taken in one sense (clockwise or anti-clockwise). Note that, from (9)

$$\Pi U_\ell \sim \exp\{ieF_P da\} \quad \Pi U + \Pi U^+ = 2\cos\{F_P da\} \; , \tag{11}$$

so that (10) approaches the continuum actions as $d\ell$ becomes small.

Since the ranges of the functional integration are finite, lattice Q.E.D. is sometimes known as compact Q.E.D.

It is easy to see that lattice Q.E.D. has monopoles, even if they are not explicitly introduced. We need simply take A, in Eq. (9), to be the monopole potential at the mid-point of the lattice line. The total rotation ΠU over the plaquette enclosing the string is $e^{2i\pi}$, so that the string does not contribute to the action. Due to the compactness of lattice Q.E.D., the string is "invisible".

To define the monopole number, we might first define the number of strings passing through a plaquette as

$$n_P = \text{Integer nearest } \frac{1}{2\pi i} \Sigma \ln U \quad . \tag{12a}$$

The number of monopoles in a cube is then defined as

$$n_c = \Sigma \, n_P \quad , \tag{12b}$$

the sum being taken over the six plaquettes surrounding the cube. The number of monopoles within a single cube is not really a significant quantity, and is in any case not unambiguously defined by (11), since the rotation angles may be charged by 2π. Nevertheless, the definitions (12) correctly identify the total monopole number.

The relation between monopoles and confinement in lattice gauge theory was first examined by Banks, Myerson and Kogut.[12] We shall give an outline which is, in detail, slightly closer to the work of Peskin.[13] For weak coupling, the main contributions to the action (10) comes from values of ΠU_ℓ near 1. The variable F_P, defined by (11), will then be close to $2\pi m_P$, where m_P is some integer, and

$$\Pi U + \Pi U^+ \sim 1 - \frac{1}{2}(F_P - 2\pi m_P)^2 \quad . \tag{13}$$

The partition function is thus approximated by:

$$Z = \int_{-\pi}^{\pi} \mathcal{D}A \left\{ \prod_P \sum_{m_P = -\infty}^{\infty} \right\} \mathrm{Exp}\left\{ - \sum_P (F_P - 2\pi m_P)^2 \right\} \tag{14a}$$

The expression on the right of (14a) is known as the Villain form of the partition function. It is a good approximation to the original partition function when g is small. Furthermore, it has the same symmetry properties as the original partition function, and one would expect it to lead to a similar phase structure.

One may change the variable A by 2π, provided one makes a compensatory change of the variables m_P:

$$2\pi m_P \rightarrow 2\pi m_P + \frac{\Delta A_\mu}{\Delta x^\nu} - \frac{\Delta A_\nu}{\Delta x^\mu} , \tag{15}$$

where $\mu\nu$ is the plane of the plaquette P. We may therefore integrate over all A, provided we sum over only one "gauge copy" of the m_P's according to the transformation (15). A suitable restriction is to take m_P to be non-zero only if P is in the 12 plane. Thus

$$Z = \int_{-\infty}^{\infty} \mathcal{D}A \left\{ \prod_{P'} \sum_{m_{P'} = -\infty}^{\infty} \right\} \mathrm{Exp}\left\{ - \sum_{P'} (F_{P'} - 2\, m_{P'})^2 \right\}, \tag{14b}$$

P' in 12 plane.

The expression on the right of (14b) is just the Dirac Lagrangian (discretized) with m_P, equal to the number of strings passing through the plaquette. The strings are taken to be in the 3 direction or, if we use a space-time lattice, the Dirac sheets are in the 03 plane. Thus compact Q.E.D. is quantitatively equivalent to non-compact Q.E.D. together with monopoles.

Banks, Kogut and Myerson estimate that their model of Q.E.D. with monopole world-lines has a phase transition, from a non-confined to a confined phase, at roughly the same value of g as that found from Monte-Carlo calculations of lattice Q.E.D.

Peskin has shown that the Villain form of the model may be rewritten as a model of point particles in four dimensions. The ordered phase is the confining phase, the disordered phase the non-confining phase.

Monopoles and non-Abelian (SU(2)) Lattice-Gauge Theories

The purpose of this part of the talk is not to discuss any further the relation between Wu-Yang monopoles, which survive in the continuum limit, and confinement. This question has not been related to the lattice approach as yet. By considering Dirac (or Z_2) monopoles, which do not survive in the continuum, physicists have been able to obtain an understanding of some otherwise puzzling transitions which are observed in lattice-gauge Monte-Carlo calculations.

The action in the SU(2) lattice-gauge theory is a straightforward generalization of the Abelian action:

$$\text{Action} = -\frac{1}{g^2} \sum_P \text{Tr} \left\{ \Pi U_\ell + \Pi U_\ell^+ \right\} , \tag{16}$$

with U in the fundamental representation. One can also define an SO(3) lattice-gauge theory with a similar action (multiplied by a factor $\frac{1}{4}$), but with the U's in the adjoint representation.

The string tension in the SU(2) theory, calculated by Monte-Carlo methods, has qualitatively the form shown in Fig. 3. The fact that

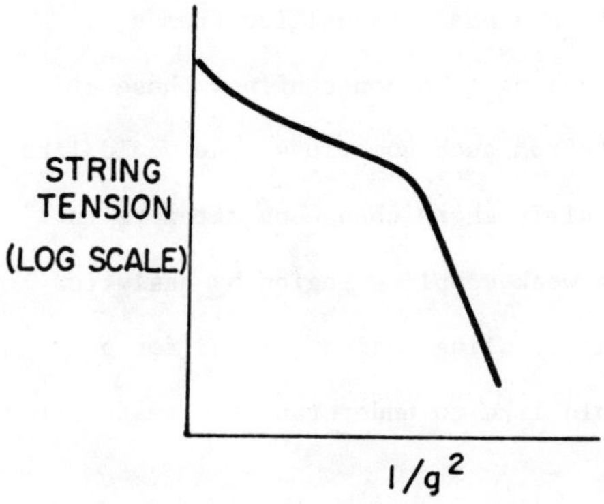

Fig. 3

Figure 3. String tension in SU(2) lattice-gauge theory

it is non-zero, even in the weak-coupling region, indicates that we have confinement in the continuum theory. It is most impressive that the slope of the weak-coupling end of the curve agrees with the asymptotic-freedom prediction to within statistical errors.

The sharp change in the slope between weak and strong couplings may be a cause for some concern. Though the change appears to be a very rapid continuous change rather than a phase transition, one may ask whether this fact is certain, in view of the statistical errors. One may similarly ask whether the non-zero slope in the weak-coupling region might not be a statistical error; one is looking for the coefficient of the area term in the logarithm of the Wilson loop, but one can only examine fairly small loops, where the term proportional to the area is small compared to other terms. One may therefore worry whether the bump in the curve might not be a phase transition from a confined phase at strong coupling to a non-confined phase at weak coupling. Quite apart from such questions, one would like to know the effect of the fairly sharp change on attempts to calculate quantities in the weak-coupling region by analytic continuation from the strong-coupling region. Thus, for a variety of reasons, we should like to understand the reasons for the rapid change.

In the SO(3) model, there is an actual phase transition between the weak- and strong-coupling regions (Halliday and Schwimmer,[14] Greensite and Lautrup[15]). One appears to have confinement on both sides of the phase transition; the remarks regarding the uncertainties of the SU(2) model are equally applicable here.

To clarify the phase structure further, Creutz and Bhanot[16] examined a combined SU(2) - SO(3) model:

$$\text{Action} = -\sum_P \text{Tr}\left\{\beta_1 \Pi U_{\ell,SU(2)} + \beta_2 \Pi U_{\ell,SO(3)}\right\} . \qquad (17)$$

The phase structure found by them is shown in Fig. 4. The

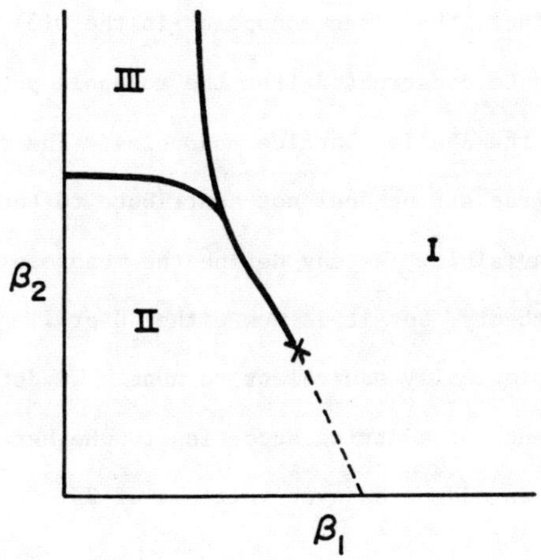

Fig. 4

Figure 4. Phase structure of the SU(2) - SO(3) lattice-gauge theory. solid lines represent first-order phase transitions, the cross a critical point. The rapid transition in the SU(2) model (horizontal axis) thus appears to be due to the nearness of an actual phase transition.

One might consider models where $\frac{\beta_2}{\beta_1}$ is fixed. For $\frac{\beta_2}{\beta_1} = \infty$, the model is pure SO(3), and we have a single phase transition. For large finite values of $\frac{\beta_2}{\beta_1}$, we have two phase transitions. As $\frac{\beta_2}{\beta_1}$ decreases, the two phase transitions approach one another and coalesce into one; at a smaller value of $\frac{\beta_2}{\beta_1}$ the phase transition disappears, leaving us with a rapid continuous transition.

To see how this structure can be understood in terms of monopoles, we first examine the SO(3) theory, which has Dirac monopoles. In fact, the Dirac monopoles in the O(3) lattice-gauge theory may be constructed from the monopole potential in the same way as the Abelian lattice monopoles. The rotation of 2π around the Dirac string does not contribute to the action; the string is invisible. We may define the monopole number as in the Abelian theory, but it is now either 0 or 1; two Dirac strings are topologically equivalent to none. We define the presence or absence of a string according to whether the rotation angle is nearer an odd or an even multiple of 2π:

$$\sigma_P = \text{Sgn Tr} \prod_P U_e \tag{18}$$

If $\sigma_P = -1$, a string passes through the plaquette. The monopole number is

$$n_c = \frac{1}{2}\left\{1 - \sigma_P\right\} \quad , \tag{19}$$

the sum being taken over the faces of the cube.

The SU(2) theory does not possess Dirac monopoles, since a rotation through 2π does contribute to the action and the string is no longer invisible.

Halliday and Schwimmer[14] use the following Villain form for the SO(3) action:

$$\text{Action} = -\frac{1}{g^2} \sum_P \left\{ |\text{Tr}\Pi U_\ell| + |\text{Tr}\Pi U_\ell^+| \right\} \quad , \tag{20a}$$

$$= -\frac{1}{g^2} \sum_P \sigma_P \, \text{Tr} \left\{ \Pi U_\ell + \Pi U_\ell^+ \right\} \quad . \tag{20b}$$

The group is SO(3) and not SU(2), since the action does not change when rotation angle changes by 2π. As in the Abelian case, the Villain form (20b) differs from the original form (16) through the variables σ_p. Now, however, σ_p is not an independent variable, but is defined by Eq. (18).

Halliday and Schwimmer do find a first-order phase transition with the action (20). The monopole density is much greater on the strong-coupling side of the phase transition than on the weak-coupling side, so that the transition appears to be due to monopole condensation. The SO(3) model, unlike the Abelian model, still has Wu-Yang monopoles, which do not disappear as we approach the continuum limit. Hence, assuming that confinement is due to monopole condensation, one might expect confinement on both sides of the phase transition; this is observed in the calculations.

Though the Dirac strings are no longer invisible in the SU(2) model, Brower, Kessler and Levine[17] suggest that we continue to define the monopole number by (19). The monopoles should now be regarded as ends of open strings rather than as independent objects. If a small admixture of SU(2) coupling is added to the SO(3) coupling, one would still expect a phase transition due to condensation of such strings. Figure 4 shows that the phase transition persists nearly, to but not quite, to the point of pure SU(2) coupling. B.K.L. compute the monopole density in the pure SU(2) model. They find that it is roughly constant on the strong-coupling side of the transition, but that it decreases rapidly on the weak-coupling side. The transition thus appears to be due to condensation of strings with monopoles at their ends.

To understand the phase structure of Fig. 4 more fully,

one must bear in mind that one could also have closed Dirac strings with no monopoles. If the density of open strings is large, one might also expect a large density of closed strings. Thus region II, the strong-coupling region, would be a region with a high density of monopoles, or open strings, and a high density of closed strings. Region I, the weak-coupling region, would not have a high density of either. In region III, the weak-coupling SO(3) region, one would expect a low density of monopoles (or open strings), but one might expect a high density of closed strings, which carry no energy in the pure SO(3) limit.

B.K.L. confirm these conclusions by adding, to the action, a chemical potential which suppresses the monopoles. The changes induced by the chemical potential are what one expects if the interpretation just given is correct.

With the work of Halliday and Schwimmer and Brower, Kessler and Levine, we now appear to have a reasonably good understanding of the transitions in lattice-gauge SU(2) and SO(3) models.

Attempts to Include Monopoles Explicitly in Continuum Gauge Theory Calculations

I shall preface the last section of the talk by an observation of Bardakci and Samuel,[18] who examined, *inter alia*, an SU(2) theory with a single scalar ϕ field in the U-gauge ($\phi^1 = \phi^2 = 0$, $\phi^3 \neq 0$). The remaining Abelian field in the U-gauge develops a Dirac-string singularity along those lines (sheets in four dimensions) where ϕ was in the 3-direction, with

$\phi^3 < 0$, before the gauge transformation. Hence the U-gauge should explicitly include monopoles at the points where $\phi = 0$.

It must be emphasized that the U-gauge can in principle be used for any phase. It is most convenient for the Georgi-Glashow phase where $\langle \phi \rangle \neq 0$, but it is not restricted to this phase. In any order of the perturbation theory of the Georgi-Glashow phase, $\phi \neq 0$ and the monopoles do not appear. They must, however, be included in an exact treatment.

For the confinement problem we are interested in a Lagrangian without a scalar field. Moreover, the U-gauge is inconvenient because of its short-distance singularities. Recent work by 't Hooft[19] suggests a procedure for treating pure Yang-Mills theories and avoiding, at least in part, the short-distance singularities of the U-gauge. The work is preliminary, and it is not at present known whether one can obtain a practical calculational scheme. The approach is very interesting, and has at least the possibility of getting to the heart of the confinement problem. We shall confine ourselves to SU(2); 't Hooft treats the general SU(N) Yang Mills theory.

Since the theory of interest has no elementary scalar field, 't Hooft uses a composite field, such as

$$\phi = \epsilon^{\alpha\beta\gamma} F^{\beta}_{\mu\nu} \nabla^2 F^{\gamma,\mu\nu} \quad . \tag{21}$$

For $N \geqslant 3$, we can omit the ∇^2 operator and replace ϵ by the symmetric coupling of three adjoint representations.

't Hooft attempts to find a gauge which resembles the Lorentz

Gauge at short distances and the U-gauge at large distances. Ghosts would appear, but would only propagate for those short distances described by the Lorentz gauge. The Gribov ambiguity might hopefully be avoided, since the Lorentz gauge is only used where the coupling is fairly weak.

A gauge which fulfills these physical requirements is the usual R-gauge

$$\frac{\partial A^{\alpha}_{\mu}}{\partial x_{\mu}} - c\varepsilon^{\alpha 3\beta} v^{3} \phi^{\beta} = 0, \quad v = \langle \phi \rangle, \quad \alpha = 1, 2 . \tag{22}$$

One may replace v^3 in (22) by ϕ^3. Alternatively, we may multiply (22) by v^3 to obtain the equation

$$v^{3} \frac{\partial A^{\alpha}_{\mu}}{\partial x_{\mu}} - c\varepsilon^{\alpha 3\beta} (v^{3})^{2} \phi^{\beta} = 0 , \tag{23}$$

following which we set $(v^3)^2$ on the right of (23) equal to a constant. We thus have the two possible gauges

$$\frac{\partial A^{\alpha}_{\mu}}{\partial x^{\mu}} - c\varepsilon^{\alpha 3\beta} \phi^{3} \phi^{\beta} = 0 , \quad \alpha = (1,2) \tag{24}$$

or

$$\varepsilon^{\alpha 3\beta} \phi^{3} \frac{\partial A^{\beta}_{\mu}}{\partial x^{\mu}} + c\phi^{\alpha} = 0 . \quad \alpha = (1,2) \tag{25}$$

If we ignore the singular nature of the operator products in (21) and (24), the first term of (24) or (25) dominates for short distances, the last for large distances. Thus these gauges do interpolate between the Lorentz gauge and the U-gauge. To maintain this property with the operator products in (21) and (24), it would appear to be necessary to spread out the products in space-time.

Gauges (24) and (25) both have monopoles since they approach the U-gauge at large distances. Each, unfortunately, has its difficulties. At large distances, the gauge condition (24) is $\phi^1 = \phi^2 = 0$, but the sign of ϕ is not fixed. We can have "phantom surfaces" where ϕ changes sign. Such surfaces are gauge artifacts and, needless to say, they complicate the calculation. Equation (25), on the other hand, may be rewritten

$$\frac{\partial A^{\beta}_{\mu}}{\partial x^{\mu}} - c(\phi^3)^{-1}\varepsilon^{\alpha 3\beta}\phi^{\beta} = 0 \quad . \tag{26}$$

The presence of the factor $(\phi^3)^{-1}$ on the right is an undesirable feature.

For these reasons, and because of the generally complicated nature of the problem, the work should at present be regarded as being in its formative stage. It is certainly a stimulating approach, and may lead to a practicable approximation scheme.

Acknowledgement

Research supported by the National Science Foundation under grant number PHY79-23251.

References

1. H.B. Nielsen and P. Olesen, Nucl. Phys. B61, 45 (1973).
2. Y. Nambu, Phys. Rev. D10, 4262 (1974).
3. S. Mandelstam, Phys. Reports 236, 245 (1976).
4. G. 't Hooft, in: High-Energy Physics, Proc. European Phys. Soc. Int. Conf., ed. A. Zichichi (Editrice Compositori, Bologna, 1976) p. 1225.
5. T.T. Wu and C.N. Yang, in: Properties of Matter under Unusual Conditions, eds. H. Mark and S. Fernbach (Interscience, New York, 1969) p. 349.
6. G. 't Hooft, Nucl. Phys. B79, 276 (1976).
7. A.M. Polyakov, Nucl. Phys. B120, 429 (1977).
8. G.K. Saviddy, Phys. Lett. 71B, 133 (1977).
9. F. Wilczek, private communication from D. Gross.
10. H. Pagels and E. Tomboulis, Nucl. Phys. B143, 485 (1978).
11. G. 't Hooft, Nucl. Phys. B138, 1 (1978).
12. T. Banks, R. Myerson and J. Kogut, Nucl. Phys. B129, 493 (1977).
13. M.E. Peskin, Ann. Phys. (N.Y.), 113, 122.
14. I. Halliday and A. Schwimmer, Phys. Lett. 101B, 327 (1981); Phys. Lett. 102B, 337 (1981).
15. J. Greensite and B. Lautrup, Phys. Rev. Lett. 47, 9 (1981).
16. G. Bhanot and M. Creutz, Brookhaven Preprint.
17. R.C. Brower, D.A. Kessler and H. Levine, Phys. Rev. Lett. 47, 621 (1981); Dynamics of SU(2) Lattice Gauge Theories (Harvard Preprint).
18. K. Bardakci and S. Samuel, Phys. Rev. D18, 2849 (1978).

19. G. 't Hooft, Topology of the Gauge Condition and New Confinement Phases in Non-Abelian Gauge Theories (California Institute of Technology Preprint).

CONFINEMENT AND MAGNETIC CONDENSATION FOR $N \to \infty$

P. Olesen

The Niels Bohr Institute, University of Copenhagen

DK-2100 Copenhagen Ø, Denmark

Abstract

We discuss why magnetic condensation is necessary for confinement. Using the Makeenko-Migdal equation we then indicate that for $N \to \infty$ (in SU(N)) one has a condensate of magnetic strings in the QCD vacuum.

1. Introduction

Many authors believe that in continuum QCD the confinement property is associated with a vacuum consisting of objects which have non-trivial topological properties. One can ask whether this belief is really right?

The answer to this question could be found if one can show that the assumed confinement (Wilson loop ~ exp(- (area)) etc.) leads/does not lead to some kind of non-trivial vacuum topology.

We shall show (in Sect. 3) that if one assumes confinement, then there must exist a certain additive flux. In the rest of this paper (Sects. 4-7) we shall then argue by means of the Makeenko-Migdal equation that (at least for large N) this implies the existence of SU(N) vortex lines (N large). Such fluxes have the property that N of them can end in "nothing":

This means that if these vortex lines condense, then they will also branch (SU(3)):

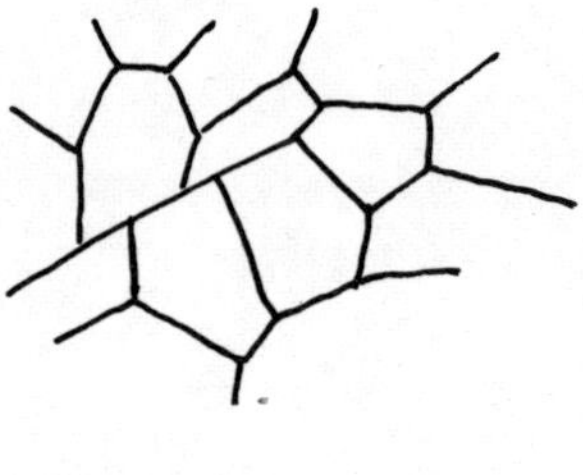

In a condensed state the "entropy" (i.e., the disorder) is thus very high. Flux tubes that end in "nothing" can in a certain sense be said to end in a monopole. Thus, in the condensate one can also describe the state as consisting effectively of monopoles. This point of view has been described by Mandelstam in his talk at this meeting. Consequently there are good reasons to believe that the vacuum state has more than one (alternative) description.

Because of the high entropy it is clear that it is only meaningful to describe fluxes in a vacuum state statistically. The usual topological language for describing monopoles and vortex lines can therefore not be applied in a straightforward manner. In sects. 2 and 3 we shall discuss this point further.

2. A simple intuitive reason why a disordered magnetic magnetic vacuum confines

We shall now give a simple intuitive reason why a disordered magnetic vacuum state confines.

In solid state physics there exists a beautiful argument[1] which shows that if you have a D-dimensional system in a random magnetic field H(x), then this is equivalent to having a D-2 dimensional system without the random field. Here H(x) satisfies

$$\langle H(x)\rangle = 0 \quad , \quad \langle H(x)^2\rangle \neq 0 \tag{2.1}$$

for any x. Further, there should be no long range correlations, so the correlation function $\langle H(x)H(y)\rangle$ is essentially a smeared δ-function $\approx \delta^D(x-y)$.

Inspired by this result as well as by some work[2] done in Copenhagen on the QCD vacuum, Ambjørn and the author then proposed[3] that the QCD vacuum should be described by a random magnetic field. It was shown by Nielsen and the author[4] and by Parisi[5] that such an assumption gives a simple intuitive reason why a disordered vacuum confines.

To see this, let us consider a random flux Φ (this is better than H, since the magnetic field is not gauge invariant in the Yang-Mills case). To produce meaningful results we have to average over a statistical ensemble, described by a distribution function $D(\Phi)$. The Wilson loop is then given by

$$W(C) = \int d\Phi \; e^{i\Phi} \; D(\Phi), \tag{2.2}$$

where C is some closed curve. Eq. (2.2) comes about because $e^{i\Phi}$ is the "value" of W in a single sample, corresponding to a flux Φ through a surface bounded by C. Randomness now means

$$\langle \Phi \rangle_A = 0, \tag{2.3}$$

where $\langle\ \rangle_A$ means average of the flux through a surface A. Further, there should be no long range correlations. To implement this we take a planar curve C and consider a flat surface (bounded by C) with area A. We now imagine that A is subdivided in many smaller areas

$A_1,\ldots,A_n$ $(n \to \infty)$, with $A = \sum_{i=1}^{n} A_i$. Since there are no long range correlations, $\langle \Phi^2 \rangle$ is additive,

$$\langle \Phi^2 \rangle_A = \sum_i \langle \Phi^2 \rangle_{A_i} \propto A, \tag{2.4}$$

because the number of subareas is of the order A.

Since the flux is independently distributed, it is natural to assume that the central limit theorem can be applied. Hence, for $n \to \infty$

$$D(\Phi) \to \frac{1}{\sqrt{\pi c A}} e^{-\Phi^2/cA}, \tag{2.5}$$

where c is a constant. Inserting eq. (2.5) in eq. (2.2) we easily obtain[4,5)]

$$W(C) \sim e^{-A} \tag{2.6}$$

Thus, we have obtained Wilson's area law from the assumption of random fluxes.*) It is, of course, a highly non-trivial problem to show that the QCD vacuum can indeed be described by a random flux distribution.

3. An additive non-Abelian flux for a confining system

From a more strict point of view the argument given above can be criticized because it introduces the rather unspecified flux Φ. It is well known that it is a

---

*) If we do not consider a flat curve C, the area in eq. (2.6) must be the minimal area spanned by C, since by analogy with the solid state result[1)] one can only expect the random fluxes to produce the leading behavior. Non-minimal surfaces give non-leading contributions.

highly non-trivial problem to find a reasonable non-Abelian flux which is additive and conserved.

We shall now show[6)] that there exists an additive flux for confing systems: <u>a necessary and sufficient condition for confinement is the existence of a flux which is independently distributed and which is additive</u>[6)].

Let us consider the operator

$$U[C] = P \exp i \oint_C A_\mu(x)\, dx_\mu , \tag{3.1}$$

which is unitary even after renormalization, at least if the curve C is smooth (see Aoyama[7)]). Thus the eigenvalues are of the form

$$\exp i\alpha_m [A_\mu(x), C] \tag{3.2}$$

with $1 \leq m \leq N$ for SU(N). In the U(1) case α_m is a magnetic flux. We shall show that even in the SU(N)-case α_m is still an additive flux, provided one has confinement. Let us introduce the Wigner spectral density,

$$\rho_C(\alpha) = \frac{1}{N} \sum_{m=1}^{N} \delta_{2\pi}(\alpha - \alpha_m[A_\mu(x), C]) \tag{3.3}$$

where $\delta_{2\pi}(x)$ is a δ-function with support for $x = 2\pi q$, $q = 0, \pm 1, \pm 2, \ldots,$ since α_m in eq. (3.2) is only defined modulo 2π. The physical quantity is the average spectral density

$$\langle \rho_C(\alpha) \rangle = \int dA_\mu\, e^{-S(A_\mu)} \frac{1}{N} \sum_{m=1}^{N} \delta_{2\pi}(\alpha - \alpha_m[A_\mu(x), C]) . \tag{3.4}$$

Using the representation

$$\delta_{2\pi}(x) = \frac{1}{2\pi} \sum_{n=-\infty}^{+\infty} e^{-in\alpha} \tag{3.5}$$

we easily obtain from eq. (3.4)

$$\langle \rho_c(\alpha) \rangle = \frac{1}{2\pi} \sum_{n=-\infty}^{+\infty} e^{-in\alpha} \langle W(C^n) \rangle, \tag{3.6}$$

where

$$W(C^n) \equiv \frac{1}{N} \langle \mathrm{Tr}\, U(C)^n \rangle$$

$$= \int_{-\pi}^{\pi} d\alpha \; e^{in\alpha} \langle \rho_c(\alpha) \rangle \tag{3.7}$$

is the Wilson loop for the curve traversed n times. In particular the Wilson loop is given by

$$W(C) = \int_{-\pi}^{\pi} d\alpha \, e^{i\alpha} \langle \rho_c(\alpha) \rangle, \tag{3.8}$$

which is of the form (2.2). Thus, we can interpret $e^{i\alpha}$ as the "value" of the Wilson loop in a statistical sample with flux distribution $\langle \rho_c(\alpha) \rangle$, i.e. $\langle \rho_c(\alpha) \rangle$ is the probability for the flux α to occur.

If one has confinement then we expect

$$W(C) \approx e^{-A} N(A), \tag{3.9}$$

where $N(A)$ represents non-leading terms (e.g. polynomials). Similarly, $W(C^n)$ is given in terms of the potential for n quarks and n antiquarks, so we expect

$$W(C^n) \approx e^{-|n|A}\, N_n(A), \tag{3.10}$$

where $N_n(A)$ are again non-leading terms and where n should not be a multiple of N. Inserting eqs. (3.9) and (3.10) in eq. (3.6) we obtain[6] as far as the leading terms are concerned

$$\langle \rho_A(\alpha) \rangle \approx \int_{-\pi}^{\pi} d\alpha_1 \ldots \int_{-\pi}^{\pi} d\alpha_n \, \langle \rho_{A_1}(\alpha_1) \rangle \ldots$$

$$\ldots \langle \rho_{A_n}(\alpha_n) \rangle \, \delta_{2\pi}(\alpha_1 + \ldots + \alpha_n - \alpha),$$

$$A = A_1 + \ldots + A_n . \tag{3.11}$$

Here the additivity of the area is crucial. A perimeter behavior does thus not satisfy eq. (3.11) in a non-trivial way. This condition implies[6] that <u>the fluxes through the smaller areas are distributed independently, subject to the condition that the flux is additive modulo</u> 2π. This is thus a consequence of confinement.

One can also go in the opposite direction*): eq. (3.11) actually implies confinement:

$$W(A) = \int_{-\pi}^{\pi} d\alpha \, e^{i\alpha} \langle \rho_A(\alpha) \rangle$$

$$\approx \int_{-\pi}^{\pi} d\alpha_1 e^{i\alpha_1} \langle \rho_{A_1}(\alpha_1) \rangle \ldots \int_{-\pi}^{\pi} d\alpha_n \, e^{i\alpha_n} \langle \rho_{A_n}(\alpha_n) \rangle$$

$$= W(A_1)\, W(A_2) \ldots W(A_n). \tag{3.12}$$

---

*) It turns out that because of periodicity one cannot apply the central limit theorem (see ref. 6).

Taking $A_1 = A_2 = \ldots = A_n \equiv A_o$ we obtain

$$W(A) \approx [W(A_o)]^{A/A_o}, \tag{3.13}$$

which holds for the leading terms, so A must be the minimal surface. We therefore see that a necessary and sufficient condition for confinement is the randomness (3.11).

The problem then arises as to how one can realize eq. (3.11) in concrete dynamical examples. We shall just give one simple example, where there exists two magnetic vortex lines which are well separated:

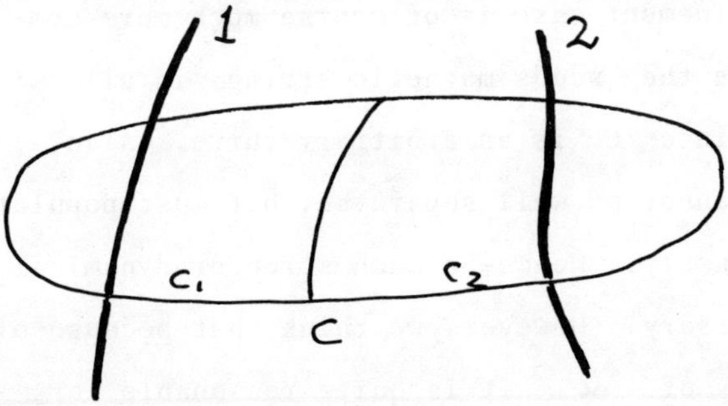

In the confinement case, the curve C is of course completely arbitrary. In the present example C is a fixed curve enclosing the flux tubes 1 and 2. The eigenvalues are degenerate for each vortex. Hence we have e.g.

$$\langle \rho_{C_1}(\alpha) \rangle = \int dA^1_\mu \, e^{-S(A^1_\mu)} \, \delta_{2\pi}(\alpha - \bar{\alpha}^1) \tag{3.14}$$

where $\bar{\alpha}^1$ is the common value of the eigenvalues for the string no. 1. For string no. 2 a similar expression is valid. Hence

$$\int_{-\pi}^{\pi} d\beta \langle \rho_{C_1}(\beta)\rangle \langle \rho_{C_2}(\alpha-\beta)\rangle$$

$$= \iint dA_\mu^1 \, dA_\mu^2 \, e^{-S(A_\mu^1) - S(A_\mu^2)} \delta_{2\pi}(\alpha - \bar{\alpha}^1 - \bar{\alpha}^2)$$

$$\approx \langle \rho_C(\alpha)\rangle \tag{3.15}$$

which is eq. (3.11) for n=2. Here we used that strings no. 1 and 2 are well separated so that their actions are additive.

The confinement case is of course much more complicated. One then needs magnetic strings in all directions, since C is an arbitrary curve. Also, the strings cannot be well separated, but must populate space very densely. Hence- a much stronger dynamical input is necessary. However, we think that because of the additivity of α it is quite reasonable to expect that only vacuum objects with non-trivial topology can do the job of providing an additive flux.

4. The Makeenko-Migdal equation

In an attempt to find a dynamical scheme which can give an indication of how additivity of the flux arises in a confining state, we shall use the Makeenko-Migdal equation[7]. Therefore, let us start by giving a short derivation of this interesting equation[*].

---

[*] The M-M-eq. has also been derived by Guerre and De Angelis[8] and by Förster[9] for the lattice case.

The derivation is formal in the sense that no attempt is made at deriving a renormalized equation. Hence a regulator is needed. It could be a lattice, but it could also be other regulators like dimensional regularization.

For calculational simplicity $A_\mu(x)$ is taken to be anti-hermitian. Again we start from

$$U[C] = P\, e^{\oint_C A_\mu(x)\, dx_\mu}, \qquad (4.1)$$

so that the Wilson expectation is given by

$$W[C] = \frac{1}{N}\langle \text{Tr}\, U[C]\rangle. \qquad (4.2)$$

We then define the variation

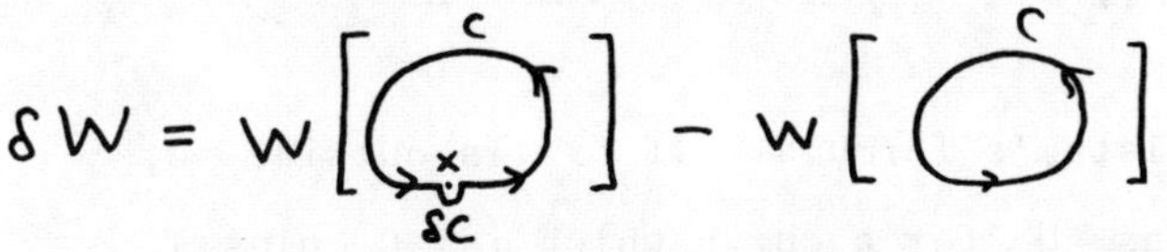

$$\delta W = W\!\left[\;\underset{\delta C}{\overset{C}{\bigcirc_x}}\;\right] - W\!\left[\;\overset{C}{\bigcirc}\;\right] \qquad (4.3)$$

or the "twisted wire" variation[*)]

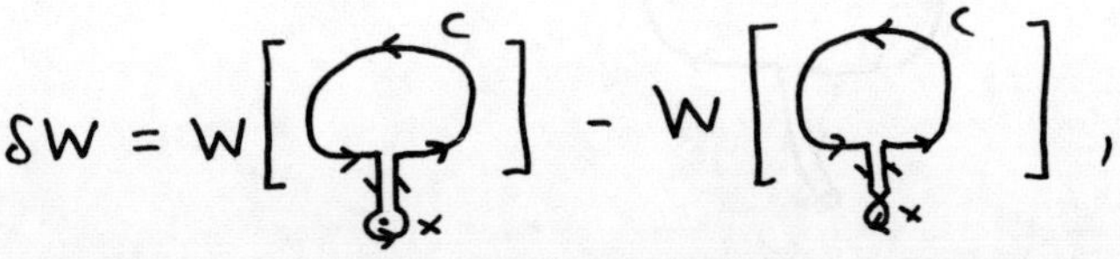

$$\delta W = W\!\left[\;\overset{C}{\bigcirc}\!\!\!\!\text{\tiny{\$x}}\;\right] - W\!\left[\;\overset{C}{\bigcirc}\!\!\!\!\text{\tiny{\$x}}\;\right], \qquad (4.4)$$

where the point x does not have to be on the curve C. With the first definition we get

---

[*)] This variation is the continuum version of the corresponding lattice variation.

$$\delta W = \frac{1}{N}\left\langle \mathrm{Tr}\left(e^{\oint_{\delta C} A_\mu dx_\mu} - 1\right) U_{x+\varepsilon \to x-\varepsilon} \right\rangle \tag{4.5}$$

where $U_{x+\varepsilon \to x-\varepsilon}$ is U for the curve C with a small piece cut out in the neighbourhood of x. With the second definition $U_{x+\varepsilon \to x-\varepsilon}$ is U for the curve C connected to the point x. Expanding we have

$$e^{\oint_{\delta C} A_\mu dx_\mu} \approx 1 + F_{\mu\nu}(x)\,\delta\sigma_{\mu\nu}(x) + O(\delta\sigma^2), \tag{4.6}$$

where $\delta\dot{\sigma}_{\mu\nu}(x)$ is an infinitesimal area element around the point x. Consequently

$$\frac{\delta W[C]}{\delta\sigma_{\mu\nu}(x)} = \frac{1}{N}\langle \mathrm{Tr}\, F_{\mu\nu}(x)\, U_{xx} \rangle, \tag{4.7}$$

which is Mandelstam's formula. If x is outside C, then U_{xx} means U for a curve which is C plus a curve connecting x to C:

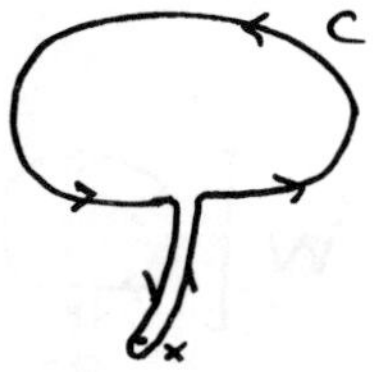

Since there is an $F_{\mu\nu}(x)$-insertion in the point x, the connecting curves do not cancel out, and the result (4.7) therefore in general depend on which connecting curve one uses.

Since x does not have to be on the curve C we can differentiate eq. (4.7) and remembering that

$$U_{xy} = P\, e^{\int_x^y A_\mu dx_\mu} \tag{4.8}$$

we obtain

$$\partial_\mu \frac{\delta W[C]}{\delta \sigma_{\mu\nu}(x)} = \frac{1}{N} \langle \mathrm{Tr}\, \nabla_\mu F_{\mu\nu}(x)\, U_{xx} \rangle . \tag{4.9}$$

So far only "kinematics" has been used. Next we use the actual form of the Yang-Mills action (i.e. dynamics):

$$\langle \mathrm{Tr}\, \nabla_\mu F_{\mu\nu}(x)\, e^{-S} \rangle$$

$$= \int dA_\mu\, \nabla_\mu F_{\mu\nu}(x)\, e^{-S}\, U_{xx} . \tag{4.10}$$

Here $\nabla_\mu F_{\mu\nu} e^{-S}$ is the functional derivative $\delta e^{-S}/\delta A_\nu$. Then, if we shift $\delta/\delta A_\nu$ by a partial functional integration we obtain*)

$$\langle \mathrm{Tr}\, \nabla_\mu F_{\mu\nu}(x)\, U_{xx} \rangle$$

$$= g_0^2 \left\langle \frac{\delta U^{ij}[C]}{\delta A_\nu^{kl}(x)} \right\rangle \left(\delta^i{}_k\, \delta^j{}_l - \frac{1}{N} \delta^{ij} \delta_{kl}\right) \tag{4.11}$$

for the group SU(N). The variational derivative of U is, however, simple because of eq. (4.1),

$$\frac{\delta U^{ij}}{\delta A_\nu^{kl}(x)} = \oint_C dy_\nu\, \delta^{(D)}(x-y)\, U^i{}_k[C_{xy}]\, U^j{}_l[C_{yx}] . \tag{4.12}$$

---

*) It should be noticed that if we had Higgs fields then this result would change by a non-trivial current (bilinear in the Higgs fields) term.

Using eqs. (4.9), (4.11), and (4.12) we obtain

$$\partial_\mu \frac{\delta W[C]}{\delta \sigma_{\mu\nu}(x)} = g_0^2 N \oint_C \delta^{(D)}(x-y)$$

$$\times \left\{ \left\langle \frac{1}{N} \mathrm{Tr}\, U[C_{xy}] \, \frac{1}{N} \mathrm{Tr}\, U[C_{yx}] \right\rangle - \frac{1}{N^2} \left\langle \frac{1}{N} \mathrm{Tr}\, U[C] \right\rangle \right\} . \tag{4.13}$$

This equation is still rather complicated. An important simplification occurs when $N \to \infty$. If one has a number of gauge invariant operators $\mathcal{O}_1 \ldots \mathcal{O}_p$, then in the limit $N \to \infty$ with $g_0^2 N$ fixed, their expectation value factorizes to any finite order of perturbation theory,

$$\langle \mathcal{O}_1 \ldots \mathcal{O}_p \rangle = \langle \mathcal{O}_1 \rangle \ldots \langle \mathcal{O}_p \rangle \left[1 + O\left(\frac{1}{N^2}\right) \right] . \tag{4.14}$$

Abstracting this property from perturbation theory (this is, of course, a highly non-trivial assumption) eq. (4.13) becomes[7,8,9]

$$\partial_\mu \frac{\delta W[C]}{\delta \sigma_{\mu\nu}(x)} = g_0^2 N \oint_C dy_\nu \, \delta^{(D)}(x-y) \, W[C_{xy}] \, W[C_{yx}] . \tag{4.15}$$

Here it was used that the δ-function in eq. (4.13) ensures that $\mathrm{Tr}U[C_{xy}]$ and $\mathrm{Tr}U[C_{yx}]$ are actually gauge invariant quantities (because $x=y$).

Eq. (4.15) is a closed equation in loop space - however, loop space is very big! A general solution of eq. (4.15) should involve information on <u>any</u> curve, no matter how weird.

It should be noticed that there is no scale in eq. (4.15), except from the regulator, which is needed to make sense out of the line integral over the δ-function. Thus, eq. (4.15) does not contain dimensional transmutation (Λ_{QCD} is not present), since it is not a renormalized equation. It is not known whether eq. (4.15) changes in an essential way under renormalization. There exists an unproven conjecture by a Berlin group (Kaschluhn, private communication) that eq. (4.15) is form invariant, but δ is replaced by a smeared δ-function which then becomes a dynamical quantity, to be computed e.g. order by order of perturbation theory (improved by the renormalization group).

Solutions of eq. (4.15) in general depend on the regulator that one uses. Such solutions must also satisfy the Bianchi identity

$$\epsilon_{\alpha\beta\gamma\delta}\, \partial_\beta \frac{\delta W[C]}{\delta \sigma_{\gamma\delta}(x)} = 0, \qquad (4.16)$$

which follows from Mandelstam's formula (4.9).

5. The spectral density from the M-M-equation

We shall now return to the spectral density $\langle \rho_c(\alpha) \rangle$ introduced in sect. 3,

$$\langle \rho_c(\alpha) \rangle = \frac{1}{2\pi} \sum_{n=-\infty}^{+\infty} e^{-in\alpha}\, W[C^n]. \qquad (5.1)$$

We shall then use the M-M-equation (4.15) to obtain information on $\langle \rho_c(\alpha) \rangle$. In 2 dimensions this has been done in much detail in the lattice case[10] and in the continuum[11]. The results are in agreement with results obtained by other methods, but the M-M-equation is usually a much easier way of obtaining these results. In D-dimensions one obtains[12] from eq. (4.15) and (5.1) the following "differential"-integral equation*)

$$\partial_\mu \frac{\delta \langle \rho_c(\alpha) \rangle}{\delta \sigma_{\mu\nu}(x)} = -g_0^2 N \oint_C dy_\nu \, \delta^{(D)}(x-y)$$

$$\times \frac{\partial}{\partial \alpha} \left\{ \langle \rho_c(\alpha) \rangle \, \mathrm{P} \int_{-\pi}^{\pi} d\beta \, \langle \rho_c(\beta) \rangle \cot \frac{\beta-\alpha}{2} \right\}$$

$$= 2 g_0^2 N \oint_C dy_\nu \, \delta^{(D)}(x-y)$$

$$\times \frac{\partial}{\partial \alpha} \left\{ \langle \rho_c(\alpha) \rangle \, \mathrm{P} \int_{-\infty}^{+\infty} d\beta \, \frac{\langle \rho_c(\beta) \rangle}{\alpha - \beta} \right\}, \qquad (5.2)$$

where the last form follows from periodicity of $\langle \rho_c(\alpha) \rangle$.

---

*) In ref. 12 it was assumed that there exists a master field. This assumption is, however, unnecessary in order to derive eq. (5.2), as one can easily see.

6. An effective action

We shall now show that eq. (5.2) has an interpretation in terms of an effective action. This will lead to information on the vacuum structure. Our arguments follow those in ref. 13, although we do not now assume the existence of a master field, as we did in ref. 13. (I thank P. Rossi for an interesting discussion of this point).

From eq. (5.1) we have

$$\int d\alpha\, \partial_\mu \frac{\delta \langle \rho_C(\alpha) \rangle}{\delta \sigma_{\mu\nu}(x)}$$

$$= \frac{i}{2\pi} \sum_{-\infty}^{+\infty} e^{-in\alpha} \frac{1}{n} \partial_\mu \frac{\delta W[C^n]}{\delta \sigma_{\mu\nu}(x)} . \tag{6.1}$$

Going back to eq. (4.9) we see that this can be written

$$\int d\alpha\, \partial_\mu \frac{\delta \langle \rho_C(\alpha) \rangle}{\delta \sigma_{\mu\nu}(x)}$$

$$= \frac{i}{2\pi N} \sum_{-\infty}^{+\infty} e^{-in\alpha} \left\langle \mathrm{Tr} \left(g_0^2 \frac{\delta S}{\delta A_\nu(x)} U(C^n_{xx}) \right) \right\rangle . \tag{6.2}$$

We now fix a gauge where $U(C_{xx})$ is diagonal, and obtain

$$\int d\alpha\, \partial_\mu \frac{\delta \langle \rho_C(\alpha) \rangle}{\delta \sigma_{\mu\nu}(x)}$$

$$= \frac{g_0^2}{2\pi N} \sum_{n=-\infty}^{\infty} \left\langle \sum_\ell \left(\frac{\delta S}{\delta A_\nu(x)} \right)_{\ell\ell} e^{in(\alpha_\ell - \alpha)} \right\rangle . \tag{6.3}$$

Using the continuous notation $(N \to \infty)$ this can be written

$$\int d\alpha \, \partial_\mu \frac{\delta \langle \rho_C(\alpha) \rangle}{\delta \sigma_{\mu\nu}(x)}$$

$$= -g_0^2 \left\langle \int_{-\pi}^{\pi} d\beta \, \rho_C(\beta) \langle \beta | \frac{\delta S}{\delta A_\nu(x)} | \beta \rangle_C \, \delta(\beta - \alpha) \right\rangle$$

$$= -g_0^2 \left\langle \rho_C(\alpha) \langle \alpha | \frac{\delta S}{\delta A_\nu(x)} | \alpha \rangle_C \right\rangle . \qquad (6.4)$$

In the basis in which $U(C_{xx})$ is diagonal we have used the notation $|\beta\rangle$ for a basis vector. From eq. (5.2) we now obtain

$$-g_0^2 \left\langle \rho_C(\alpha) \langle \alpha | \frac{\delta S}{\delta A_\nu(x)} | \alpha \rangle_C \right\rangle$$

$$= i g_0^2 N \oint dy_\nu \, \delta^{(D)}(x-y) \langle \rho_C(\alpha) \rangle$$

$$\times \int_{-\pi}^{\pi} d\beta \langle \rho_C(\beta) \rangle \cot \frac{\alpha - \beta}{2}$$

$$+ \text{const.}, \qquad (6.5)$$

where const. is independent of α. Thus, const. can depend on other curves than C.

Eq. (6.5) looks like an average equation of motion, since $\delta S / \delta A_\nu$ is contained on the left hand side. We shall now find a more direct interpretation of eq. (6.5) by going to loop space.

We thus imagine that space is somehow spanned by curves. How this should be done in details is not known at present. The main problem is to ensure the group property

$$U(C_1) \, U(C_2) = U(C_1 \cup C_2), \qquad (6.6)$$

where C_1 and C_2 are open curves that join smoothly in one point. We shall, however, assume that ordinary space can be replaced by the space of all curves with some appropriate measure. Further, for simplicity we assume that space can be spanned in such a way that through each point in space there is only one curve.

We then look for the QCD-action S expressed in terms of $U(C)$ instead of $A_\mu(x)$, since we can imagine that $A_\mu(x)$ is a functional of $U(C)$ for all curves. The guiding principle in the construction of S is that eq. (6.5) should be a consequence.

Since $A_\mu(x)$ is regarded as a functional of $U(C)$ we have

$$\frac{\delta S}{\delta A_\mu^{kn}(x)} = \sum_{C'} \frac{\delta S}{\delta U^{lm}(C')} \frac{\delta U^{lm}(C')}{\delta A_\mu^{kn}(x)}$$

$$= \oint_C dy_\mu \, \delta^{(D)}(x-y) \, U^l{}_k(C) \, \delta^m{}_n \frac{\delta S}{\delta U^{lm}(C)}$$

$$+ C\text{-independent terms.} \tag{6.7}$$

In the last step we assumed that C is a simple curve without self-intersection (presumably self-intersecting curves have a small measure in the space of all curves). In a representation where $U(C)$ is diagonal we then obtain

$$\langle\alpha| \frac{\delta S}{\delta i A_\mu(x)} |\alpha\rangle_C = - \oint_C dy_\mu \, \delta^{(D)}(x-y) \left(\frac{\partial S}{\partial \alpha}\right)_C$$

$$+ C\text{-independent terms.} \tag{6.8}$$

We now insert eq. (6.8) in eq. (6.5). Taking into account only those terms that refer to the curve C, we have

$$\left\langle \rho_c(\alpha) \left(\frac{\partial S}{\partial \alpha}\right)_c \right\rangle$$

$$= N \langle \rho_c(\alpha) \rangle \int_{-\pi}^{\pi} d\beta \langle \rho_c(\beta) \rangle \cot \frac{\alpha - \beta}{2} . \qquad (6.7)$$

Since the left hand side is the expectation value of the product of two gauge invariant quantities we can use factorization for $N \to \infty$ and obtain

$$\left\langle \left(\frac{\partial S}{\partial \alpha}\right)_c \right\rangle = 2N \int_{-\infty}^{+\infty} d\beta \, \frac{\langle \rho_c(\beta) \rangle}{\alpha - \beta} . \qquad (6.8)$$

If one leaves out the average < >, eq. (6.8) agrees with equations of motion obtained by Brezin, Itzykson, Parisi, Zuber[14] for the $N \to \infty$ 0-dimensional case and by Gross and Witten[15] for the 2-dimensional lattice case. In those cases there exists effectively a master field. However, in our case this cannot be deduced from eq. (6.8).

Eq. (6.8) has been derived on the assumption that space can be spanned by curves in such a way that only one curve passes through each point of space. In this case eq. (6.8) becomes a constraint on the effective action in loop space.

7. Consequences of confinement

We shall now investigate the form that S must have according to eq. (6.8) in the case of confinement. One of the very interesting features of the M-M-equation is that it can accomodate the area law. For large, smooth curves one has the solution[16,17]

$$W(C) \sim e^{-aA-bL} , \qquad a = \text{const.}, \; b = \text{const.} \tag{7.1}$$

where A is the area of the minimal surface and L is the perimeter. The perimeter dependence can be interpreted as a rudimentary "dynamical Higgs" phase (if a=0). From eqs. (7.1) and (3.6) we thus have asymptotically

$$\langle \rho_c(\alpha) \rangle \approx \frac{1}{2\pi} \Big[1 + 2 e^{-aA-bL} \cos\alpha + O(e^{-2aA-2bL}) \Big]. \tag{7.2}$$

Inserting eq. (7.2) in the equation of motion (6.8) we obtain

$$\left\langle \left(\frac{\partial S}{\partial \alpha} \right)_c \right\rangle \approx 2N e^{-aA-bL} \sin\alpha \tag{7.3}$$

Eq. (7.3) implies that the effective action is of the form

$$S_c \approx -2 e^{-aA-bL} N \sum_i \cos\alpha_i + \text{const.}$$

$$\approx -2 e^{-aA-bL} N^2 \int_{-\pi}^{\pi} d\alpha \langle \rho_c(\alpha) \rangle \cos\alpha + \text{const.} \tag{7.4}$$

With this action the equation of motion (7.3) can be derived from an effective partition function $(N \to \infty)$

$$Z^{eff} \approx \text{const.} \int dU(C)$$
$$\times \exp\left\{ N \sum_{C} \frac{1}{g_{eff}(C)^2} \operatorname{Tr} U(C) + \dots \right\}, \tag{7.5}$$

with

$$g_{eff}(C)^2 \approx e^{aA + bL} \rightarrow \infty . \tag{7.6}$$

Disregarding problems on how to define the appropriate measure in loop space, eq. (7.5) is strongly reminiscent of Wilson's lattice action in the strong coupling limit. We emphasize that the validity of eq. (7.5) is based on the continuum equation of motion (6.8) following from the M-M-equation, so eq. (7.5) is an unrenormalized continuum equation.

For $N \rightarrow \infty$ we thus have the behavior

$$Z^{eff} \sim \prod_{C} e^{-\frac{1}{2} S_C \mu} \tag{7.7}$$

where μ is a measure for curves. We shall now estimate the vacuum energy by enclosing the universe in a box, which is spanned by curves:

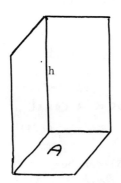

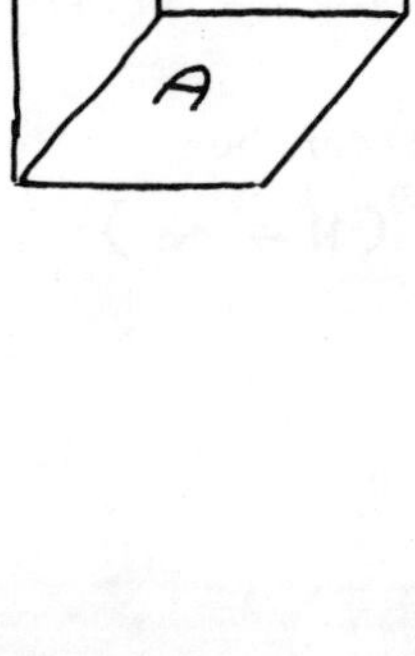

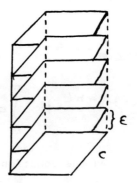

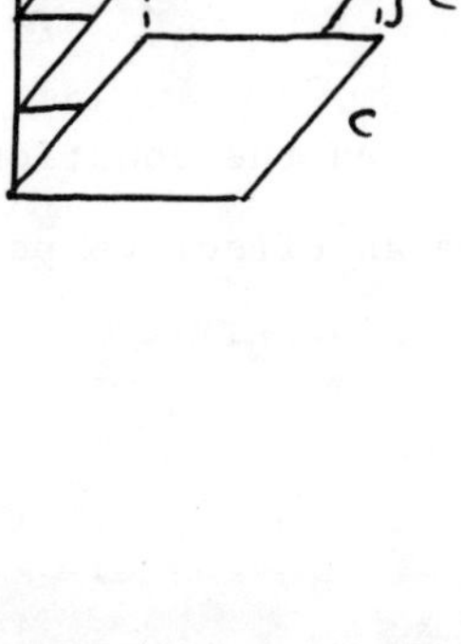

We then have from eq. (7.7)

$$E_{VAC} \approx \text{const. } N h e^{-aA-bL} \sum_m \cos \alpha_m + \text{const.}$$

$$\approx \text{const. } N^2 h e^{-aA-bL} \int_{-\pi}^{\pi} d\alpha \langle \rho_c(\alpha) \rangle \cos\alpha + \text{const.} \tag{7.8}$$

The vacuum energy is thus lowest in the confining state. The rudimentary "dynamical Higgs" ($W \sim e^{-bL}$) appears unstable (for $N \to \infty$). Thus, in the framework of the QCD M-M-equation, a "Higgs phase" (in some rudimentary sense) is not actually forbidden, but it has to be disregarded because of lack of stability.

The eigenvalues α_m can be computed. For $N \to \infty$ one has

$$\langle \rho_c(\alpha) \rangle = \frac{d\left(\frac{m}{N}\right)}{d\alpha}, \quad -\frac{N}{2} \le m \le \frac{N}{2}. \tag{7.9}$$

From eq. (7.2) we thus have

$$\alpha_m \approx \frac{2\pi m}{N} + O\left(e^{-aA-bL}\right). \tag{7.10}$$

Thus, the eigenvalues of $U(C)$ are $e^{i\alpha_m} \approx e^{\frac{2\pi i m}{N}}$.
Writing

$$E_{VAC} = \sum_m E_{VAC}(m) + \text{const.}, \tag{7.11}$$

we have

$$E_{VAC}(m) \approx -N^2 e^{-aA-bL} h \cos \frac{2\pi m}{N} + \cdots \tag{7.12}$$

Eq. (7.11) shows that the vacuum is a superposition of states each of which is characterized by the magnetic quantum number m.

Eq. (7.12) agrees with the expression obtained by 't Hooft[18] for the vacuum energy with total magnetic twist m in a confining state. In ref. 18 eq. (7.12) has been derived by use of electric <—> magnetic duality, which has not been used in the M-M-equation.

We take the agreement between eq. (7.12) and 't Hooft's expression as an indication that the additive flux α needed for confinement can only be realized by means of a vacuum filled with topologically non-trivial objects (magnetic vortex lines or, alternatively, monopoles). In the M-M-equation no non-trivial topology has been put in, but if one considers the confining state, such objects nevertheless show up, as we have seen.

For further discussion of the vacuum as a superposition of states with the magnetic quantum number m we refer to ref. 13.

Acknowledgements

I thank G. Parisi and P. Rossi for some interesting remarks.

References

1 - Y. Imry and S.K. Ma, Phys.Rev.Letters 35 (1975) 1399;

- A. Aharony, Y. Imry, and S.K. Ma, Phys.Rev.Letters 37 (1976) 1364;

- K.B. Efetov and A.I. Larkin, Sov.Phys. JETP 45 (1977) 1236;

- A.P. Young, J.Phys. C10 (1977) 1257;

- G. Parisi and N. Sourlas, Phys.Rev.Letters 43 (1979) 744.

2 - For a review see e.g. P. Olesen, Physica Scripta 23 (1981) 1000;

- H.B. Nielsen, in "Particle Physics" (Eds. I. Andric, I. Dadic and N. Zorko, North-Holland Publ.Co., Amsterdam 1981).

3 - J. Ambjørn and P. Olesen, Nucl.Phys. B170 [FS1] (1980) 60.

4 - H.B. Nielsen and P. Olesen, Niels Bohr Institute preprint NBI-HE-79-45 (unpublished) (1979).

5 - G. Parisi, rapporteurs talk at the Wisconsin Conf. (1980).

6 - P. Olesen, Nucl.Phys. B200 [FS4] (1982) 381.

7 - Yu.M. Makeenko and A.A. Migdal, Phys.Letters 88B (1979) 135.

8 - D. Angelis anf F. Guerre, Nuovo Cim. Lett. to be published.

9 - D. Förster, Phys.Letters 87B (1979) 87;
Nucl.Phys. B170 (1980) 107.

10 - G. Pafutti and P. Rossi, Phys.Lett. 92B (1980) 321.

11 - V.A. Kazakov and I.K. Kostov, Nucl.Phys. B176 (1980) 199;
- B. Durhuus and P. Olesen, Nucl.Phys. B184 (1981) 461;
- P. Rossi, M.I.T. preprint (1980).

12 - B. Durhuus and P. Olesen, Nucl.Phys. B184 (1981) 406.

13 - P. Olesen, Nucl.Phys. B184 (1981) 429.

14 - E. Brezin, C. Itzykson, G. Parisi, and J.B. Zuber, Comm.Math.Phys. 59 (1978) 35.

15 - D. Gross and E. Witten, Phys.Rev. D21 (1980) 446.

16 - Yu.M. Makeenko and A.A. Migdal, Phys.Letters 97B (1980) 253; Moscow preprint ITEP-170, 1980.

17 - P. Olesen and J.L. Petersen, Nucl.Phys. B181 (1981) 157.

18 - G. 't Hooft, Nucl.Phys. B138 (1978) 1; B153 (1979) 141.

MONOPOLES IN THE PRESENT AND EARLY UNIVERSE

T.W.B. Kibble

Blackett Laboratory
Imperial College, London.UK.

ABSTRACT.

This is a review of the physics of monopoles produced at the phase transition commonly supposed to have occurred in the very early history of the universe associated with grand unification. Estimates of the initial monopole density are examined and compared with observational limits from a variety of sources. Attempts to avoid the contradiction by invoking first-order transitions with extreme supercooling are discussed in detail, and other escape routes, for example involving multiple phase transitions, more briefly.

INTRODUCTION.

Fifty years ago when Dirac first discussed magnetic monopoles, non-Abelian gauge theories had not been invented. It is of course these models that have given a new lease of life to the idea. In particular in grand unified theories (GUTs) it seems almost inevitable that monopoles can exist. If we also accept the Hot Big Bang theory of the early universe, then it is hard to avoid the production of large numbers of monopoles at very early times.

I shall begin by reviewing briefly current ideas about the early history of the universe. (For further discussion and references see Kibble 1980). Then I shall discuss estimates of the initial density of monopoles, first of all making for simplicity the rather unrealistic assumption that the universe undergoes a second-order phase transition. In reality, for reasons I will return to, it is expected to be first-order, but the simple treatment should still apply to the case of a weak first-order transition. Next I shall turn to the observational limits on the monopole density in the recent universe, and demonstrate the apparent contradiction which constitutes the cosmic monopole problem.

In the remainder of the talk, I shall discuss possible ways out of this impasse. Several authors have suggested that it can be avoided if the transition is strongly first-order. I shall discuss this possibility at some length, before turning more briefly to other escape routes based on scenarios with multiple phase transitions or in which monopoles become confined.

Monopoles in Spontaneously Broken Gauge Theories. There have been discussions at this meeting of several different kinds of monopoles, and it may therefore be well to emphasize at the outset which kind I intend to discuss. I shall be talking exclusively about the monopoles that appear in spontaneously broken gauge theories, of which the 't Hooft-Polyakov monopole ('tHooft 1974, Polyakov 1974) is the paradigm.

Let us consider a theory with gauge group G, broken by the Higgs mechanism in which a scalar field ϕ acquires a vacuum expectation value. (Most of the discussion would apply equally to a dynamical symmetry-breaking mechanism in which ϕ is a composite field). The vacuum expectation value $\langle\phi\rangle$ is constrained, at the tree level, to minimize the Higgs potential $U(\phi)$, a quartic polynomial. If ϕ_o is a point on the minimum surface M of U, and H is the corresponding isotropy subgroup of G,

$$H = \{\Omega \varepsilon G: \Omega\phi_o = \phi_o\}$$

then M, whose points label the degenerate vacuum states of the theory, may be identified with the quotient space G/H, the space of left cosets of H in G.

The group G is a simple or semisimple Lie group, so that its second homotopy group $\pi_2(G)$ is trivial, and in most models is simply-connected, so that also $\pi_1(G)$ is trivial. If it is not, we can replace G by its simply connected covering group $\tilde{G}$, and H by its inverse image $\tilde{H}$ under the canonical projection of $\tilde{G}$ onto G. Then

$$M = G/H = \tilde{G}/\tilde{H}.$$

The topologically stable monopoles are then classified by the second homotopy group of M, namely

$$\pi_2(M) = \pi_1(\tilde{H}).$$

In particular, if H contains a U(1) factor, then $\pi_2(M)$ contains a corresponding factor Z.

It follows that if we start with a semisimple grand-unification group G and end with a group containing an unbroken U(1) subgroup- as we must to accommodate electromagnetism - then stable monopoles are inevitable. In the simplest GUTs, G is SU(5) or SO(10) and the symmetry breaking

$$G \to SU(3) \times SU(2) \times U(1)$$

occurs at an energy of the order of the mass m_X of the superheavy lepto-quark gauge bosons, typically in the range 10^{14} to 10^{15} GeV. At this energy the strong, weak and electromagnetic interactions would all have the same strength, characterized by the grand unified coupling constant g, with

$$\alpha = g^2/4\pi \simeq 0.025.$$

In this case, we expect stable monopoles of mass

$$m_m \simeq m_X/\alpha \sim 10^{16} \text{ GeV}.$$

Z_2 monopoles. It should be remarked that although most of the discussion in this field has focussed exclusively on these U(1) monopoles, there could also be others. The group $\pi_1(H)$ may contain finite factors such as Z_2, for example in the breaking of SU(3) to SO(3). The corresponding monopoles would carry not an additive magnetic charge but a mod-2 quantum number. However their existence would not greatly affect most of the subsequent discussion.

THE EARLY UNIVERSE.

I shall assume without much discussion that the early universe was isotropic and homogeneous and in a state of thermal equilibrium. These assumptions are a natural extrapolation backwards of what we know of the present universe, especially the isotropy of the cosmic microwave background radiation.

Robertson-Walker metric. Isotropy and homogeneity imply that the universe may be described by the Robertson-Walker metric,

$$ds^2 = dt^2 - R^2(t)\, d\sigma^2$$

where $d\sigma^2$ is a spatial metric of constant curvature K. The time development of the scaling function R(t) is governed by Einstein's equation

$$\left(\frac{\dot{R}}{R}\right)^2 = \frac{8\pi}{3m_{P\ell}^2}\rho - \frac{K}{R^2} + \Lambda$$

where $m_{P\ell}$ is the Planck mass

$$m_{P\ell} = G^{-\frac{1}{2}} \simeq 1.2 \times 10^{19}\,\text{GeV},$$

ρ is the density and Λ is the cosmological constant.

At very early times, we may assume (for reasons I will explain in a moment) that the universe is filled with an ideal gas at a temperature T much larger than any of the masses. In that case,

$$\rho = N_* \frac{\pi^2}{30} T^4$$

where N_* is the total effective number of helicity states of different particle species (the number of boson states plus 7/8 of the number of fermion states). In a typical GUT, N_* is initially of order 160.

If the expansion of the universe can be regarded as adiabatic, then

$$sR^3 = \text{constant}$$

where s is the entropy density

$$s = N_* \frac{2\pi^2}{45} T^3.$$

It follows that so long as N_* is constant $T \propto 1/R$. Over long periods this law is only approximate, because as T falls below various masses the corresponding particles effectively disappear and the effective N_* decreases - to about 100 after the first grand unification phase transition, and eventually down to about 4 or 5 in the present universe.

During any period where N_* is effectively constant, we have therefore $\rho \propto R^{-4}$. Hence at small values of R the other terms on the right hand side of the Einstein equation (K/R^2 and Λ) are relatively unimportant and may be ignored. Then the equation simplifies to

$$\left(\frac{\dot{R}}{R}\right)^2 = \frac{8\pi}{3m_{P\ell}^2} \rho$$

which yields $R \propto t^{\frac{1}{2}}$.

A useful relation between time and temperature which follows from this is

$$tT^2 = \left(\frac{45}{16\pi^3 N_*}\right)^{\frac{1}{2}} m_{P\ell} \simeq \frac{0.3}{\sqrt{N_*}} m_{P\ell} .$$

Justification of ideal gas approximation. At first sight the ideal-gas approximation for a system at densities vastly greater than nuclear might seem rather suspect. What makes it at least plausible is asymptotic freedom.

At the very high energies involved, all interactions are indeed weak, since $\alpha \ll 1$. Moreover it is easy to estimate the mean free path λ. We have $\lambda = 1/n\sigma$, where the number density n is given by

$$n = N_*' \frac{\xi(3)}{\pi^2} T^3$$

(N_*' differs from N_* only in the weighting of fermion states) and a typical cross-section σ may be taken to be

$$\sigma \sim \alpha^2/T^2.$$

Thus

$$\lambda \sim \frac{10}{N_*' \alpha^2 T} \sim \frac{100}{T}$$

which is certainly large compared to the average interparticle spacing ~1/3T. This helps to justify the ideal-gas approximation.

<u>Assumption of thermal equilibrium.</u> Unfortunately it is also true that initially when $T \gtrsim 10^{-3}\ m_{P\ell}, \lambda \gg t$. This means that the universe has had no time to reach equilibrium.

In fact the initial thermal equilibrium is very hard to understand. This is closely related to the Horizon Problem emphasized by Guth (1981): why is the universe homogeneous on scales vastly greater than the horizon distance (the limit of previous causal contact, which at time t is of order 2t)? Probably an answer to this question can be found only by going back before the Planck time, when gravity becomes as strong as the other interactions. This we can do only when we have a proper theory of quantum gravity. I shall not attempt to resolve this problem here but simply accept initial thermal equilibrium as a hypothesis.

<u>Flatness Problem.</u> I shall also not discuss further this related problem, but mention it here for completeness.

As is well known, if the cosmological constant Λ is zero the universe is closed (K>0) or open (K≤0) according as $\rho > \rho_c$ or $\rho \lesssim \rho_c$ where ρ_c is the critical density

$$\rho_c = \frac{3m_{P\ell}^2 \; H^2}{8\pi} \quad , \quad H = \frac{\dot{R}}{R} .$$

Present observational limits based on measurements of the Hubble constant H and decelleration parameter q are roughly

$$.02\rho_c < \rho < 2\rho_c .$$

However if we follow ρ and ρ_c back in time these limits narrow, until at the Planck time

$$\left(\frac{|\rho_c - \rho|}{\rho} \right)_{P\ell} \lesssim 10^{-58}$$

Why should ρ then have been so close to the critical value ρ_c? At present we do not know. Nor similarly do we know why the cosmological constant should be observationally so very small:

$$\frac{|\Lambda|}{m_{P\ell}^2} \lesssim 10^{-122} .$$

INITIAL MONOPOLE DENSITY.

For simplicity let me start by considering a second-order phase transition. Above the critical temperature T_c the order parameter $\langle\phi\rangle$ is zero. For $T < T_c$ it has the typical form

$$|\langle\phi\rangle| \simeq \eta \left(1 - \frac{T^2}{T_c^2}\right)^{\frac{1}{2}}$$

where η is a constant of the same order of magnitude as T_c.

The masses of the lepto-quark gauge bosons X and Higgs bosons H are proportiomal to $|<\phi>|$:

$$m_X \sim g|<\phi>|$$
$$m_H \sim h|<\phi>|$$

where g is the gauge constant and h^2 the quartic Higgs coupling. (Of course in general there is more than one such constant. These are to be regarded only as order-of-magnitude relations). The inverse quantities

$$\xi = m_H^{-1} \quad , \quad \lambda = m_X^{-1}$$

are the two length scales familiar in superconductivity theory, the correlation length and penetration depth.

<u>The Ginzburg temperature.</u> Now let us consider the situation just after the phase transition, when T has fallen slightly below T_c. The typical shape of the effective potential V (ϕ) - which replaces U(ϕ) at finite temperature - is then as shown in Fig. 1. Here ΔV is the difference in free energy density between the symmetric maximum at $\phi=0$ and the asymmetric minimum. Typically it is of order

$$\Delta V \sim h^2 \ |<\phi>|^4 .$$

So long as this central hump is fairly low, thermal fluctuations back and forth across it will be common. The relevant condition is that

$$\frac{4\pi}{3} \ \xi^3 \ \Delta V \lesssim T.$$

The temperature at which equality holds is the Ginzburg temperature T_G (Ginzburg 1960). Once T has fallen below T_G, $<\phi>$ may fluctuate near the potential-minimum surface M but is unlikely to fluctuate back to the top of the hump (except on a very small spatial scale that is irrelevant for our purposes). From this point on, $<\phi>$ must lie near M, but the choice of one point on M rather than another is random, and must be expected to be different in different regions of space. Monopoles are trapped regions within which $<\phi>$ has to pass through 0, where there are topological obstructions to extending $<\phi>$ continuously without leaving the minimum surface M. Thus for $T < T_G$ the monopoles are effectively frozen in.

For weak coupling, T_G lies not far below T_c. As an estimate of the length scale of the topological structures formed we may take the correlation length ξ_G at the Ginzburg temperature, which (Kibble 1976) is of order

$$\xi_G \sim 1/h^2 T_c .$$

The simplest way to visualize the structure is to imagine that at points separated by distances of order ξ_G, $<\phi>$ randomly chooses a point on M, choices at different points being uncorrelated. Then from each point a domain of roughly constant $<\phi>$ spreads out. Where the domains meet, $<\phi>$ varies smoothly from one value to the next. Occasionally at the corners where four domains meet monopoles are trapped. If the probability of this happening is p, then the initial monopole density should be (Einhorn 1980)

$$n_m \sim p\xi_G^{-3} \sim ph^6 T_c^{\,3} .$$

The number p is related to the geometry of the gauge group, and is typically of order 1/8.

Monopole-to-entropy ratio. Of more interest than n_m itself is the ratio

$$r = n_m/T^3$$

which is roughly constant during adiabatic expansion, except for the effects of annihilation. It is better in fact, as pointed out by Einhorn and Sato (1981), to use the monopole-to-entropy ratio

$$\hat{r} = \frac{n_m}{s} = \frac{45}{2\pi^2 N_*} r.$$

At the time of monopole production in the very early universe $\hat{r} \sim 0.02r$, while in the recent universe $\hat{r} \sim 0.5r$. Thus the difference though not large could be significant.

The above estimate for n_m based on the correlation length at the Ginzburg temperature gives

$$\hat{r}_{in} \sim ph^6 \sim 10^{-8\pm?}$$

There is considerable uncertainty about this figure, because of uncertainty in the value of h^2 and the precise criterion for defining T_G, but as we shall see this is really irrelevant, because the final value $\hat{r}_{fin}$ will be almost independent of $\hat{r}_{in}$ so long as $\hat{r}_{in} \gtrsim 10^{-10}$.

Causality limit. The above estimate depends on various, possibly questionable, assumptions about the monopole production mechanism. It is therefore of interest to find a limit that is independent of many of these details (Einhorn 1980). Whatever dynamical processes are involved, an upper limit on the length scale over which correlations can have been established is provided by the causal horizon distance at $T=T_G$, namely

$$2t = \left(\frac{45}{4\pi^3 N_*} \right)^{\frac{1}{2}} \frac{m_{P\ell}}{T_G^2} .$$

This yields a lower limit on the monopole density, namely

$$n_m \gtrsim \frac{p}{\frac{4\pi}{3} (2t)^3}$$

whence

$$\hat{r}_{in} \gtrsim 2.5p\sqrt{N_*} \left(\frac{T_G}{m_{P\ell}} \right)^3 \sim 10^{-12}.$$

It might be argued that a solution to the horizon problem could allow correlations to be established over much greater distances, and hence overcome this limit. However, this is unlikely because whatever correlations might exist at the Planck time would be destroyed by thermal fluctuations well before the time of the phase transition.

Special case of large m_H. It has been argued by Bais and Rudaz (1980) that the initial monopole density might be reduced in special circumstances where the monopole mass at the Ginzburg temperature $m_m(T_G)$ is large compared to T_G. If the monopole density is given by a thermal equilibrium distribution at $T = T_G$, then we should expect

$$\hat{r}_{in} \simeq \frac{45}{2\pi^2 N_*} \left[\frac{m_m(T_G)}{2\pi T_G} \right]^{3/2} e^{-m_m(T_G)/T_G}$$

The ratio in the exponent depends on the ratio h/g of the coupling constants or equivalently on m_H/m_X. Bais and Rudaz showed that typically

$$\frac{m_m(T_G)}{T_G} \sim 30 \frac{m_H}{m_X} .$$

This would give a very small value of $\hat{r}_{in}$ if $m_H \gtrsim 2.5m_X$. This may be uncomfortably large but perhaps not impossibly so.

However this argument has been strongly criticized Einhorn (1980), who queried the assumption of thermal equilibrium, and by Linde (1980a). Linde pointed out that when $m_X^{-1} >> m_H^{-1} = \xi$ then in a monopole-antimonopole pair separated by a distance of order ξ there will be a large cancellation between the long-range parts of the magnetic field which provide the dominant contribution to the monopole mass. Hence the energy of the pair is much less than $2m_m(T_G)$, in fact of order T_G. Thus there is no exponential suppression of pairs.

In these circumstances we should expect that at T_G there is a large number of monopoles and antimonopoles, typically separated by distances of order ξ_G. Presumably many of these would annihilate rapidly, before becoming completely separated. To estimate how many survive we need to estimate the likely excess of monopoles over antimonopoles within some volume large compared to ξ_G. Possibly a modified form of the Bais-Rudaz argument could be used to do this, but it would be necessary to compute an effective monopole mass in the presence of a dense plasma of monopoles and antimonopoles.

In any event, as Einhorn pointed out, it is difficult to see how one could thus avoid the limit based on the causal horizon distance.

OBSERVATIONAL LIMITS ON $\hat{r}$.

Once produced monopoles disappear (in most models) only by annihilation with antimonopoles, as discussed by Preskill (1979) and Zel'dovich and Khlopov (1976).

Monopole annihilation. Preskill studied the process of diffusion of monopoles towards antimonopoles through the plasma of light charged particles followed by capture in Bohr orbits and finally annihilation. Translated to $\hat{r}$ his equation for the rate of change with temperature is

$$\frac{d\hat{r}}{dT} = \left(\frac{16\pi^3 N_*}{45}\right)^{\frac{1}{2}} \delta \frac{m_{P\ell}}{T^2} \hat{r}^2$$

where

$$\frac{1}{\delta} = \frac{\zeta(3)}{\pi^2} \sum_s \frac{q_s^2}{4\pi} \sim \frac{1}{13}$$

The sum here is over all species of charged particles. This equation holds so long as the monopole mean free path is short compared to the capture distance (at which the Coulomb energy matches the thermal energy), i.e. down to a final temperature

$$T_f = \frac{2\delta^2}{\alpha_m^4} m_m ,$$

where

$$\alpha_m = \frac{g_m^2}{4\pi} \sim \frac{1}{\alpha} .$$

Note that

$$T_f \sim 10^{-4} m_m \sim 10^{12} \text{ GeV}.$$

For $T < T_f$ the probability of annihilation is almost vanishingly small, and $\hat{r}$ is essentially constant. The final conclusion may be summarized thus :

If $\hat{r}_m \gtrsim 10^{-10}$, then $\hat{r}_{fin} \sim 10^{-10}$.

If $\hat{r}_{in} \lesssim 10^{-10}$, then $\hat{r}_{fin} \sim \hat{r}_{in}$.

(The limit here is somewhat higher than Preskill's because of different choices for the parameters). Clearly annihilation is not very effective in reducing $\hat{r}$.

Limits on the present or recent monopole density can be derived from a number of different types of observational data.

1. <u>Present mass density.</u> From measurements of the Hubble constant and deceleration parameter we know that the present total mass density of the universe cannot much exceed the critical density. In particular the total contribution of monopoles cannot do so. Thus

$$n_m(3K)\, m_m \lesssim \rho_c \sim 10^{-122}\, m_{P\ell}^4$$

This yields the limit

$$\hat{r}_{now} \lesssim 10^{-24}.$$

It might be argued that since the formation of galaxies and stars there has been an opportunity for monopoles to accumulate in special places such as the cores of massive stars where perhaps they have annihilated. It is not clear how such a mechanism could operate but even if it does there is another limit which predates the era of galaxies, and to which I now turn.

2. <u>Mass density at time of helium synthesis.</u> The currently accepted scenario for synthesis of primordial helium is very sensitive to the mass density of the universe at the relevant time, when $T \sim 1\text{MeV}$. A larger density would mean faster expansion. This in turn would mean that the neutron-to-proton ratio freezes out at a higher temperature (because at the freeze-out point the relevant reaction rate equals the expansion rate). Since most of the neutrons end up incorporated in helium, this would raise the primordial helium abundance.

The observational situation is somewhat confused. Some observations seem to be in conflict with the standard theory, even without monopoles or any other extra contributions. Thus it is difficult to come to firm conclusions. However it seems reasonable to suppose that the total contribution of monopoles to the mass density cannot much exceed that of light particles. This means

$$n_m\,(1\text{ MeV})\,m_m \lesssim N_* \frac{\pi^2}{30}\,T^4$$

and hence

$$\hat{r}\,(1\text{ MeV}) \lesssim \frac{3}{4}\,\frac{T}{m_m} \sim 10^{-19}.$$

3. <u>Energy balance in galactic magnetic fields.</u> One of the best limits, due to Parker (1970, Bludman and Ruderman 1976), comes from the observation that there are magnetic fields in galaxies, typically of a few microgauss. These fields are sufficient to accelerate monopoles to energies of about 10^{11} GeV, corresponding to velocities

$$\frac{v}{c} \sim 10^{-2} \text{ to } 10^{-3}.$$

These magnetic fields are presumably maintained by the dynamo effect of differentially rotating gas in the galaxy. Their survival to the present time places an upper limit on the monopole density. If there were too many monopoles, the energy of the magnetic fields would be dissipated in accelerating them.

Based on this argument, Parker found an upper limit of 10^{-26} on the monopole-to-baryon ratio in the galaxy. However he was considering lighter monopoles that would reach relativistic velocities. The corresponding limit for $v/c \sim 10^{-3}$ is

$$n_m/n_B \lesssim 10^{-23}.$$

Since the baryon-to-entropy ratio is $n_B/s \sim 10^{-10}$, we have

$$\hat{r} \lesssim 10^{-33}.$$

Lazarides, Shafi and Walsh (1981) argue that this process may have been accelerating monopoles out of the galaxy throughout its history and hence obtain a limit on the primordial monopole-to-baryon ratio in the galaxy

$$n_m/n_B \lesssim 10^{-20\pm1}$$

or

$$\hat{r} \lesssim 10^{-30\pm1}$$

This seems fairly conservative. These authors also find corresponding limits on the monopole flux. The flux of monopoles accelerated in our galaxy is

$$f_{gal} \lesssim 10^{-3}\ m^{-2}\ yr^{-1}.$$

In addition, there would be an isotropic component of energetic monopoles accelerated in other galaxies, but this would be much smaller,

$$f_{iso} \lesssim 10^{-5\pm1}\ m^{-2}\ yr^{-1}.$$

Recently, another limit of a similar kind has been obtained by Drukier (1981) based on the magnetic field of magnetic dwarf stars. I shall not discuss this limit here because it will be presented in the following talk.

4. <u>Searches in cosmic rays.</u> It is also possible to search more directly for monopoles, either in cosmic rays or bound in matter. A review of the limits obtained has recently been presented by Longo (1981), and some of the evidence will be discussed in a later contribution to this meeting, by G. Giacomelli (1981).

Ionization by monopoles is a rapidly varying function of velocity. Essentially they will ionize if $v/c \gtrsim 3\times10^{-3}$, corresponding to $m_m \lesssim 10^{16}$ GeV. Monopoles with velocities just above this limit will be lightly ionizing and should be detected in quark searches. Thus for monopoles with $v/c \gtrsim 5\times10^{-3}$ we can conclude (Jones 1977) that the flux f is limited by $f \lesssim 300\ m^{-2}\ yr^{-1}$.

Faster monopoles would be easier to detect. The best limit is that of Kinoshita and Price (1981) who give

$$f \lesssim 0.4\ m^{-2}\ yr^{-1}$$

for monopoles with $v/c \gtrsim 0.02$. (See also Fleischer et. al. 1971).

These limits are not very restrictive when compared with those discussed above.

5. Searches in matter. Widely varied samples of matter have been checked for the presence of monopoles with negative results. For example in searches in lunar material and meteorites (Eberhard et al. 1971, Ross et al. 1973) a limit

$$\frac{n_m}{n_B} \lesssim 3\times10^{-28}$$

was obtained.

If this matter were a representative sample of the matter in the universe as a whole this would correspond to a very low limit, $\hat{r} \lesssim 10^{-38}$. However, these measurements are only sensitive to ligher monopoles, with masses $\lesssim 5 \times 10^{14}$ GeV. If such monopoles were present in the material from which the solar system condensed they would have become trapped in the matter and should still be there. Heavier monopoles on the other hand would be pulled free by gravity and would now be found only at the centre of the earth or other large bodies.

This limit is not therefore very significant for monopoles in the most familiar GUTs, but it would become very significant in alternative models with lighter monopoles.

FIRST-ORDER TRANSITIONS.

I now turn to various possible ways of avoiding the paradox that the monopoles predicted by GUTs are nowhere to be found. Many suggested escape routes are based on the assumption that the symmetry-breaking phase transitions are not second-order but first-order.

Reasons for first-order transitions. In fact there are several grounds for believing that the transitions must in fact be first-order.

Higher-order corrections to the effective potential can be important when the Higgs coupling is very small, i.e. $h^2 \sim g^4$ (Linde 1976). Then the gauge-particle one-loop corrections are comparable to the bare $U(\phi)$, and yield an effective $V(\phi)$ with a metastable minimum at $\phi=0$. (See Fig. 2).

Similarly when $U(\phi)$ contains no quadratic terms, the one-loop corrections are crucial. This is the mechanism of Coleman and Weinberg (1973), which has been invoked in the cosmic monopole context by Cook and Mahanthappa (1981) and by Billoire and Tamvakis (1981).

Even at tree level, $U(\phi)$ often contains cubic terms. For example, if $G = SU(5)$ and ϕ belongs to the adjoint representation, then (Guth and Tye 1980, Guth and Weinberg 1981, Kennedy, Lazarides and Shafi 1981)

$$U(\phi) = -\tfrac{1}{2}\mu^2 \mathrm{tr}\phi^2 + \tfrac{1}{4}a\,(\mathrm{tr}\phi^2)^2 + \tfrac{1}{2}\,b\,\mathrm{tr}\phi^4 + \tfrac{1}{3}\,c\,\mathrm{tr}\phi^3.$$

Such cubic terms always lead to a first-order transition.

In any event there will be cubic terms, proportional to m^3T, in the temperature-dependent corrections to the effective potential. It is thus inevitable that the transition will in reality be first-order. What is in question is whether it is <u>strongly</u> first-order.

Supercooling. When the temperature falls below the critical temperature T_c, no transition occurs and T continues to fall with the system in what is now a "false vacuum", separated by a barrier from the true vacuum, as in Fig. 2. The transition actually occurs by tunnelling through the barrier (or by thermal fluctuation over it), leading to formation of bubbles of the new phase which then expand until they coalesce and the transition is complete.

The probability of bubble formation per unit space-time volume, f, is typically (Coleman 1977, Callan and Coleman 1977) of the form

$$f = Ae^{-B}$$

where B is the Euclidean action for the "bounce" solution and A is a ratio of functional determinants.

In application to the early universe we must take account of the universal expansion. Guth and Tye (1980) calculated the fraction p of space in the new phase as a function of time. The result can most easily be expressed in terms of the "conformal" time variable

$$\tau = \int^t \frac{dt}{R(t)} ,$$

and is

$$p(\tau) = \exp - \frac{4\pi}{3} \int_0^\tau d\tau'(\tau-\tau')^3 R^4(\tau') f(\tau')$$

The implications of this result have been studied particularly by Einhorn, Stein and Toussaint (1980) and by Einhorn and Sato (1981).

Exponential expansion. If the transition is strongly first-order, i.e. with a large barrier between false and true vacuum states, there will be extreme supercooling. As T falls, the contribution to the density from light particles will fall like T^4, but there will remain a roughly constant vacuum energy density

$$\rho_{vac} = \Delta V$$

equal to the difference in V between the two minima. Eventually, ρ_{vac} will come to dominate ρ. When this happens the universe will start expanding exponentially,

$$R \propto e^{Ht} ,$$

with

$$H^2 = \frac{8\pi}{3m_{P\ell}^2} \rho_{vac} .$$

A reasonable order-of-magnitude estimate of H is

$$H \sim \frac{m_H}{m_{P\ell}} T_c \sim 10^{11} \text{ GeV}.$$

This exponential expansion makes it difficult to complete the transition,(Einhorn, Stein and Toussaint, 1980, Guth and Weinberg 1981). Essentially there are two possibulities. Either $f >> H^4$ in which case the transition is completed rather quickly, or $f \lesssim H^4$ in which case it is never completed at all. The first case corresponds to a weakly first-order transition which is not significantly different from a second-order one. In the second, there are always regions of old phase present because the expanding bubbles never catch up with the overall exponential expansion. (Even if $p(\tau)$ eventually tends to unity the bubbles may fail to coalesce, i.e. the new phase does not percolate. See Guth and Weinberg 1981). This scenario would predict unacceptably big large-scale inhomogeneity.

Second critical temperature. Apparently what we need to escape from this dilemma is a bubble-nucleation probability f that is very small initially but then suddenly becomes large, i.e.

$$f << H^4 \quad \text{for} \quad T_c > T > T_{c2}$$

$$f \gg H^4 \quad \text{for} \quad T < T_{c2}$$

with $T_{c2} << T_c$.

The simplest case in which this would occur is that of a classical instability. At T_{c2} the metastable false vacuum state becomes actually unstable. (See Fig. 3). Then we have supercooling almost down to T_{c2}, followed by the simultaneous nucleation of many bubbles and rapid completion of the transition.

The first-order transition releases a large latent heat, the vacuum energy ρ_{vac}, which reheats the universe to a temperature T_r not far below to T_c. The large entropy generation dilutes the baryon-to-entropy ratio. Thus to retain the successful explanation of baryon asymmetry as due to decays of X or H bosons we must ensure that at least $T_r \gtrsim m_H$, (or of course m_X) so that baryon number can be re-generated.

Unfortunately this superficially attractive scenario does not really help to solve the cosmic monopole problem. Within each bubble the symmetry-breaking direction is chosen randomly, and presumably independently in different bubbles. If so, the expected monopole density is

$$n_m \simeq pn_b \simeq p\ell^{-3}$$

where n_b is the number density of monopoles, and ℓ a typical distance between bubble nuclei. Now let us suppose for simplicity that f is zero above T_{c2} and constant below it. Then we easily find that

$$\ell \sim f^{-\frac{1}{4}} \quad .$$

The requirement that the transition be completed is $f \gg H^4$. Hence $\ell \ll H$, and it follows that

$$\hat{r}_{in} = \frac{n_m}{S(T_r)} \gg \frac{pH^3}{N_* \frac{2\pi^2}{45} T_r^{\,3}} \sim \frac{p}{N_*} \left(\frac{m_H \, T_c}{m_{P\ell} T_r} \right)^3 \quad .$$

Remarkably this is independent of T_{c2}, though it does depend on the temperature T_r after reheating. Typically we have

$$\hat{r}_{in} \gtrsim 10^{-18}$$

unless m_H is very small ($< 10^{13}$ GeV, say) in which case the exponential expansion rate is also small. Moreover this is only a lower limit on $\hat{r}_{in}$. In many cases, it would be as large as in a second-order transition.

It seems therefore that exponential expansion followed by a classical instability cannot solve the cosmic monopole problem. However I shall return to this point shortly, to discuss a recent proposal by Linde.

Premature transition. Another reason for doubting whether supercooling and exponential expansion can solve our problem is that prolonged supercooling may actually be impossible. Three separate lines of argument suggest that there may be a limit to the possible extent of supercooling. At least two of these are apparently unrelated, but all three suggest similar limits.

The first argument is due to Abbott (1981) who pointed out that in the exponential-expansion era there is a large scalar curvature

$$R = -12H^2.$$

Now for any scalar field ϕ we may expect to find an effective curvature-dependent interaction term,

$$\tfrac{1}{2}b\, R\phi^2,$$

which will contribute a squared mass for ϕ of order

$$m^2 \sim 12bH^2.$$

If $b > 0$ this has the effect of preventing the symmetry breaking while if $b < 0$ it will induce the transition, typically at a temperature of order H.

More recently Hut and Klinkhamer (1981) pointed out that exponential expansion implies the existence of event horizons (beyond which future causal contact is impossible so long as the expansion lasts), at a distance of order $1/H$. Consequently there is no longer an observer-independent definition of the vacuum or particle states. There will be mixing of positive and negative frequencies for modes with wave-length exceeding $1/H$, i.e. for frequencies $<H$. Hut and Klinkhamer argue that when T falls below H this mode mixing will induce the transition.

The third argument due to Horibe and Hosoya (1981) may be a variant of the second. They point out that the event horizons imply particle creation, as in the Hawking effect, and by analogy suggest that it will have a blackbody spectrum at the Hawking temperature $T_H \simeq H/2\pi$.

Taken together the three arguments make it very plausible that supercooling cannot proceed below a temperature of order $H \sim 10^{11}$ GeV.

There is one further, unrelated point, made by Sher (1981). Even if supercooling can continue beyond 10^{11} GeV, it may not go as far as is often envisaged, because of the temperature-dependence of the effective coupling constant. Indeed Sher finds that the effect of including this dependence is to raise the temperature at which the transition occurs in the Coleman-Weinberg case from as low as 1 GeV to 2×10^{10} GeV. Clearly this dependence cannot be ignored.

Gentle first-order transitions. Linde (1981) has recently proposed an interesting new variant on the supercooling theme. I should like to call this a "gentle" first-order transition, but I must emphasize that it is far from "weak". In fact, Linde envisages a transition occuring after extreme supercooling, when the true vacuum has an energy far below the false one (as in Fig. 2).

Linde argues that ϕ would not jump suddenly to the minimum ϕ_0 but rather to a point close to ϕ_1, where the free energy is equal to that of the false vacuum. Then the subsequent evolution from ϕ_1 to ϕ_0 would be slow, determined by an appropriate solution of the classical field equations. Initially we should expect

$$\phi(t) \propto e^{mt}$$

where

$$-m^2 \simeq \frac{d^2V}{d\phi^2} \quad .$$

In order of magnitude, $m \sim T_{c2}$, so the time required for ϕ to grow is of order $1/T_{c2}$.

Initially, there is still a large vacuum energy, almost as large as before, so that exponential expansion continues for a time of order $1/T_{c2}$ after the transition. During this time the universe expands by a factor

$$e^{H/T_{c2}} \sim e^{25000} .$$

Clearly the radius of a bubble then vastly exceeds the radius of the universe. In fact our visible universe lies entirely within a single bubble. This automatically solves the cosmic monopole problem, since the monopole density is of the same order as the bubble density. It also solves the horizon problem, since clearly the interior of one bubble is causally connected. In addition, Linde claims that it solves the flatness problem.

It is true of course that there will be far distant regions of the universe still in the old false-vacuum phase but they will be so far away as to have no observable influence. (This is arguably a rather unaesthetic feature of the model).

An interesting aspect of the model is that baryon number may be regenerated even without reheating to m_H, by the decay of the classical Higgs field ϕ which behaves in this respect just like Higgs bosons.

One problem of course is that the transition might be terminated prematurely at $T \sim H$, as I discussed earlier. Linde claims that the model can be designed to avoid this problem. Also, as has been pointed out by E. Weinberg (1981), it seems very doubtful whether there would be enough entropy within the bubble (rather than localized near its walls) to yield a reasonable final temperature. Nonetheless the model is an intriguing one.

MULTIPLE TRANSITIONS.

Hitherto I have assumed that only a single phase transition is relevant, but there are many models in which several transitions occur in succession. For example in the simplest SU(5) model, the parameters may be chosen so that instead of breaking directly to $SU(3) \times SU(2) \times U(1)$ it breaks first to $SU(4) \times U(1)$.

Fate of monopoles at successive transitions. The general case of a symmetry breaking scheme

$$G \to H_1 \to H_2$$

has been discussed in some detail by Bais (1981). If G is semisimple and both H_1 and H_2 contain U(1) factors then monopoles can exist in both H_1 and H_2 phases. The interesting question is what happens to the monopoles of the H_1 phase at the second phase transition. In general the answer to this question is quite complicated, especially when - as in the example I quoted - H_2 is not a subgroup of H_1.

There is a considerable range of possibilities. The H_1 monopoles may survive unchanged or decay to new H_2 monopoles that are often lighter; they may decay to the vacuum or become joined by relatively light flux tubes of confined (non-Abelian) magnetic flux, or finally they may become unstable to radial expansion. This last is an intriguing possibility discovered by Steinhardt (1981). Essentially what happens is that the cores of the H_1 monopoles make a smooth transition to the H_2 phase and then expand to form the new vacuum state. Meanwhile the regions of old vacuum are squeezed to form new monopoles.

Which of these possibilities actually occurs depends on the detailed topological relationships between the symmetry groups. However it seems to be generally true that a non-negligible fraction of the original H_1 monopoles either survives or decays to form new stable monopoles, so that it is hard in this way to escape the cosmic monopole problem.

Monopole-free intermediate phase. One scenario that does seem to work has been suggested by Langacker and Pi (1980). It makes use of an earlier observation by S. Weinberg (1974) that symmetry may sometimes be lower on the high-temperature side of a phase transition. A well known example is Rochelle salt (sodium potassium tartrate).

Langacker and Pi devised a model involving three Higgs doublets with the symmetry-breaking scheme

$$\begin{aligned} SU(5) &\to SU(3) \times SU(2) \times U(1) \\ &\to SU(3) \to SU(3) \times U(1) \end{aligned}$$

Monopoles are formed at the first transition but disappear completely at the second, because there is no longer an unbroken U(1) factor in the symmetry group. Further monopoles may be formed in the final transition but their number should be quite small. Indeed when the symmetry is restored rather than being broken there is no reason to expect the initially random distribution of points on the potential-minimum surface M that previously made monopole formation inevitable. Heavy monopoles would be suppressed by a large exponential factor.

This is a rather inelegant solution to the cosmic monopole problem, requiring a complicated ad hoc set of Higgs fields, but it does appear to be viable.

MAGNETIC MASS.

Another possible escape route has been proposed by Linde (1981), based on a study of the infrared behaviour of non-Abelian gauge theories at finite temperature.

To calculate Green's functions at finite temperature T we have to use Feynman rules with a discrete energy variable :

$$\int \frac{d^4k}{(2\pi)^4} \rightarrow T \sum_n \int \frac{d^3\underset{\sim}{k}}{(2\pi)^3}$$

where

$$\omega_n = \begin{cases} 2n\pi T & \text{for bosons} \\ (2n+1)\,\pi T & \text{for fermions.} \end{cases}$$

In a massless theory the infrared divergences come from the n=0 boson terms. Thus if we are interested only in infrared behaviour we may replace our original four-dimensional theory with an effective three-dimensional gauge theory with a (dimensional) coupling constant g^2T (Gross, Pisarski and Yaffe 1981). This theory has power-law infrared divergences, showing that perturbation theory must break down at some point.

If we examine the gauge field propagator near k=0 in the Feynman gauge we find that the time component has a natural cut-off:

$$D_{oo}(\omega = 0,\underset{\sim}{k}) \sim \frac{1}{\underset{\sim}{k}^2 + g^2T^2} \,.$$

Here gT has been called the "electric mass". Non-Abelian electric charges at finite temperature are screened with a screening length 1/g T. On the other hand for the spatial components we find

$$D_{ij}(\omega=0,\underset{\sim}{k}) \sim \frac{1}{\underset{\sim}{k}^2} \delta_{ij} .$$

Higher-order contributions are increasingly infrared divergent, possibly signifying the appearance of a non-perturbative "magnetic mass" which would replace $1/\underset{\sim}{k}^2$ here by $1/(\underset{\sim}{k}^2+m_{mag}^2)$. (The actual situation is slightly more complicated because of transversality requirements).

If we assume the existence of a magnetic mass and attempt a self-consistent calculation we find a series of the form

$$m_{mag}^2 = m_{mag}^2 \sum_L C_L \left(\frac{g^2 T}{m_{mag}} \right)^{2L}$$

which clearly suggests that m_{mag} is of order g^2T. Recent lattice calculations give some support to this idea (Billoire et al. 1981).

Linde (1981) suggests that such a magnetic mass would lead to a non-Abelian analogue of the Meissner effect, with magnetic flux confined to flux tubes with a thickness of order $1/g^2T$ and a string tension of order g^2T^2. If this interpretation is correct, monopoles should be confined in the sense that monopoles and antimonopoles are joined by flux tubes that pull them together and lead to rapid annihilation.

It is not clear however whether this interpretation is correct. The generation of a magnetic mass is not the only way of avoiding power-law infrared divergences. It would be sufficient to soften the $1/\underset{\sim}{k}^2$ behaviour of the propagator, say to $1/|\underset{\sim}{k}|$, leaving the magnetic force long-range. Moreover, even if a magnetic mass is generated it is not clear that it would lead to confinement rather than merely screening.

CONCLUSIONS.

Magnetic monopoles are an almost inevitable concomitant of grand unification in the early universe. Their initial density corresponds to a monopole-to-entropy ratio $\hat{r}_{in}$ that is unlikely to be less than 10^{-12}. This ratio cannot be much reduced by annihilation (below about 10^{-10}).

On the other hand the ratio $\hat{r}$ could not have been larger than 10^{-19} at the time of helium synthesis and must now be less than about 10^{-30} - a limit derived primarily from considerations of the energy balance of galactic magnetic fields. If the monopole mass is much less an even more severe limit exists derived from non-observation of monopoles in matter.

Various possible escape routes from the cosmic monopole problem exist. It is conceivable that a solution of the horizon problem might limit monopole production, or that for very large m_H the number produced may be exponentially suppressed. Severe supercooling followed by a "gentle" fist-order transition might do the trick, if not prematurely terminated and if the interior of the bubble is not too cold. In the context of multiple transitions the problem can be avoided but so far only in rather artificial and unattractive models. Finally there is the possibility of confinement by a magnetic mass.

At present none of these alternatives seems wholly satisfactory. The cosmic monopole problem remains a problem.

ACKNOWLEDGEMENTS.

I am indebted to many colleagues for useful comments, and especially to Dr. F.A. Bais and Dr. E. Weinberg for pointing out after this talk several points that I had overlooked.

NOTE ADDED.

There is another suggested escape route which came to my attention after this talk had been prepared, and which deserves a mention.

About a year ago Goldman, Kolb and Toussaint (1981) discussed the possibility that gravitational clumping of monopoles might sufficiently enhance the annihilation rate to escape the cosmic monopole problem. They concluded that it could not.

However this scenario has recently been reexamined by Dicus and Teplitz (1981). They envisage a period when the universe was monopole-dominated from about 10^6 GeV down to 1 MeV, during the latter part of which monopoles clump together. They then argue that monopole annihilation could maintain the temperature at about 1 MeV for a considerable period. The far out-of-equilibrium annihilation would generate a great deal of entropy and hence much reduce the monopole-to-entropy ratio. Of course by the same token it would reduce the baryon-to-entropy ratio. It seems difficult to achieve an adequate reduction of monopoles without losing the baryons too. Nevertheless this is an interesting scenario which if correct would have profound implications for cosmology, particularly nucleosynthesis and galaxy formation.

REFERENCES

ABBOTT, L.F. (1981) Nucl. Phys. B185, 233

BAIS, F.A., (1981) Physics Letters 98B, 437

BAIS, F.A. and RUDAZ, S. (1980) CERN preprint, TH-2885-CERN.

BILLOIRE, A and TAMVAKIS, K. (1981) CERN preprint, TH-3019-CERN.

BLUDMAN, S.A. and RUDERMAN, M.A. (1976) Phys. Rev. Letters 36, 840.

CALLAN, C.G. and COLEMAN, S. (1977) Phys. Rev. D16, 1762.

COLEMAN, S. (1977) Phys. Rev. D15, 2929.

COLEMAN, S. and WEINBERG, E. (1973) Phys. Rev. D7, 1888.

COOK, G.P. and MAHANTHAPPA, K.T. (1981) Phys. Rev. D23, 1321.

DRUKIER, A.K. (1981) Following contribution to this meeting.

EBERHARD, P.H., ROSS, R.R., ALVAREZ, L.W. and WATT, R.D. (1971). Phys. Rev. D4, 3260.

EINHORN, M.B. (1980) NORDITA preprint, NORDITA-80/24; Proceedings, Europhysics Conference on Unification of the Fundamental Interactions, Erice 1980.

EINHORN, M.B., STEIN, D.L. and TOUSSAINT, D. (1980). Phys. Rev. D21, 3295.

FLEISCHER, R.L., HART, H.R., NICHOLS, G.E. and PRICE, P.B. (1971) Phys. Rev. D4, 24.

GIACOMELLI, G. (1981). Lecture later in these proceedings.

GINZBURG, V.L. (1960) Fiz. Tverdogo Tela 2, 2031 (Sov.Phys.-Solid State 2, 1824).

GROSS, D.J., PISARSKI, R.D. and YAFFE, L.G. (1981). Rev. Mod. Phys. 53, 43.

GUTH, A.H. (1981) Phys. Rev. D23, 347.

GUTH, A.H. and TYE, S.H. (1980). Phys. Rev.Letters 44, 631.

GUTH, A.H. and WEINBERG, E., (1981) Phys. Rev. D23, 876.

HORIBE, M. and HOSOYA, A. (1981). Osaka University preprint OU-HET 41.

HUT, P. and KLINKHAMER, F.R. (1981). Physics Letters 104B, 439.

JONES, L.W. (1977). Rev. Mod. Phys. 49, 717, exp. pp. 735-6.

KENNEDY, A., LAZARIDES, G and SHAFI, Q. (1981). Phys. Letters 79B, 239.

KIBBLE, T.W.B. (1976). J. Phys. A9, 1387.

KIBBLE, T.W.B. (1980) Physics Reports 67, 183.

KINOSHITA, K. and PRICE, P.B. (1981), Phys. Rev. D24, 1707.

LANGACKER, P. and PI, S.-Y. (1980). Phys. Rev. Letters 45, 1.

LAZARIDES, G., SHAFI, Q and WALSH, T.F. (1981) Physics Letters 100B, 21.

LINDE, A.D. (1976) Zh. Eksp. Teor. Fiz. Pis'ma Red. 23, 73 (JETP Letters 23,64)

LINDE, A.D. (1980a) Lebedev Physical Institute Report no.125 (BI-TP80/20), published in part in Linde (1980b).

LINDE, A.D. (1980b) Physics Letters 96B, 293.

LINDE, A.D. (1981) Lebedev Physical Institute report no. 229.

LONGO, M.J. (1981) University of Michigan preprint, UM HE 81-33.

PARKER, E. (1970). Astrophys. J. 160, 383

POLYAKOV, A.M. (1974). Zh. Eksp. Teor. Fiz. Pis'ma Red. 20, 430 (JETP Letters 20,194

PRESKILL, J.P. (1979) Phys. Rev. Letters 43, 1365.

ROSS, R.R., EBERHARD., P.H., ALVAREZ, L.W. and WATT, R.D. (1973). Phys. Rev. D8, 698.

SHER, M.A. (1981) Phys. Rev. D24, 1699.

STEINHARDT, P.J. (1981) Phys. Rev. D24, 842.

'tHOOFT, G. (1974) Nucl. Phys. B79, 276.

WEINBERG, E. (1981) Private communication.

WEINBERG, S. (1974) Phys. Rev. D9, 3357.

ZEL'DOVICH, Ya. B. and KHLOPOV, M.Y. (1976). Phys. Letters 79B, 239.

ADDENDUM

DICUS, D.A. and TEPLITZ, V.L. (1981) University of Texas preprint DOE-ER-03992-458.

GOLDMAN, T. KOLB, E . and TOUSSAINT, D. (1981). Phys. Rev. D23, 867.

BILLOIRE, A., LAZARIDES, G. and SHAFI, Q. (1981) Phys. Letters 103B, 450.

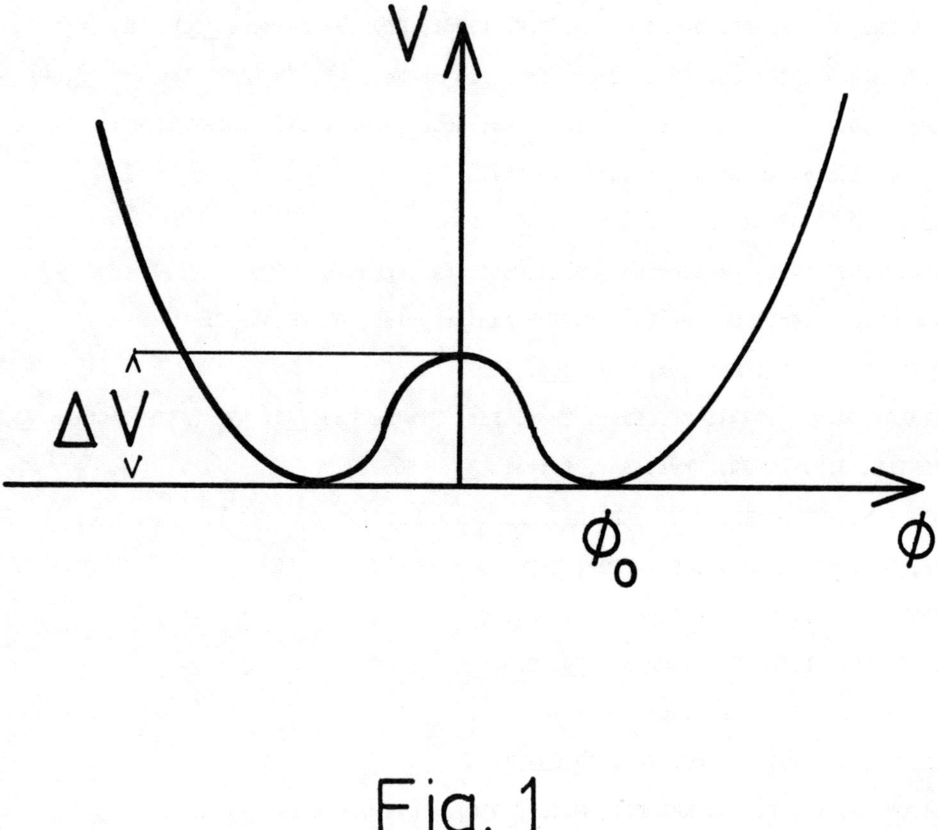

Fig. 1

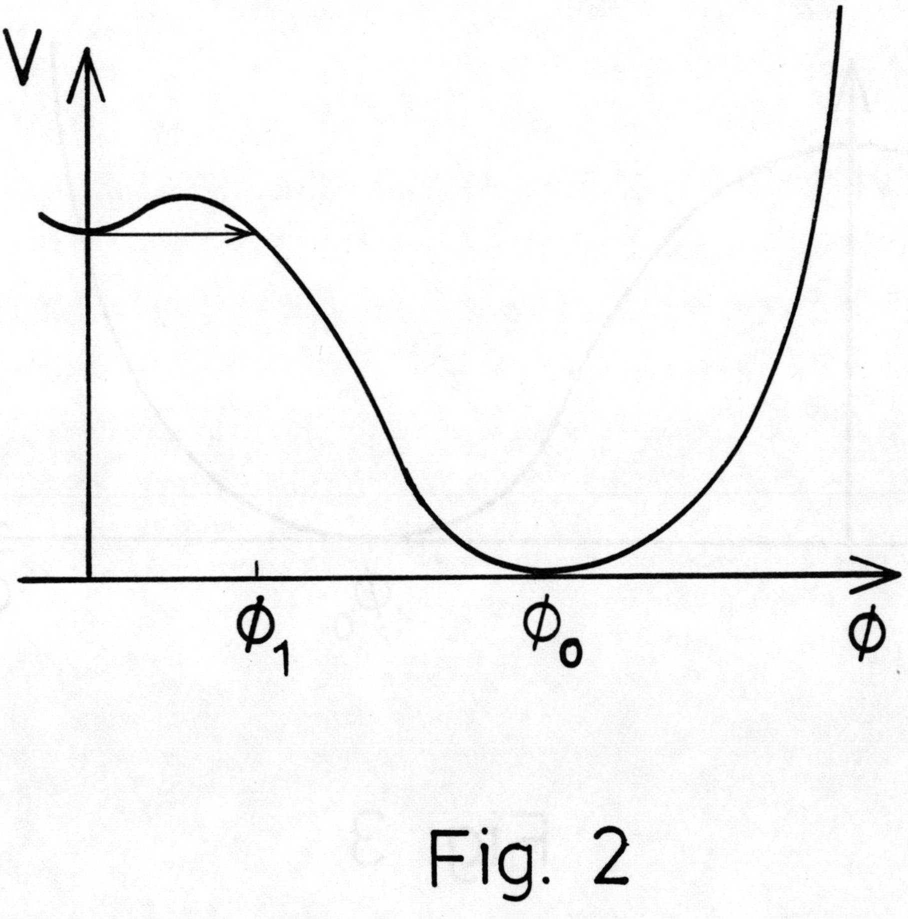

Fig. 2

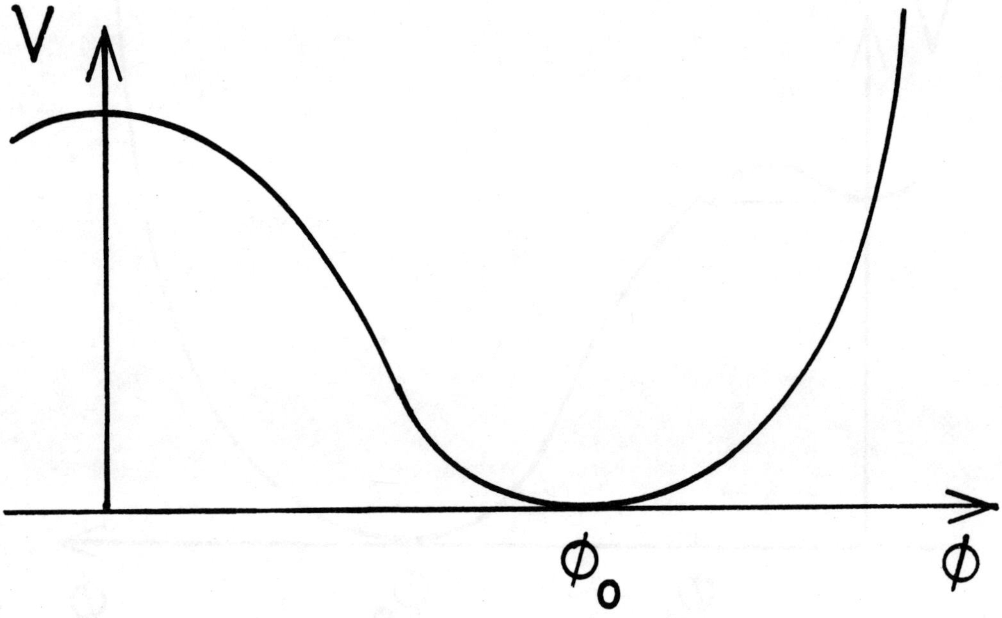

Fig. 3

REVIEW OF THE EXPERIMENTAL STATUS (PAST AND FUTURE) OF MONOPOLE SEARCHES

G. Giacomelli

Istituto di Fisica dell'Università di Bologna.Italy.

and

Istituto Nazionale di Fisica Nucleare, Sezione di Bologna,Italy.

1. INTRODUCTION

Magnetic monopoles were introduced by P.A.M. Dirac in 1931[1] in order to explain the quantization of the electric charge. In fact Dirac has shown that if one magnetic charge exists the quantization of angular momentum would be violated unless there existed a quantum of electric charge and of magnetic charge. Therefore in this reasoning the quantization of electric charge follows from the existence of at least one free magnetic charge. Dirac established also the basic relation between the elementary electric charge, e, and the magnetic charge, g:

$$g = ng_o = \frac{1}{2}\frac{\hbar c}{e} n \simeq \frac{137}{2} en \tag{1}$$

where g_o is the smallest magnetic charge and n is an integer which in the original proposal could assume the values n=1,2,3...... According to Schwinger there are difficulties in Dirac's formulation unless n is an even integer[2].

The hypothesis of the existence of magnetic monopoles solves also a number of problems and brings in other attractive features.

For instance the existence of magnetic charges and of magnetic currents symmetrizes in form the Maxwell's equations[*].

Moreover Schwinger[2] has shown that an explanation of the zero-mass for the photon follows naturally from the existence of both electric and magnetic charges.

These types of reasoning were the basis for the introduction of what we may now call the "classical magnetic monopoles", characterized by a large magnetic charge and by a relatively low mass (say of the order of a few GeV).

More recently Grand Unified Theories (GUT) of weak, electromagnetic and strong interactions, with symmetries which are broken at distances of the order 10^{-28} cm, predict the existence of magnetic monopoles[3,4,5]. The monopoles are produced at the phase transition (at the time) when a gauge group

---

(*) However the symmetry would not be perfect since the smallest magnetic charge is predicted to be much larger than the smallest electric charge, eq. (1).

breaks spontaneously leaving a subgroup, which contains $U(1)_{EM}$. The monopole mass (m_M) is connected to the vector boson mass m, which defines the unification scale

$$m_M \simeq \frac{2\sqrt{2}\pi m}{G^2} \tag{2}$$

where G^2 is the dimentionless coupling constant. Assuming $m=m_x \simeq 10^{14}$ GeV and $G^2=0.02$, one obtains a minimum monopole mass of the order of $m_M \simeq 10^{16}$ GeV $\simeq$ $\simeq$ 0.02 μg. This is an enormous mass; therefore magnetic monopoles cannot be produced at any existing (and even at any concivable) accelerator. They could only be made in the first instants of the universe, at times of the order of 10^{-35} seconds after the Big Bang.

Cosmic, or GUT, monopoles of lowest mass are expected to be stable, since magnetic charge should be conserved like electric charge. Therefore the original monopoles produced at the early stage of the universe should still be around as cosmic relics, whose kinetic energy has been strongly affected by their travel hystory through galactic magnetic fields.

Monopole masses of different orders of magnitude have been hypothesized in different models. For instance assuming a mass relation of the type (2), using the nominal W-boson mass and the electromagnetic coupling constant, one obtains $m_M \sim 4 \cdot 137 \cdot m_W \sim 10^4$ GeV [3].

In this talk I shall review the experimental searches for magnetic monopole, starting with "classical" monopoles and then proceeding to GUT monopoles. I shall not discuss here the tachyon monopole, that is the possibility that magnetic monopoles could be tachyons [6], nor the dyons [7] which could possess both electric and magnetic charges.

Section 2 deals with the general properties of monopoles, properties based on the Dirac relation. Section 3 is devoted to the searches for classical monopoles. Section 4 describes the "history" of GUT monopoles after the production in the Big Bang [8]. In Section 5 is discussed the relevance of previous experiments to the search for GUT poles. The experiments which are presently searching for GUT poles are described in Section 6. The outlook for future experiments is given in Section 7.

2. PROPERTIES OF MAGNETIC MONOPOLES

We shall summarize briefly the main features of magnetic monopoles which can be obtained from relation (1). If n=1 and the elementary charge is that of the electron(*), one has:

(a) Magnetic charge $g_o = \frac{1}{2}\frac{\hbar c}{e} = \frac{137}{2} e = 3.29\times10^{-8}$ cgs units (3)

(b) Coupling constant $\frac{g_o^2}{\hbar c} = \frac{e^2}{\hbar c}\left(\frac{g_o}{e}\right)^2 = \frac{1}{137}\left(\frac{137}{2}\right)^2 = 34.25$ (4)

(c) Energy acquired in a magnetic field B $W(eV) = g\,H\,l = 2.06\times10^4 B(gauss)\,l(cm)\,n$ (5)

Thus, because of the large g-value, monopoles acquire large energies in even modest magnetic fields.

(d) Ionization energy loss $\left(\frac{dE}{dx}\right)_g = \left(\frac{dE}{dx}\right)_q \left(\frac{g_o}{e}\right)^2 (n\beta)^2$ (6)

where

$$\left(\frac{g_o}{e}\right)^2 = \left(\frac{137}{2}\right)^2 = 4700 \quad (7)$$

Relation (6) connects the ionization energy loss of a magnetic charge with that of an electric charge with the same β. In relation (6) are missing some terms due to the interaction of the traveling magnetic monopole with the magnetic dipoles of the electrons or Zeeman ionization(8,9). These terms have a lower β-dependence and therefore become important at low values of β. Fig. 1 shows the ionization energy loss for monopoles computed on the basis of eq. (6) plus other quasi static terms(8). Relativistic monopoles ionize 4700 times the value of minimum ionizing particles. The ionization is equal to minimum ionizing particles for $\beta \simeq 1.4\times10^{-3}$.

(*) If the elementary electric charge would be that of quarks, with charge 1/3, one would clearly have an elementary magnetic charge 3 times larger and ionization losses 9 times larger (eq. (6)). A similar situation arises if $n > 1$

(e) Energy loss in a conductor - A monopole moving in a conductor produces a time-varying magnetic field, therefore a time-varying electromotive force and eddy currents. The effective interaction with the conduction electrons is possible only when the pole is moving more slowly than the electrons in the atom. The energy loss is approximately given by[9]:

$$-\frac{dE}{dx} \simeq \frac{4\pi^2 n g^2 e^2}{m_e c v_o} \beta \tag{8}$$

where $v_o \simeq 10^8$ cm/s and n is the number of conduction electrons per cm^3. For copper one has:

$$\frac{dE}{dx} \simeq 2x10^2 \beta (GeV/cm) \tag{9}$$

which saturates at losses of the order of 1 GeV/cm for $\beta \gtrsim$ few10^{-3}. This is the dominant loss at low β. In an insulator the losses are small[10].

(f) Losses by hysteresis in ferromagnetic materials are small ($\sim$0.1 MeV/c).

(g) Magnetic monopoles may be trapped in bulk paramagnetic and ferromagnetic materials by an image force, which in ferromagnetic materials may reach the value of 10 eV/Å (11 eV/Å in iron, 3.5 in magnetite)[11].

3. SEARCHES FOR CLASSICAL MONOPOLES

In the early 1970's the "classical" monopole was considered to be a member of the family of "well known undiscovered objects". Searches were made at every new higher energy accelerator, in cosmic rays and in bulk matter[12,13].

Rough estimates of the mass of these monopoles were tried assuming that the classical electron radius, r_e, is equal to the classical monopole radius, r_g:

$$m_g = \frac{g^2}{e^2} \cdot m_e = (\frac{137}{2} n)^2 m_e \tag{10}$$

If n=e=1 it follows that $m_g \simeq 2.4$ GeV; if n=2, e=1, $m_g \simeq 10$ GeV. If quarks with fractional charge exist, then $m_g = (\frac{137}{2} x3)^2 m_e \simeq 22$ GeV for n=1 and $m_g \simeq 86$ GeV for n=2, etc. These masses would be within reach of many present high energy accelerators. One thus thought that monopoles could be produced in high energy collisions, the simplest production reaction being of the type:

$$p+p \longrightarrow p+p+g+\bar{g} \tag{11}$$

where $\bar{g}$ is an antimonopole. In the context of GUT monopoles one may now see how reliable these estimates were!

I refer to ref. 12 for a complete list of references and to ref. 13 for a more thorough coverage of the experimental searches for classical monopoles. In this section only the main lines of searches will be briefly mentioned. It must also be recalled that the main method of detection is based on the high energy loss of fast monopoles.

3.1 Accelerator searches

Broadly speaking the methods employed for searches for free magnetic poles may be classified into two groups:

(i) Direct detection of monopoles, immediately after their production in high energy collisions.

(ii) Indirect searches, where monopoles are searched for a long time after their production. A broad class of experiments could be classified as indirect. They involve accelerators, cosmic rays and bulk matter experiments.

3.1.1 - Examples of direct searches are the experiments performed at the CERN-ISR with plastic detectors[14,15] (*). A set of thin plastic sheets of Nitro

---

(*) Plastic detectors act as threshold devices with not well defined thresholds. Approximate thresholds for monopoles are: Nitrocellulose: $\beta n > 0.02$; Lexan (Makrofol E): $\beta n > 0.3$; Mica: $\beta > 2$; CR 39: $\beta n > 0.003$.

cellulose and of Makrofol E surronded an intersection region of the ISR. Magnetic monopoles produced in pp collisions should have crossed the ISR vacuum wall and some of the plastic sheets. Heavily ionizing monopoles should have left a sort of dislocations along their paths in the plastic sheets. When properly developed, the sheets should show holes along monopole tracks (the tracks should have been roughly perpendicular to the sheets). The holes were searched for using standard optical microscopes.

3.1.2 - Examples of indirect searches at high energy accelerators are those which have been performed at the CERN-ISR, Fermilab, CERN-SPS and at other lower energy accelerators, using ferromagnetic materials[16-18]. For instance in the experiment of ref. 18 the 400 GeV protons from the SPS interacted (before reaching a beam dump) in a series oftargets made of compacted ferromagnetic tungsten powder. The poles produced in high energy pp and pn (and also πN) collisions should lose quickly their energy and be brought to rest inside the target, where they are assumed to be bound. More specifically, in this experiment the monopoles should be trapped in one of the small pieces of ferromagnetic tungsten. This should avoid the possibility of monopole--antimonopole annihilations. Later on the pieces of the material were placed in front of a pulsed solenoid, capable of giving a magnetic field of more than 200 Kgauss. This should have been large enough to extract and accelerate the monopoles, which should have been detected in nuclear emulsions and in plastic sheets.

In this sort of experiments one can in principle obtain very good cross section upper limits, since one can integrate the production over long time inervals. But there are clearly several hypotheses on the behaviour of monopoles in matter. In fact each experimental group takes special precautions, like segmentation of the targets, stripper foils to dislodge the paramagnetic molecules which may attach to a monopole, etc.

Fig. 2 summarizes schematically, as a function of the magnetic charge, the production cross section upper limits (at the 95% C.L.) in pN collisions[13,18]. Fig. 3. summarizes the same limits as a function of monopole mass. Solid lines refer to "direct" measurements; dashed lines to "indirect" measurements at high energy accelerators; dotted lines refer to "indirect" cosmic rays experiments.

3.2 Cosmic ray searches

Searches for a flux Φ of fast magnetic monopoles were made using counters, plastic detectors and nuclear emulsions.

3.2.1 - Searches with electronic detectors. Most of these searches were aimed at detecting lowly ionizing quarks at sea level and at mountain altitude[13,19]. The informations on magnetic poles is only indirect, from a reanalysis of the data. The experimental upper limits are modest, $\Phi < 3\times10^{-2}\,cm^{-2}\,yr^{-1}$ [13,19].

3.2.2 - Searches with emulsion and plastic detectors. In 1975 was reported a monopole candidate from a high altitude, balloon-born stack of plastic detectors, nuclear emulsions and a Cherenkov detector[20]. The detector had an area of 18 m^2, was quite elaborate (35 layers of lexan and 3 of emulsion) and was flown for 15 days. The main purpose of the experiment was the search for heavy nuclei, with $20 \leqslant Z \leqslant 83$, in the cosmic radiation.

After a long debate the authors concluded that they had an unusual event, which could be:

(i) a supermassive particle with $\beta \simeq 0.4$, $Z \simeq 95$ and $m > 10^3$ GeV;

(ii) a fast antinucleus with $Z/\beta \simeq -110$, $76 \leqslant |Z| \leqslant 96$; the antinucleus fragmented and lost one or two charges;

(iii) a very fast nucleus with $Z \simeq 112$, $\beta \geqslant 0.99$.

Because of inconsistencies in the various detector readings, the authors finally excluded a monopole[*].

From this exposure, and from two subsequent ones, they obtained an upper limit $\Phi < 10^{-4}\,cm^{-2}\,yr^{-1}$.

3.2.3 - Searches for ancient tracks with traversed samples of mica and obsidian[21,22]. It has to be remembered that mica is full of tracks from

---

(*) But they added that the event could also be compatible with a monopole with $\beta \simeq 0.4$, n=2, $m > 10^{11}$ GeV; they said that "such a large mass is not excluded by theory; but is perhaps offensive".

natural radioactivity. As far as monopoles these detectors have high thresholds ($\beta n > 2$). Within this limitation the authors reported a flux upper limit $\Phi < 10^{-11} cm^{-2} yr^{-1}$ in mica and $\Phi < 10^{-9} cm^{-2} yr^{-1}$ in obsidian. These good limits are obtained because the rocks had ages of approximately $2x10^{8}$ years. The area scanned (with optical microscopes) was 380 cm^2.

3.2.4 - Magnetic poles from outer space or produced by cosmic rays at the top of the atmosphere could be broughy to rest if their energy is not extremely high. If they have large masses, they would then drift slowly in the earth magnetic field.

In a number of experiments the lines of force of the magnetic field of a solenoid were mostly supplied by the earth magnetic field. Thus the drifting poles would be sucked by the lines of force of the solenoid, accelerated and detected by counters or emulsions[(23,24)]. The estimated upper limit found corresponds roughly to $\Phi < 10^{-6} cm^{-2} yr^{-1}$.

3.3 Searches in bulk matter

3.3.1 - An experiment from Berkeley used as detector a superconducting coil in which an electromotive force, and thus a current charge, should be induced by a magnetically charged particle present in a sample which was moved through the coil[(25)]. Using multiple traversals of the sample they achieved the proper sensitivity. As samples they used about 20 Kg of lunar material, magnetite from earth mines and 2 Kg of meteorites. They placed a limit of less than $2x10^{-4}$ monopoles per gram of lunar material. This gave the best limit in terms of flux of monopoles, assuming a constant flux over the long time during which the moon remained unaltered. If poles had low masses and not too high energies, the estimated flux was $< 3x10^{-9}$ poles $cm^{-2} yr^{-1}$. The flux limit becomes much smaller for monopoles with higher kinetic energies, $\Phi < 3x10^{-18} T^2 (GeV^2)$, and drops to negligible values for $T > 10^8$ GeV. Assuming that monopoles are produced by cosmic rays one obtained the cross section upper limits shown in Figs. 2 and 3.

3.3.2 - Another group searched for monopoles in magnetite (from a surface mine) and from ferromanganese nodules (from deep ocean sediments) using a layout similar to the one described in 3.1.2 [26]. The poles should have been extracted, accelerated and sent towards a detector by a large magnetic field (pulsed or continuous). The detectors consisted of plastic sheets of lexan and nitrocellulose.

One has to recall now that a field of 60 Kgauss is needed to extract poles with n=1,2,3; moreover the acceleration yields poles with sufficient velocities to produce ionization only if the pole masses are less than 10^4 GeV.

The experiment used 7.7 Kg of material, having an age of approximately 1.6×10^7 years. Thus they obtained the limit $\Phi<10^{-10}\,cm^{-2}\,yr^{-1}$ assuming trapping in the sample.

3.4 Multi-γ events

Five peculiar photon showers were found in nuclear plates in high-altitude cosmic-ray exposures [27]. They are characterized by a very energetic narrow cone of tens of γ-rays, without any incident charged particle. The total energy in the photons is of the order of 10^{11} eV. The radial spread of photons (10^{-3}-10^{-4} rad) suggests a c.m. velocity corresponding to $\gamma \gtrsim 10^3$. The energies of the photons in the overall c.m. system are very small, orders of magnitude too low to have π^0 decays as their source.

One of the possible explanations of these events could be the following: a high energy γ-ray, with energy of the order of 10^{12} eV, produces in the plate a monopole-antimonopole pair by a mechanism similar to e^+e^- production. The monopole-antimonopole pair then suffers bremstrahlung and annihilation producing the final multi-γ events. The 5 events correspond to a significantly large production cross section. Experiments performed at the ISR and at Fermilab failed to observe them (by at least two orders of magnitude) [28,29].

3.5 Discussion

Classical monopoles have been searched in many ways, using many different techniques (What has been reported here is only a sample of these searches). The experiments yielded null results. The significance of these is shown in terms of production cross sections in Figs. 2 and 3 and in terms of flux limits in Table 1.

4. THE HISTORY OF RELIC G.U.T. MONOPOLES

Most Grand Unified Theories predict, using the "standard" model of the Big Bang, a large production of magnetic monopoles at $t \sim 10^{-35}$ sec [8,30-33]. The predicted ratio of the number of monopoles ($r=n_M/T^3 \sim 10^{-10}$) is so large that the monopole mass density would exceed the mass density of the universe by orders of magnitude. Thus theoreticians tried to find ways to reduce this number.

After production the monopoles are expected to lose linetic energy and to eventially reach termal velocities, with $\beta \sim 10^{-10}$ and kinetic energies $T \sim 1$ MeV, at the matter-dominated area ($t \sim 10^{11}$ sec) [8,34].

At the cosmic time $t \sim 10^{17}$ seconds the matter started to condense into galaxies: at the same time galactic magnetic fields should have developed. These fields act as monopole accelerators.

Few of the original monopoles should have been lost from birth to the time of formation of the galaxies. Possible losses could be due to monopole--antimonopole annihilation, which could play a role only in the early universe when matter density was large; its effect is expected to be important only if the number of monopoles was really very large [34]. It should not be able to reduce the ratio r below 10^{-10}. At the time of the helium synthesis r was estimated to be $\sim 10^{-19}$.

Magnetic monopoles inside the galaxies should be accelerated preferentially in the plane of the galaxy, by magnetic fields of the order of $B \simeq 5 \times 10^{-6}$ gauss acting over distances comparable to the radii of the galaxies ($r \sim 5 \times 10^{22}$ cm), spiralling outward. Monopoles would thus be ejected with velocities[8]:

$$\beta_{ejected} \simeq \left(\frac{2\,g\,B\,r}{m_M c^2}\right) \simeq 3x10^{-2} \left(\frac{m_M}{10^{16}\,GeV}\right)^{1/2} \qquad (12)$$

Monopoles of $m_M=10^{16}$ GeV would have $\beta \simeq 3x10^{-2}$ and kinetic energies $T \simeq 5x10^{12}$ GeV. Higher energies could be achieved over longer distances and higher field regions. But energy losses in ionized gas clouds and in monopole-photon (3°K) collisions would bring the kinetic energy downward towards the value quoted.

Thus the monopoles would spiral outward in the galaxies and after a time of 10^6-10^7 years would be ejected. (The escape velocity from the galaxy is $\beta=10^{-3}$)(*). The ejected monopoles would give rise to an isotropic flux of "high energy" monopoles. One may compute an upper limit on the number of these monopoles in the following way[8,35]. The regeneration time of the galactic magnetic field from the dynamo effect is of the order of $\tau \sim 10^8$ years, which is longer than the average time required by a monopole to escape from the galaxy. Thus if the number of escaping poles is too large, they would extract energy from the field faster than it could be replenished and the field would be destroyed. One may assume that the energy in all the poles should be equal to the energy stored in the galactic fields. From this equality one obtains for the number of monopoles per unit volume

$$n_M \leqslant \frac{Bt}{8\pi g r \tau} \sim 10^{-20}\,cm^{-3} \qquad (13)$$

from which one obtains an upper limit for the flux of $\Phi < 3.10^{-9}\,cm^{-2}\,yr^{-1}$. This number has large uncertainties; it may be raised upward by about two-three orders of magnitude[34] (The flux of primordial extragalactic poles of high energy is not restricted by the above limitation).

These relatively fast poles would have had the time to encounter many galaxies, where they could be accelerated or deaccelerated. On the average this would have no net change of energy in the field of the galaxies.

(*) The escape velocity from the galaxy may be simply estimated from the relation $G\, m_M\, M_{gal}/r^2 = m_M v^2$; one has $\beta = v/c = (G\, M_{gal}/r)^{1/2}/c \simeq 10^{-3}$ and $T=5x10^9$ GeV (we assumed $M_{gal} \simeq 4x10^{44}$ g and $r \simeq 3x10^{22}$ cm).

Besides the flux of extragalactic poles, there should also be a flux of poles with velocities equal or smaller than the escape velocity. To these poles the bound (13) does not apply and one should expect a larger number(*).

In conclusion from the above reasonings we may expect:

(i) an isotropic flux of monopoles of extragalactic origin, which have $\beta \sim 10^{-2}$ and $T \sim 10^{12}$ GeV, with a flux of 10^{-9}-10^{-6} cm^{-2} yr^{-1}.

(ii) a non-isotropic flux of monopoles from our galaxy, with $\beta \sim 10^{-3}$ and $T \sim 10^{10}$ GeV, and a flux of 10^{-4}-10^{-6} cm^{-2} yr^{-1}.

These monopoles have very large ranges in matter.

5. RELEVANCE OF "CLASSICAL" MONOPOLE SEARCHES TO GUT MONOPOLES

In this Section will be discussed the relevance of previous searches for "classical" monopoles as far as the superheavy monopoles. Let us recall first some properties:

(a) pole mass: $m_M = 10^{16}$ GeV.

(b) most poles are expected to have: $\beta \sim 10^{-3}$, $T \sim 10^{10}$ GeV.

(c) $\beta \sim 10^{-2}$, $T \sim 10^{12}$ GeV.

(d) poles with $\beta \sim 10^{-4}$ may be stopped in earth.

(e) poles with $\beta \sim 10^{-3}$ may be stopped in stars.

(f) kinetic energy acquired in earth gravitational field = 1.2 GeV/m.

(g) earth escape velocity = 11 Km/sec (corresponding to a pole with $T=10^{8}$ GeV).

(h) gravitational binding to earth 0.1 eV/Å.

(i) no inelastic collisions with nucleons, because the available energy in a pole-nucleon collision is less than 1 MeV.

Table 1 gives a summary of the searches, which will now be commented:

(1) <u>At accelerators</u> - Clearly these searches are relevant only for the production of poles with $m_M < 30$ GeV.

(2) <u>Measurements of poles as a flux of the cosmic radiation</u> - It has to be

(*) These limits are uncertain because of the many doubtfull hypotheses made (f.i. the poor knowledge of the magnetic fields at the time of the formation of the galaxies). One has also neglected possible effects of magnetic stars(36,37).

remembered that in most previous experiments one did not think that monopoles could reach the earth.

(a) Experiments performed with counters (more generally with electronics devices) were tuned to fast particle and were therefore insensitive to slow particles. Experiments of this type may play instead an important role in the future.

(b) The experiment performed at high altitude using lexan plus emulsion detectors had a global threshold of $\beta n > 0.3$ (fixed by lexan). It would therefore be OK only for high velocities and/or large n-values.

(c) The experiment which looked for fossil tracks in mica and obsidian had a high threshold, $\beta n > 2$; one can thus repeat more strongly the same comment made for the search 2b.

(d) Heavy poles would fall through the earth and cannot be found in the atmosphere; thus the search for poles drifting in the atmosphere is not relevant.

(3) Searches in bulk matter

(a) The use of the coil detector where a voltage (and thus a current) is induced by a pole moving with the sample is a good method. On the other hand it is improbable that heavy poles are stopped at the surface of the moon at the surface of the earth or in meteorites. These last two cases could nevertheless be important for future searches (using lots of material). For the moon, one has also to remember that the lunar material was taken to the earth, experiencing high ($\sim 10^2$g) deceleration. Monopoles trapped in all materials, but ferromagnetic, would have been lost. There would be a partial loss also of monopoles trapped in iron meteorites (at least in that parts which melts)(*).

(b) The search performed trying to extract with a strong magnetic field poles from magnetite and ferromanganese is not relevant because the velocities acquired by the poles would not be sufficient to ionize (and the detectors used had high thresholds).

In conclusion: none of the previous searches was really relevant to the question of the existence of massive poles(+). Only the search with the superconductive coil could give an upper limit (of the order of 0.1-1 poles $cm^{-2}yr^{-1}$). One has also learned that heavy poles are "delicate" objects, which may be lost by small accelerations, like when turning a sample upside down.

The limits for possible monopoles with $m_M \sim 10^4$ GeV could instead be appreciable ($\Phi < 10^{-9} cm^{-2} yr^{-1}$).

(*) Since all elements heavier than hydrogen and helium were sinthesized inside the stars, it is unlikely that meteorites would originally be very reach in monopoles. They would have to pick up monopoles (with low efficiency) in their travel. Furthermore monopoles in meteorites may get lost when they impact the earth: they experimence decelerations of $\sim 10^3$ times the acceleration of gravity on the earth and parts of the meteorites melt. Therefore monopoles in non ferromagnetic materials (bound with less than about 1 eV/Å) and from the melted parts escape. A heavy monopole with a velocity $v = v_{escape} \simeq 11$ Km/sec $\Rightarrow \beta = 3 \times 10^{-5}$ has a kinetic energy of $\sim 10^8$ GeV; if it gets free it would have a range comparable to the diameter of the earth!

(+) Considerations about monopoles inside the earth and the moon have been made by a number of authors(38-40). The earth and the moon have passed through a molten state. Therefore any heavy monopole would have rapidly migrated to the center, making very small oscillations around the potential minimum. Many monopoles would have annihilated, unless there was a relatively large magnetic field to keep monopoles separated from antimonopoles (if the field is ~ 10 gauss, it separates the poles by less than 1 cm). The present geothermal heat bulance of the earth limits the possible annihilation rate to $\sim 10^4$ poles/sec in the whole earth.

6. PRESENT SEARCHES FOR GUT MONOPOLES

In 1980 and in 1981 several informal discussions were held in many universities and laboratories about possible searches for GUT poles[10,38,41-45]. Two basic types of searches emerged:

(i) direct search for a flux of penetrating cosmic monopoles with low velocities ($10^{-3} < \beta < 10^{-1}$). In particular scintillation counters layout were considered (sensitive to $\beta > 3\times10^{-3}$);

(ii) bulk matter searches, using large quantities of ferromagnetic material and a superconducting coil.

At present only a few small-scale experiments of type (i) are being performed. They either use existing apparatus[45] or prototypes of future larger layouts[44].

The Bologna search uses the existing apparatus of a cosmic ray station located on the roof of the physics building[46]. The apparatus (Fig. 4) consists of 16 scintillation counters, each 1 m^2 in surface and 8 cm thick. Twelve counters are placed on the floor, while the other four are located two meters above. Between the two layers is located a lead absorber, 10 cm thick. The threshold on each counter is set at ~25 times minimum ionization. The trigger logic requires the firing of one and only one counter from each of the top and bottom layers. When a trigger happens, the two pulses are photographed on an oscilloscope. The final selection of a monopole candidate would be based on the two pulse heights and on the time of flight. No candidate has been recorded in about 6 months of running. The effective area times solid angle of the apparatus is 5 m^2 sr. The detector is capable of recording monopoles coming from above and from below (for these there is no background). The present upper limit is $\Phi < 6\times10^{-5}$ poles $cm^{-2}\,yr^{-1}\,sr^{-1}$; it covers the $0.007 < \beta < 0.6$ range as indicated in Fig. 5.

The Tokyo layout (see Fig. 6)[44] is much more elaborate. It is sensitive to very low ionization (for instance from slow moving quarks) and is meant to be a prototype for a much larger layout. It has a total area times solid angle of $S\Omega = 1\ m^2 sr$. At present the authors quote an upper limit of

$\Phi < 6\times 10^{-4}\,cm^{-2}\,yr^{-1}\,sr^{-1}$.

The Columbia apparatus[41] is presumably similar to the Tokyo one.

7. OUTLOOK

Searches for "classical" monopoles have been proposed also for the new generation of accelerators: an experiment[47] has been approved for the CERN $\bar{p}p$ collider (270 $\bar{p}$ against 270 p). In various reports of summer studies are described possible experiments at Isabelle[48] (pp collisions at $\sqrt{s}$=700-800 GeV), at the Fermilab $\bar{p}p$ collider ($\sqrt{s}$=1600-2000 GeV) and at LEP (e^+e^- collisions at $\sqrt{s}$=100 GeV). If performed, these experiments will be pertinent to monopole masses up to 800 GeV; the cross section upper bound will be limited by luminosities to about $10^{-34}\,cm^2$ at the $\bar{p}p$ colliders and $10^{-37}\,cm^{-2}$ at Isabelle.

In order to achieve good sensitivities in the search of a cosmic flux of ionizing GUT monopoles, following the method used by the small layout described in the previous section, one will have to use large apparatus, covering at least hundreds of square meters. In order to reduce the background these equipments will have to be located underground, like those used to study the decay of the proton. But while the experiments for proton instability must be compact (in order to reduce edge effects), those required for the GUT monopole search must have the largest surface area (thus they are "topologically" different!).

The Tokyo group is presently designing an experiment with a total surface area of 100 m^2 and Ω=10 sr. It is planned to have three layers pf scintillation counter hodoscopes, interleaved with drift chambers and absorber materials. The proposed site is the Kamioka mine, in central Japan, located at a depth of 2700 meters of water equivalent underground[44].

At CERN and at Fermilab there are suggestions of using the counters of the existing neutrino detectors (positioning them correctly).

From a series of discussions at the University of Winsconsin (at Madison) and at the University of Michigan (at Ann Arbor) emerged the interest in performing experiments of the bulk matter type using ferromagnetic materials and a superconducting loop. They would be similar to the Berkeley experiment, but performed on a large scale to offset the small probability that a monopole stops at the surface of the earth or on meteorites(*).

It is in any case clear that future searches will involve major efforts. In order to fully justify these efforts it would be important that the theoretical situation becomes more definite(+) and that limits based on cosmological ideas and on galactic magnetic fields become more sound.

(*) GUT monopoles are so massive that one can conceive even a "Pisa Tower" experiment: the sample is dropped from the tower and is stopped abruptly on the ground, with a deceleration of the order of 10^6 times the acceleration of gravity. Monopoles would thus be freed, would move with kinetic energies of the order of 100 GeV and would have a range of several hundred meters. The detection could be via one or two superconducting loops and via the light seen in a trasparent body. (Alternatively the sample at the top of the tower could be melted: monopoles would be freed and fall down).

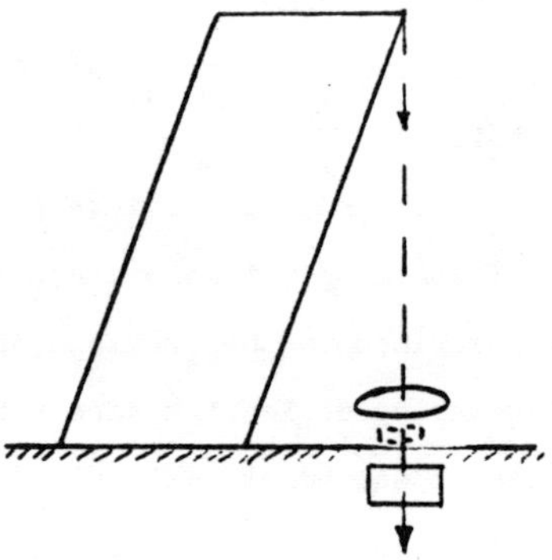

(+) It would also be interesting to have a better consensus on the problem of dressed monopoles. According to Yang a monopole may consist of a small object, around which there is neutral cloud of virtual particles (e^+e^-, $p\bar{p}$, etc), which extends up to distances of the order of the electron classical radius. Moreover a monopole in matter may tear apart some nuclei, because of the different force excersized by the monopole on neutrons and on protons. Thus a monopole may also be surronded by a cloud of debris up to distances of the order of 1 Å. In this context it is not entirely clear if at least some monopoles cannot have an effective binding energy larger than the values discussed in this report.

Acknowledgements

I would like to acknowledge many colleagues for duscissions and for sending material before publication. In particular I would like to thank Drs. E. Amaldi, G. Baroni, P. Capiluppi, R.A. Carrigan Jr., D. Cline, S.L. Glashow, M. Koshiba, M.J. Longo, G. Mandrioli, P. Musset, A.M. Rossi, A. Salam and C.N. Yang. The skillfull help of Mrs. F. Santucci is gratefully acknowledged.

References

1. P.A.M. Dirac, Proc. Roy. Soc. 133 (1931) 60.
(The theory of magnetic poles), Phys. Rev. 74 (1948) 817.
(The monopole concept) Int. J. Theor. Phys. 17 (1978) 235.
2. J. Schwinger (Magnetic charge and quantum field theory) Phys. Rev. 144 (1966) 1087.
(Electric and magnetic charge renormalization) Phys. Rev. 151 (1966) 1048.
(Sources and magnetic charge) Phys. Rev. 173 (1968) 1536.
3. G.'t Hooft (Magnetic monopoles in unified gauge theories) Nucl. Phys. B79 (1974) 276.
4. A. Polyakov (Particle spectrum in quantum field theory) JETP Lett. 20 (1974) 194.
5. E. Bogomolny (The stability of classical solutions) Sov. Journ. Nucl. Phys. 24 (1976) 449.
6. For discussions on the Tachyon magnetic monopole see: D.F. Bartlett et al. (Search for tachyon monopoles) Phys. Rev. D6 (1972) 1817.
(Search for tachyon monopoles in cosmic rays) Phys. Rev. D18 (1978) 2253.
E. Recami et al. (Tachyons, monopoles and related topics) North Holland, Amsterdam, (1978).
7. See for example: D'Adda et al. (Supersymmetric magnetic monopoles and dyons) Phys. Lett. B76 (1978) 298.
8. G. Lazarides, Q. Shafi and T.F. Walsh (Superheavy magnetic monopole hunts) TH 3008-CERN (1980), Phys. Lett. B100 (1981) 21.
9. D.M. Scott, Cambridge DAMTP 80/2 (1980). V.P. Martemianov et al., JETP 35 (1972) 20.
10. M.J. Longo (On the experimental observation of very massive magnetic monopoles) Report UM HE 89-39 (1980).
11. E. Goto, Progr. Theor. Phys. 30 (1963) 700.
12. R.A. Carrigan Jr. (Magnetic monopole bibliography 1973-76) FERMILAB-77/42 (1977).
R.E. Craven et al. (Magnetic monopole bibliography for 1977-80) FERMILAB-81/37 (1981).
13. G. Giacomelli (Searches for missing particles) Invited paper at the 1978 Singapore meeting on "Frontiers of Physics", proceedings of the Conference (1978).
14. G. Giacomelli et al. (Search for magnetic monopoles at the CERN-ISR with plastic detectors) Nuovo Cimento 28A (1975) 21.
15. H. Hoffmann et al. (A new search for magnetic monopoles at the CERN-ISR with plastic detectors) Nuovo Cimento Lett. 23 (1978) 357.
16. R.A. Carrigan Jr., F.A. Nezrick and B.P. Strauss (Search for magnetic monopole production by 300 GeV protons) Phys. Rev. D8 (1973) 3717.
(Extension of FNAL monopole search to 400 GeV), Phys. Rev. D10 (1974) 3867.

17. R.A. Carrigan Jr., B.P. Strauss and G. Giacomelli (Search for magnetic monopoles at the CERN-ISR) Phys. Rev. D17 (1978) 1754.

18. J. Barkov et al. (Magnetic monopole search at the SPS) Bologna-CERN-URSS collaboration, Expt. WA38.

19. Particle data group (Review of particle properties) Rev. Mod. Phys. 52 (1980) 1.

20. P.B. Price et al. (Evidence for detection of a moving magnetic monopole) Phys. Rev. Lett. 35 (1975) 487.
(Further measurements and reassessment of the magnetic monopole candidate) Phys. Rev. D18 (1978) 1382.

21. R.L. Fleischer et al. (Search for tracks of massive, multiply charged magnetic poles) Phys. Rev. 184 (1969) 1398.

22. H.H. Kolm et al. (Search for magnetic monopoles) Phys. Rev. D4 (1971) 1285.

23. W.C. Carithers, R. Stefanski and R.K. Adair (Search for heavy magnetic monopoles) Phys. Rev. 149 (1966) 1070.

24. M. Fidecaro et al. (Search for magnetic monopoles) Nuovo Cimento 22 (1961) 657.

25. R.R. Ross, P.H. Eberhard, L.W. Alvarez and R.D. Watt (Search for magnetic monopoles in lunar material using an electromagnetic detector) Phys. Rev. D8 (1973) 698. Phys. Rev. D4 (1971) 3260.

26. R.L. Fleischer et al. (Search for multiply charged Dirac magnetic poles) Phys. Rev. 137 (1969) 2029.
(Search for magnetic monopoles in deep ocean deposits) Phys. Rev. 184 (1969) 1393.

27. M. Schein, D.M. Haskin and R.G. Glasser (Narrow showers of pure photons at 100.000 feet) Phys. Rev. 95 (1954) 855. Phys. Rev. 99 (1955) 643.

28. G.B. Collins et al. (Unexplained multiphoton phenomen) Phys. Rev. D8 (1973) 982.
D.L. Burke et al. (Search for anomalous multiphoton production at 100-300 GeV) Phys. Lett. B60 (1975) 113.
E. Amaldi, private communication.

29. M.A. Ruderman and D. Zwanziger (Magnetic poles and energetic showers in cosmic rays) Phys. Rev. Lett. 22 (1969) 146.

30. S.L. Glashow (Particle physics away from the high energy frontier) Lectures at the 1980 Scottish Summer School, St. Andrews, Scotland (1980).

31. J. Ellis (Grand Unified Theories) Lectures at the 1980 Scottish Summer School, St. Andrews, Scotland (1980).

32. J.P. Preskill (Cosmological production of superheavy magnetic monopoles) Phys. Rev. Lett. 43 (1979) 1365.

33. M.B. Einhorn et al., UM HE 80-1 (1980), Phys. Rev. D21 (1980) 3295.

34. T. Kibble (Monopoles in the early universe) This meeting.

35. E.N. Parker (Cosmic magnetic fields) Clarendon Press (1979). Astrophys. Journ. 160 (1970) 383.

36. S.A. Bludman and M.A. Ruderman (Theoretical limits on interstellar magnetic poles set by nearby magnetic fields) Phys. Rev. Lett. 36 (1976) 840.

37. A.K. Drukier (New limits on the density of superheavy magnetic monopoles in the interstellar space) This meeting.
(The creation of magnetic monopoles in outer gaps pulsars) Astrophys. and Space Science 74 (1981) 245.

38. R.A. Carrigan Jr. (Down to earth speculations on grand unification magnetic monopoles) FERMILAB-Pub-80/58-EXP (1980).

39. R.A. Carrigan Jr. (Grand unification magnetic monopoles inside the earth) Nature 288 (1980) 348.

40. S. Nussinov (Superheavy monopoles and baryon asymmetry) Report MPI-PAE/PTh 58/81 (1981).

41. J.D. Ullmann et al. (Experimental search for Planck-mass particles) Bull. Ann. Phys. Soc. 25 (1980) 524.

42. P. Musset et al., Meeting on superheavy monopoles, CERN, March 20th, 1981.

43. D. Cline et al., Monopole seminars held at the University of Winsconsin at Madison, (1980-81).

44. M. Koshiba, private communication
T. Mashimo, K. Kawagoe and M. Koshiba (Sarch for monopoles and quarks of non relativistic velocities) Univ. of Tokyo report (1981).

45. R. Bonarelli, P. Capiluppi, I. D'Antone, G. Giacomelli, G. Mandrioli, C. Merli, A.M. Rossi and P. Serra-Lugaresi (An experimental search for cosmic monopoles) Univ. of Bologna, preprint IFUB 82/1 (1982).

46. M. Galli et al. (Cosmic ray monitor) (1970).

47. B. Aubert al al. (Search for magnetic monopoles at the CERN $\bar{p}p$ collider) Experiment UA3, Annecy-CERN collaboration.

48. G. Giacomelli and G. Kantardjian (Magnetic monopole searches at Isabelle) Proceedings of the 1981 Isabelle Summer Workshop (1981).
Reference added in proof

49. J.D. Ullmann (Limits of showly moving very massive particles coveying electric or magnetic charge) Phys. Rev. Lett. 47 (1981) 289.

Table 1 - Relevance of "classical" monopole searches to GUT monopoles: summary table.

	SEARCH TYPE	β n	FLUX (cm^{-2} yr^{-1})	RELEVANCE	REASON
1	At accelerators			No	Energies too low
2	Cosmic ray fluxes				
	Counters	β>.3	$<10^{-2}$	No	Time of flight - OK for future
	Lexan	>.3	$<10^{-4}$	May be	OK for fast poles and large n
	Tracks in mica	> 2	$<10^{-11}$	May be	OK for fast poles and large n
	Drifting poles		$<10^{-6}$	No	Improbable and not detectable
3	Bulk matter				
	Lunar material } coil			Doubtful	Capture low; lost when coming to earth
	Ferromagnetic } coil { meteorite		~ 1	Possible	Small probability - Need mass production
	Ferromagnetic } coil { earth			Possible	Small probability - Need mass production
	Ferromagnetic (solenoid)		-	No	Not enough acceleration - Lexan ⟶ too high threshold
4	Multi-γ			No	Energies too low?

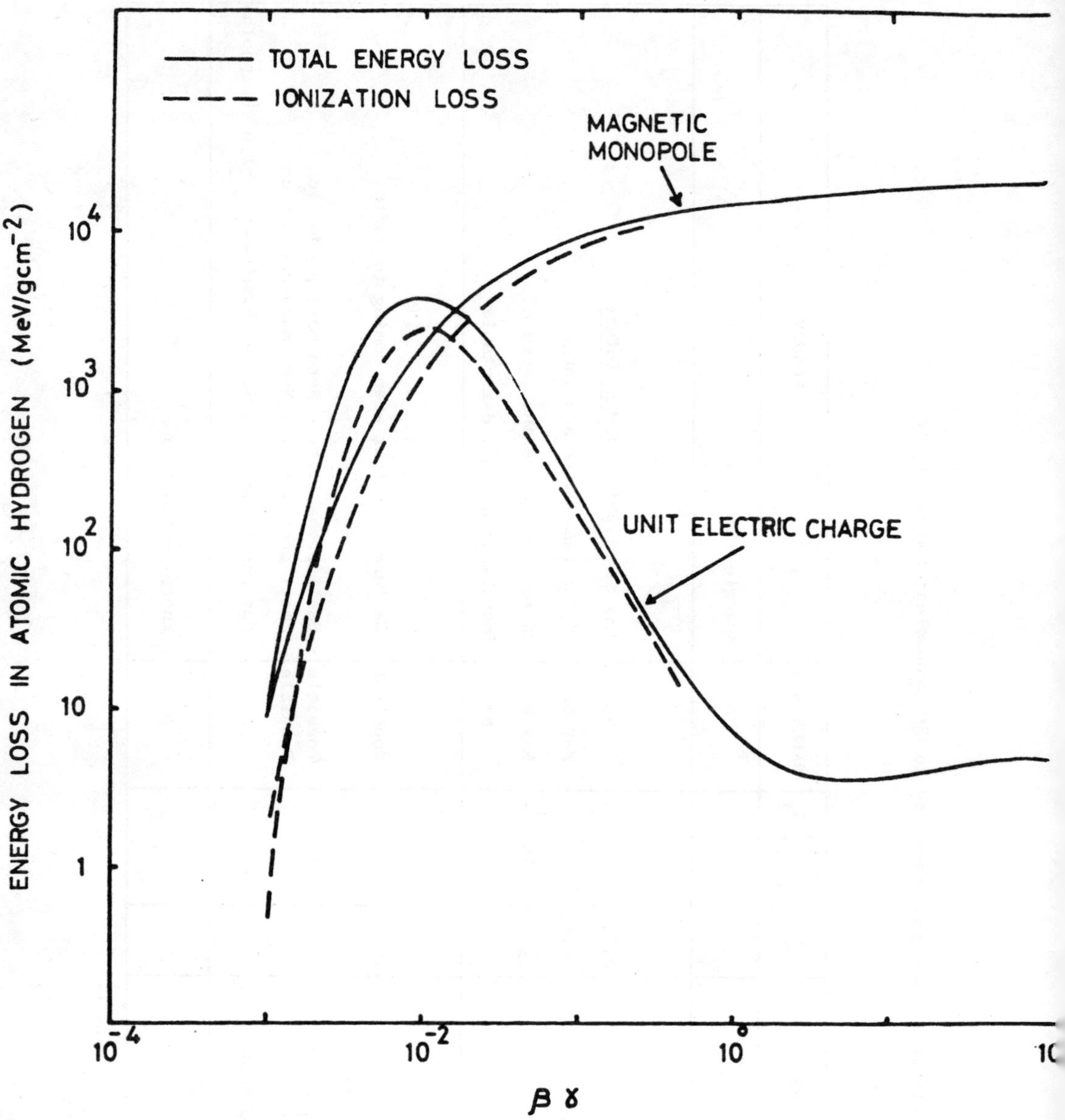

Fig. 1 – Total energy loss and ionization loss of unit magnetic monopoles and of unit electric charge in atomic hydrogen[8]. For $\beta < 10^{-3}$ the energy loss of poles is appreciable only in conductors (see Section 2e)).

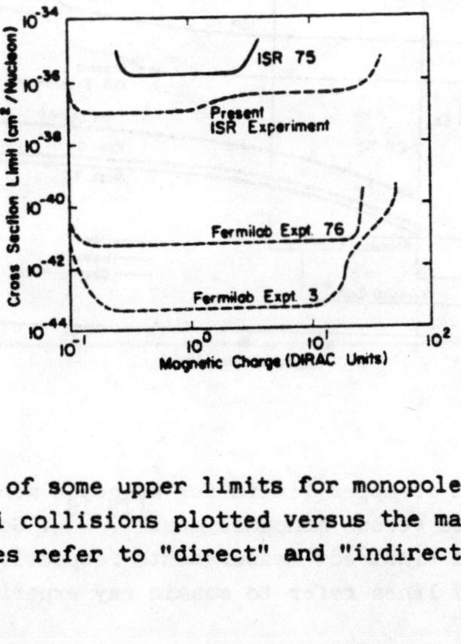

Fig. 2 - Compilation of some upper limits for monopole production (at the 95% C.L.) in p-nuclei collisions plotted versus the magnetic charge [13,18]. Solid and dashed lines refer to "direct" and "indirect" measurements.

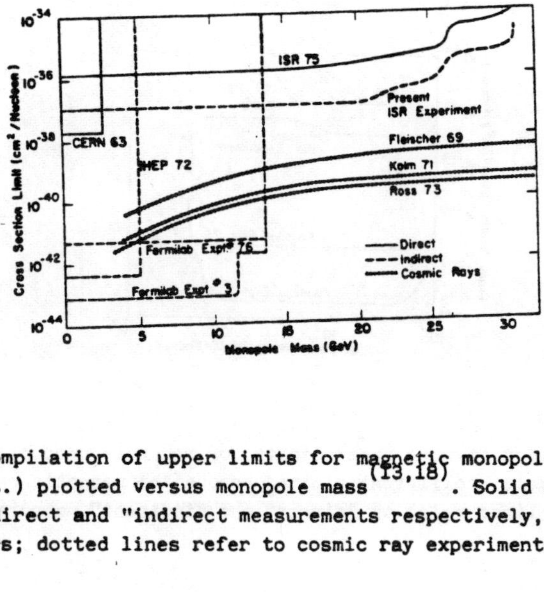

Fig. 3 - Compilation of upper limits for magnetic monopole production (at the 95% C.L.) plotted versus monopole mass [13,18]. Solid and dashed lines refer to "direct and "indirect measurements respectively, at high energy accelerators; dotted lines refer to cosmic ray experiments.

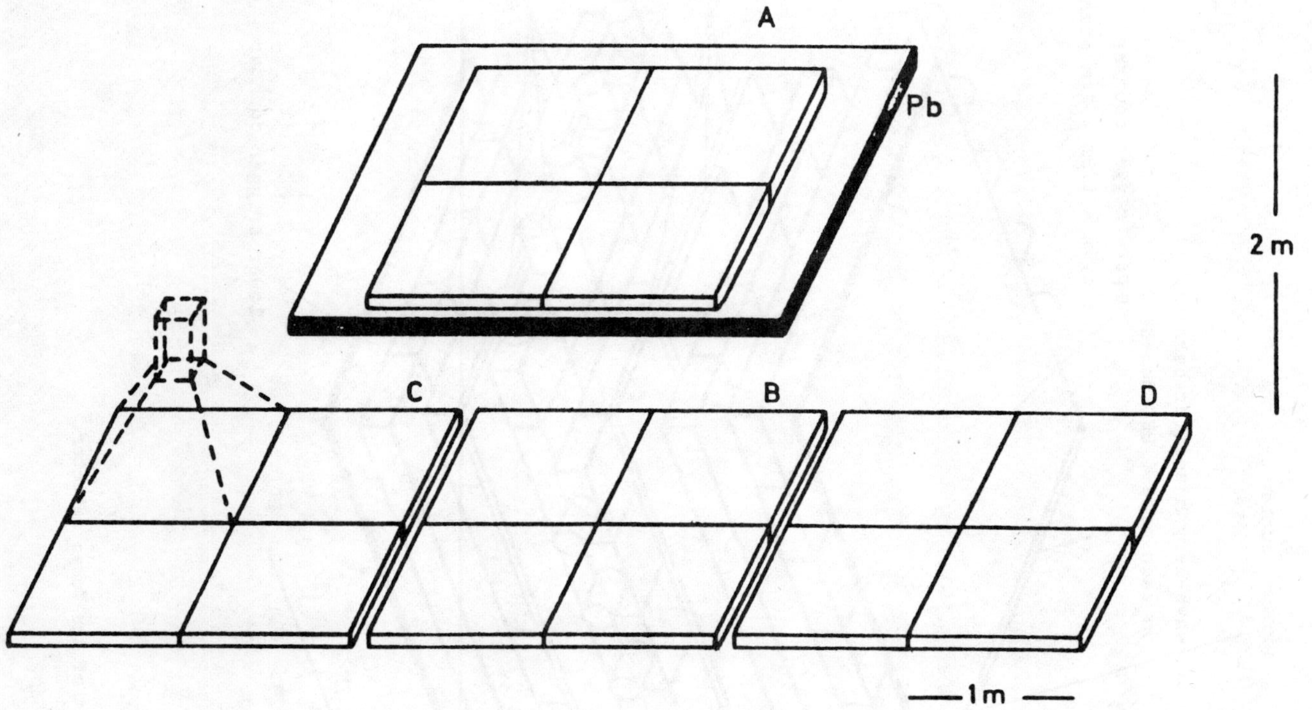

Fig. 4 - Layout of the Bologna cosmic ray monitor.

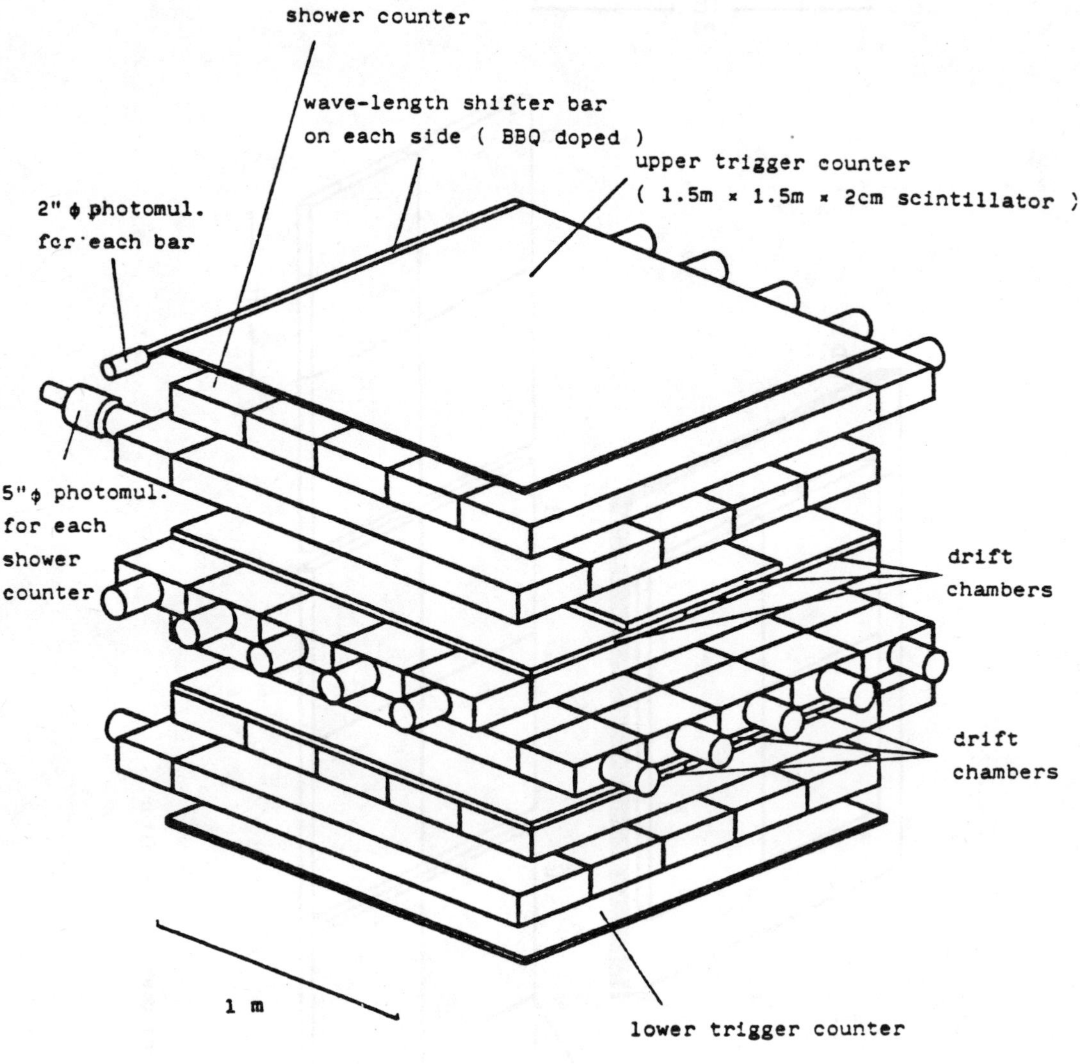

Fig. 5 - Layout of the Tokyo detector.

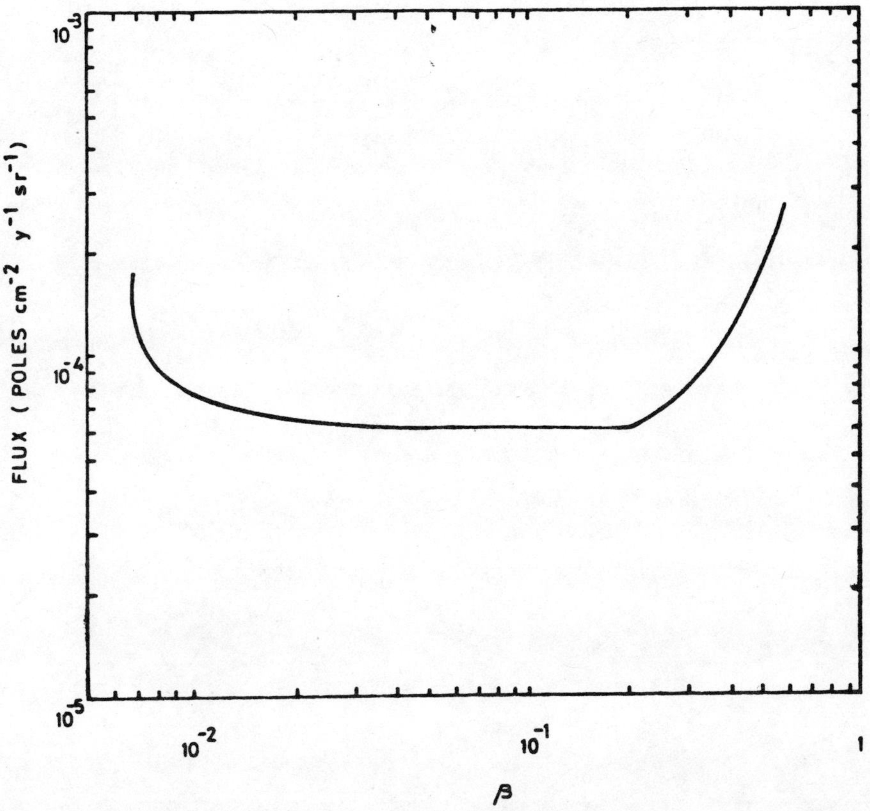

Fig. 6 - Upper limits (at the 90% C.L.) for a flux of massive GUT monopoles plotted versus their β=v/c.

NEW LIMITS ON THE DENSITY OF SUPERHEAVY MAGNETIC MONOPOLES IN THE INTERSTELLAR SPACE

A.K. Drukier

Dept. of Nuclear Medicine, Klinikum Recht der Iser, TU Munich, Ismaningerstrasse 22, 8000 Munich, Federal Republic of Germany.

With the advent of grand unified field theories -GUT's (ref.1) increasing consideration has been given to the relationship between elementary particle physics and early cosmology (ref.2).One of the predictions is the existence of superheavy monopoles, $M \approx 10^{16}$ GeV/C . The problem of monopole production and their subsequent annihilation was analyzed (ref.3,4,5) and it was found that the relic monopoles would exceed present upper limits by over 10 orders of magnitude.However ,these limits are not well established.Thus,we believe that effort to obtain new astrophysical limits is worthwhile.

The strongest limits on the density of magnetic monopoles are obtained from the persistance of the interstellar magnetic fields(Parker ,ref.6).If the monopole density is sufficiently high, monopoles would destroy galactic magnetic fields by acceleration in those fields,and

$$n_m < 10^{-26}/cm^3 \qquad (1)$$

This argument is valid only for relativistic monopoles.However, the monopole energy after acceleration by the interstellar magnetic field of a few microgauss over the length of $L=10^{22}$ cm would be 10^{11} Gev and the corresponding velocity $\beta \approx 10^{-3}$.To obtain this velocity,the superheavy monopoles should be accelerated during at least 10^{8} years. Two effects should be considered:

a) trapping by magnetic stars ;
b) forming dust particles by attraction of interstellar gas and eventually the absorbtion in ferromagnetic dust,e.g. iron meteorites.

Both processes may stop or diminish the acceleration by interstellar magnetic field,i.e. they can suppress by many orders of magnitude the limit obtained by Parker.

In the author's opinion,very interesting limits are obtained from studies of astronomical objects with strong,relict fields,e.g. magnetic white dwarfs and neutron stars.

First,it seems fairly certain that the observed fields of collapsed objects are relicts from the time of star formation, and not the products of current dynamo action.The magnetic white dwarf's lifetime may be some 10^{9} years ,i.e. they are among the oldest objects in the Galaxy.Using arguments similar to those presented by Parker, a rather stringent limit of 10^{-23} monopoles/nucleon is obtained for the case of magnetic white dwarf at the epoch of its creation.The information that already some 10^{9} years ago ,the density of magnetic monopoles has been small is interesting for Big Bang theories.

Second,from the study of binary systems in which there is an accretion of matter on the magnetic white dwarf,we have obtained a limit on the ratio of numbers of magnetic monopoles and nucleons, namely

$$r_m < 10^{-21} \text{ monopoles/nucleon} \quad (2)$$

Finally, magnetic white dwarfs can accrete nonrelativistic monopoles from the interstellar medium. The persistence of magnetic field of white dwarfs gives:

$$n_m < 10^{-27} \text{ monopoles/cm} \quad (3)$$

The derivation of this last limits is given in what follows.

A few percent of all white dwarfs are strongly magnetic (ref.7). In some magnetic white dwarfs the fields are so strong that the spectral features are very weak. In these objects the magnetic field is identified by the elliptical polarization of the optical continuum, and it is about 10^8 Gauss. Most of the magnetic white dwarfs are rotating very slowly, i.e. the rotation periods are longer than a few years. Other measurable properties of magnetic white dwarfs do not seem remarkably different from white dwarfs in general. The fact that they are not especially hot means that the time scale for the magnetic field decay is comparable with white dwarf life time, i.e. some 10^9 years.

Let us consider a magnetic white dwarf with radius 10^9 cm and a magnetic field of 10^8 Gauss. Assuming homogeneous magnetization of the star's interior, the total magnetic energy of the white dwarf is about 2×10^{42} erg. However, both monopoles and antimonopoles are accelerated by the star magnetic field and take off an energy of about 10 eV, each. The persistance of magnetic field of white dwarfs means that

$$N \times 10 \quad ev << 10 \quad erg \quad (4)$$

i.e. during the white dwarf's life time, say 10^9 years, no more than 10^{33} monopoles were accelerated on/off the star.

Let's assume that the superheavy monopoles are non-relativistic. The magnetic white dwarfs have non-negligible proper motion, say a few tens kilometers per second, and during their lifetime they traverse some 10^{22} cm. During all this time they accrete interstellar matter. In following we assume the density of 1 nucleon per cubic centimeter. The relevant cross-sections for accreation of superheavy, non-relativistic monopoles are:

- the cross-section of the magnetosphere of the magnetic white dwarf;
- the gravitational sphere of influence given by $R_g \approx MxG/v^2 = 10^{14}$ cm where $M \approx 10^{33}$ g and $v = 30$ km/sec are the mass and average velocity of the magnetic white dwarf, respectively.

The extension of the magnetosphere of magnetic white dwarf is limited by the light radius of the rotating magnetic star, $R_L = cT/2 = 10^{14}$ cm for rotation period T=1 day. The rotating periods of magnetic white dwarfs are very long, i.e. $R > 10^{16}$ cm. However, the strength of the dipole field at distance R from the star with surface field $H = 10^8$ Gauss is only a few milligauss. In the following, we assume that the superheavy monopoles are magnetically caught by the star when $H > 100$ Gauss, i.e. the magnetic cross-section is 10^{24} cm^2.

It should be pointed out that the mechanisms of gravitational and magnetic accretion are complementary. The magnetic attraction of the white dwarf is strongest when the monopole is "naked". On the other hand, the gravitational influence of the star is strongest when the monopoles are imbedded in the dusts which are expected to have a mass of a few milligrams and the velocities of $\beta < 10^{-5}$. This velocity is smaller than the proper motion of the star itself, i.e. the relation $R_g \approx M_s \times G/v^2$ is valid for superheavy monopoles imbedded in dust. At a field of a few hundreds Gauss the monopoles are stripped off the nucleons and magnetic acceleration becomes the dominating process.

Thus, the limits on the density of superheavy, non-relativistic monopoles in the interstellar matter are

$$r_m < 10^{-27} \text{ monopoles/nucleon} \qquad (5)$$

Taking into account that Parker's limit (1) is valid for "naked" monopoles and the limit (5) for monopoles imbedded in dust, we have

$$r_m < 10^{-26} - 10^{-27} \text{ monopoles/nucleon} \qquad (6)$$

This limit is valid for monopoles trapped in galaxies. However, it is possible that magnetic fields much stronger than a few microgauss existed in the early Universe. Then even the superheavy monopoles are relativistic and distributed homogeneously all over the Universe. In this case the bound obtained (ref.5) from the Hubble constant should be used:

$$r_m < 10^{-15} \text{ monopoles/nucleon} \qquad (7)$$

ACKNOWLEDGMENTS.

It is a pleasure to thanks L.Stodolsky, S.Nussinov, and H-C.Thomas for discussions.

REFERENCES

1) H.Georgi, S.L.Glashow, Phys.Rev.Lett 32 (1974) 438
H.Georgi, H.R.Quinn, S.Weinberg, Phys.Rev.Lett 33 (1974) 451
2) A.Dolgov, Ya.B.Zeldovitch, Rev.Mod.Phys 53 (1981) 1
3) G.t' Hooft, Nucl.Phys. B79 (1974) 276
A.M.Polyakov, JETP Lett. 20 (1974) 94
4) Ya.B.Zeldovitch, M.Y.Klopov, Phys.Lett 79B (1979) 239
5) J.P.Preskill, Phys.Rev.Lett. 43 (1979) 1365
6) G.V.Domogatsky, I.M.Zeleznych, Yad.Fiz. 10 (1969) 1238
E.N.Parker, Astropys.J. 160 (1970) 410
7) J.R.Angel, Ann.Rev.Astron.Astrophys. 16 (1978) 487
ibid., Astrophys.J. 216 (1977)
J.Liebert, Ann.Rev.Astron.Astrophys. 18 (1980)

TOPOLOGY OF QUANTIZED FLUX FIELDS

Herbert Jehle

Sektion Physik, Universität München, München 2, Federal Republic of Germany.

Hermann Weyl, already in the 1920's, recognized that the fundamental law of conservation of electric charge is implied by the gauge invariance of electromagnetism. Furthermore, in the wave-mechanical representation of a field particle motion, a phase factor $\exp(i\vartheta)$ of the ψ function permits a gauge function ϑ which needs to be single valued only modulo 2π. Considering the line singularities of a grad ϑ field, implied by that multivaluedness of ϑ, Fritz London recognized (cf. also Dirac, Proc. Roy. Soc. A133) the physical significance of these singularities of the gauge field, i.e. closed lines which play the role of quantized flux loops (closed, as long as one does not make the additional hypothesis of magnetic monopoles). London was able, on that basis, to explain the main features of superconductivity.

Obviously, singularities of the magnetic field have to be taken seriously, and one has to recognize their physical significance. Instead of considering, however, a kind of approach which regards flux quantization only as a condition imposed on the currents generating the magnetic field, the present project considers all magnetic fields to be defined in terms of quantized flux as the primary entity; the Maxwell-Lorentz field equations then give the electric charge and current distribution in terms of the "quantized flux fields".

How is that quantized flux field to be determined, and how may we go about building a consistent theory or at least a model of particles on that basis?

A magnetic flux field may be defined by means of a statistical distribution of alternative forms (in space time) which a quantized flux loop may adopt, i.e. superpositions with probability amplitudes in a manner similar to Feynman's superposition of alternative path histories. Those flux "loop forms" resemble the Faraday lines of a magnetic dipole. The probability amplitude superposition is to be chosen so as to satisfy the homogeneous M-L field equations for the ensuing field in the region outside the source, a source which also incorporates the properties of a spin $\frac{1}{2}$ particle. The electric field is then given with the motion of the magnetic singularity line distribution. The relativistically invariant definition of the magnetic vector potential which implies quantized flux, implies at the same time an electric potential which is related to the magnetic as the Coulomb potential of electric charge e is related to the magnetic potential of a Bohr magneton dipole.

London's concepts had been given acceptance only slowly until, in 1961, Deaver and Fairbank, and Doll and Näbauer showed the experimental evidence of quantized flux.

The task was then to carry through this programme, with respect to particle physics. We have then seen that quantum mechanical models of particles, on the basis of so-defined electromagnetic fields, permitted a consistent formulation of structures, of intrinsic quantum numbers (conservation laws) and of the interactions of particles.

Instead of the customary interpretation of electron, muon, neutrino and of quarks as point-like structures, the present project thus considers topologically specified structures of magnetic (with electric) fields, to be responsible for the properties of the particles and their interactions. This description by means of structuralized "particles" is complementary to the point-source model for leptons and for quarks (partons) which is so successfully used in QED. The present description is based on time-averages over the relativistic Zitterbewegung, with the corresponding formulation of the all important fluctuations.

It should be clearly pointed out that there is no use starting with a development of the topological aspects of such a model of particles and their interactions, before showing the consistency of that electromagnetic model with proper quantum-mechanical properties. With this model which is based on statistical distributions of flux loop forms, being a model complementary to QED, many interesting results of QED may not be obtainable, but it permits:

i) to use only electromagnetic fields (and, of course, their probability amplitude distributions) to account for the field structures of particles;

ii) to interpret the intrinsic quantum numbers, and to interpret "electromagnetic", "strong" and "weak" interactions topologically, i.e. to render the colour and flavour classification of particles in terms of topological concepts;

iii) to give an understanding for the inter-relationship of $e^2/\hbar c$ and the electron to muon (and also to tauon) mass ratios, and also to give estimates for the distribution of masses of the diverse types of mesons and baryons.

The topological basis for such a programme lies in the famous work of H. Seifert and Wm. Threlfall who showed that (in the present language) fields of reasonably defined topological structure, formed from closed flux loop forms, should be toroidal fields.

Whereas the lepton may be defined in terms of a statistical distribution of <u>one</u> single magnetic closed flux loop, an ordinary (not high mass) meson may be defined in terms of <u>two</u> co-axial flux loops (i.e. flux loop form manifolds) which occupy the two parts of three space which are subdivided by a toroidal surface. Correspondingly, an ordinary baryon may be characterized by <u>three</u> co-axial flux loop form manifolds.

Whilst the usual theories of particles are based on point charges (whose statistics is only brought to function properly through the introduction of the phenomenological concept of colour), the structuralized quark fields are not in conflict with the Pauli principle (statistics) because they are localizable objects. Also, the problem of quark confinement resolves itself because of the topological inter-linkage of the two or three quark fields.

The well understood dumb bell model of the J/ψ meson is represented by two spectator quarks (one a quark, the other an antiquark), quark fields which lie side-by-side and which are linked together by a gluon magnetic field; these three structures sharing their polar symmetry axis. That gluon field may, however, become a pair of valence quark, valence antiquark, leading to the subsequent formation of a pair of ordinary mesons when the valence q, $\bar{q}$ joined with their respective spectator $\bar{q}$, q (cf. the three sequential figures). A two-pronged jet may be a dumb bell structure with its central, somewhat parallel magnetic field lines stretched at length; or it might be that the two aforementioned newly-formed mesons simply fly apart. Around the central, somewhat parallel magnetic gluon field region, also the possibility of formation of a new q, $\bar{q}$ pair arises.with the gluon field breaking off between those q, $\bar{q}$ (resulting in two dumb bells) or even the gluon structure forming one big ring on which various q, $\bar{q}$ are linked.

The task is now to give a detailed description of the diverse interactions of particles, to identify the intermediary structures of the processes, and to obtain a consistent description of the conservation laws and of the respective intrinsic quantum numbers.

Correspondingly, the non-Abelian gauge theories permit to calculate propagators and vertex functions through a judicious choice of Lagrangians which incorporate a phenomenological formulation of the influence of topology (in particular of chiralities) of the "particles" on their interaction.

I enjoyed the hospitality and support of the Universities of Münich, Uppsala and Amsterdam, and of the Deutsche Forschungsgemeinschaft and of the Research Corporation. I am also deeply indebted for friendship, suggestions and critical discussion to many colleagues, in particular, to Professor G. Süssmann and to Professor Raj Wilson.

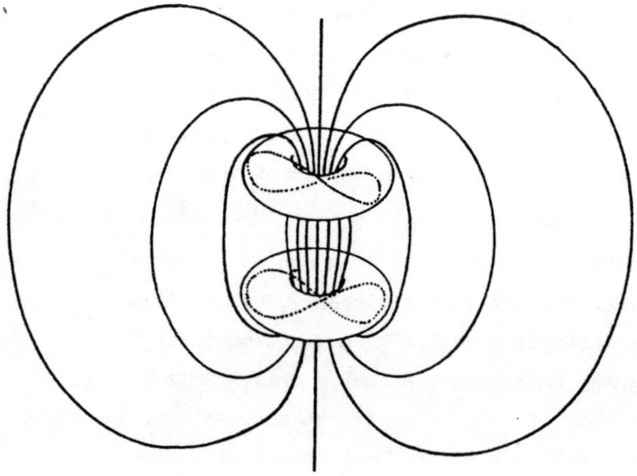

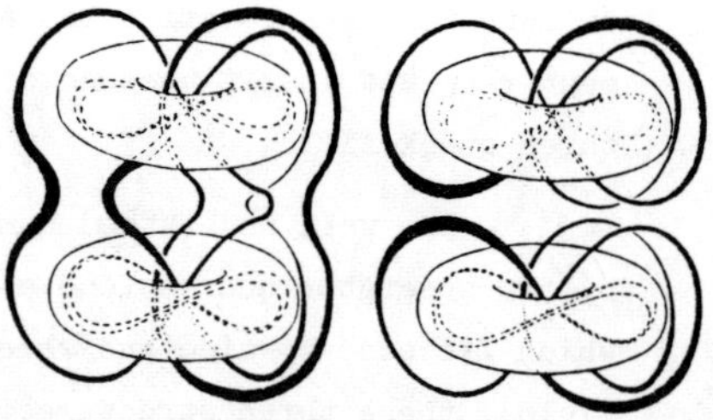

V: SURVEY AND SUMMARY

Summary Talk

* * * * * * *

L. O'Raifeartaigh

Dublin Institute for Advanced Studies, 10 Burlington Road, Dublin 4, Ireland.

---

1 Introduction.

This conference has been characterized not only by its scope and diversity, but by the unusually high level of the invited talks, the contributions, and of the discussions following them. Hence I think that I speak for all present when I offer our thanks to the organizers for ensuring this success, and also for the smooth running of the conference.

In order to summarize the talks in a unified manner, I have tried to group those that lie in the same general areas together, and to relate the different areas, as shown in Fig. 1. In this figure the general areas covered are enclosed in boxes. In the present section I consider the relationships between the boxes, and in the subsequent sections I consider the contents of the individual boxes.

The 1931 paper of Dirac, which first introduced the quantization of the product of electric and magnetic charges eg is, of course, the seminal paper for the conference. However, the reference to Maxwell and Faraday just above it is to recall that some of the themes of the conference, notably the concepts of duality and of topology (recall the vector potential and Faraday's lines of force) have even older origins.

A tremendous boost to the study of monopoles was, of course, given by the discovery in 1974 that non-abelian gauge-theories with spontaneous symmetry breakdown predicted the existence of monopoles of extended size and finite mass, and so this important development and its consequences are connected to Dirac's paper by a double line. On the other hand, a number of papers at the conference could have been presented even if the finite-mass monopole development had never taken place. As the natural (abelian) development of the Dirac theory these papers are placed in the box below the Dirac paper.

The discovery of finite-mass monopoles has had enormous repercussions on our thinking concerning unified gauge theories, and three important areas were included in the themes of the conference, as indicated in Fig. 1. First there is the positive possibility of the monopoles explaining confinement. Second there is the puzzle of non-abundance of monopoles in cosmology. Finally, there is the mathematical development in which sophisticated techniques of algebraic geometry have been used to solve the problem of constructing solutions to the monopole field equations. Unfortunately, so far this latter effort has been most successful in a case which is probably unphysical, namely the case in which the Higgs potential is zero. (This is the meaning of the little man in Fig. 1 – he is not taking a shower, but is looking for his latch key where there is light instead of where the key probably is – lower down in Fig. 1!).

In the last analysis of course, Physics is an experimental science and the conference would not be complete without experimental considerations. The experimental talks are shown in the lowest box, supporting the theoretical boxes, since theory should naturally be supported by experiment. The sad fact that must be reported, however, is that direct support, in the shape of an observed monopole, is still lacking.

Finally some of the papers at the conference related monopole theory to neighbouring areas of research such as instantons and supersymmetry. These papers are not depicted in Fig. 1 but are discussed in section 6.

2. QED Dirac Theory: First and Second Quantization.

In this section we consider the developments that are independent of the finite-mass monopoles i.e. that would have taken place even if unified gauge theories had not been discovered. At the first-quantized level we had first of all the very elegant talk by Yang, who gave a simple intuitive derivation of the Lipkin-Peshkin-Weissberger paradox (non-closure of the Jacobi identity in the presence of Dirac monopoles) and then went on to the main part of his talk which was the extension of the conventional Sturm-Lioville theory to systems with Dirac monopoles. The talk by Chan on the derivation of the equations of motion from topology was of considerable interest (and raised most questions in the discussion time). Some of the more mathematical points of this paper were later taken up by Mi Chou. The topology of classical monopoles was also stressed by Jehle (who in fact was stressing this aspect long before the concept of topological charges became popular in the seventies) by Horvathy, who generalized the Bohm-Aharonov effect to SU(n), and by Quiros, who discussed the embedding of Dirac monopoles into SU(2) monopoles.

At the second-quantized level the main question is whether QED can, in fact, be second-quantized consistently in the presence of Dirac monopoles, and in the main talk on this subject Zwanziger argued that the Dirac quantization condition $eg=2\pi n\hbar$ is both necessary and sufficient for consistency. He also pointed out that the 1931 theory appears to be remarkably durable in that it has withstood a number of tests, such as second-quantization, without requiring modification. A problem that does arise in second-quantization is the question of renormalization of e and g. This question cannot be investigated perturbatively for both e and g on account of the quantization condition. An old question here is whether eg or e/g remain unrenormalized. A number of groups presented results on this renormalization question at the conference, but unfortunately (at the time of writing) the results on the eg or e/g question do not agree. So this is a problem that remains to be resolved.

3. Monopoles in Unified Gauge Theory: Confinement.

We turn now to monopoles which are not of the original Dirac type but arise as extended, singularity-free, solutions of the Yang-Mills-Higgs field equations, with topological magnetic charge. Ironically these monopoles dispense with the two concepts, electromagnetic duality and charge quantization, which were the foundation of the Dirac theory, because in non-abelian theories the electric and magnetic fields are not quite dual (recall that $\vec{B}$ can be constructed from the vector-potential $\vec{A}$, but $\vec{E}$ requires both A_0 and $\vec{A}$ in the commutator term) and, as pointed out by Georgi and Glashow in their SU(5) grand-unification paper, the electric charge is automatically quantized for a semi-simple gauge group. However, the duality concept may return as gauge-field-monopole duality, as discussed by Olive, and, even though e is self-quantized, one still obtains the Dirac relation $g=nhe^{-1}$ for the topological magnetic charge.

One of the outstanding problems, of course, is the quantization of theories with topologically charged sectors, and programs for quantization were presented by Osborn and Manton. Some initial semi-classical steps in the program were presented explicitly by each of these authors. A quite different approach to the problem, introducing the idea of adiabatic invariants, was presented by Bogomolny.

An immediate question that arises in connection with topological charges is whether they are connected with quark confinement. For some time now it has been argued that monopoles may cause confinement of coloured charges in the same way that superconducting vortices could cause confinement of Dirac monopoles (by forcing the lines of force to be one-dimensional). The argument for this (so-called dual Meissner) effect was presented by Mandelstam who, supported it with results from QED, SO(3) and SU(2) lattice computations. A somewhat different approach to the same problem was adopted by Oleson, who considered the Migdal-Makeenko equation for Wilson loops in the 1/N limit, and showed that in this limit one may obtain the central Z_n vortex lines proposed by 'tHooft and Polyakov as a mechanism for confinement.

4. Explicit Monopole Solutions.

The explicit construction of monopole solutions to the Yang-Mills-Higgs field equations was possibly the theme of greatest interest and discussion during the conference, because the past eighteen months have seen some spectacular developments in this area. It is true that the developments have been in the special, and possibly unphysical, case in which the Higgs potential is set equal to zero after the spontaneous symmetry breakdown (and in which the Higgs field is in the adjoint representation of the gauge group) but the progress has been so rapid, and relates to so many other branches of mathematics and physics — algebraic geometry, soliton-theory, inverse-scattering problems, instantons — that the interest in this field has been widespread.

The background for all the later talks on this subject was given by Atiyah in the opening lecture of the conference. He pointed out that the non-linear systems (KdV, Sine-Gordon and other soliton systems, and latterly the instantons and monopoles) are those for which the non-linear equations arise as integrability conditions for linear systems. He then went on to review some recent work of Hitchin on monopoles. In this work Hitchin shows that the SU(2) monopoles are characterized by an algebraic (spectral) curve in the space of lines in R_3, and that this structure underlies the construction of SU(2) monopole solutions.

There have actually been three approaches to the construction of monopole solutions over the past eighteen months, and all three were presented at the conference. The first approach was due to Ward and was initially for two SU(2) monopoles, both superimposed and separated. The generalization to n superimposed monopoles was discussed by Rossi, and the final generalization to n arbitrarily placed monopoles, with 4n-1 parameters, was presented by Corrigan. The results of Hitchin actually suggest the solutions presented by Corrigan may form a complete set for SU(2). A second approach, using Bäcklund transformations for the superimposed case, and inverse scattering methods for the separated case, was presented by the group from the Eötvös University. This group also presented graphs of the energy distribution for n=2...5. These graphs give a nice intuitive picture of the monopole configurations, and of their rather unusual axial symmetry properties. The final approach to the explicit construction of solutions is due to Nahm and is the monopole analogue of the ADHM constructionf for instantons. A central role in Nahm's approach is played by the equations

$$\frac{dT_\alpha}{dt} = \varepsilon_{\alpha\beta\gamma}[T_\beta, T_\gamma]$$

where the T_α are nxn matrices. It turns out that the Lax spectrum for these equations is closely connected with the spectral curve of Hitchin mentioned above.

The generalization of the above results to groups larger than SU(2) was also considered. Explicit generalizations of his own approach to SU(3) have been constructed by Ward. The generalization of the Eötvös method, was considered by the Eötvös group itself and by Bais. The approach of Nahm, on the other hand, though less explicit than the others, lends itself more readily to generalization — indeed it makes little distinction between different groups. Other work on higher groups included work on symmetries by Cho, the extension of the index theorem (the prediction of 4n-1 parameters for SU(2) to SU(3) by Weinberg, and the construction of spherically symmetric solutions for higher groups by Olive and Ganoulis. Interesting features of the latter paper were an unexpected connection with Toda-lattices and the existence of a natural symmetry-breaking hierarchy for the exceptional groups.

Some outstanding problems that came up in the discussions were (i) the intrepretation of the 4n-1 parameters and the form of the parameter space for SU(2), (ii) the question of stability of the static n-monopole system and (iii) the question of whether the monopoles behave like solitons when they are no longer static. Some initial information on (i) was provided in the talk by Rouhani, and on (iii) by the talk of Manton, in which the motion of slow-moving monopoles was shown to be geodesic.

5. Experiment and Cosmology.

In any conference on monopoles the question that must ultimately be faced is that of their experimental observation and abundance. There were three invited talks on this subject at the conference (two on experiment by Giacomelli and Drukier respectively, and one on cosmology, by Kibble) and these talks were so well-presented and self-contained that there is very little that I can add here.

The basic problem with monopoles is that they seem to be unavoidable in grand-unified theories, because all that is required for their existence is a non-trivial map of the sphere at infinity into the (gauge-group) / (unbroken symmetry group) i.e. $S_2 \to G/H$. But if the gauge-group is simple (there is one overall coupling constant) and the unbroken symmetry group is SU(n)xU(1) where SU(n) and U(1) are the colour and EM groups respectively, then non-trivial maps always exist (technically $\pi_2(G/SU(n)xU(1)) \neq 0$). It is true that the usual arguments leading to this result are classical, but they are also topological, and it is difficult to believe that the topological effects could completely disappear at the quantum level.

Given that monopoles should exist, the question is why they are not seen. The possible terrestial searches have been described in detail in Giacomelli's talk, and it is some comfort to recall his statement that most previous searches for relatively light monopoles are not relevant for monopoles with the large masses that are predicted by grand unified theories. So the terrestial search for monopoles is only beginning.

The cosmological abundance of monopoles is also a problem. Cosmology and Thermodynamics predict an abundancy rate $r \gtrsim 10^{-2}$, where r = number of monopoles/ entropy. But as discussed in detail by Kibble and Drukier, experimental estimates, based on present densities, densities at helium synthesis, the energy balance in galactic magnetic fields and so on, produce an upper bound of at most $r \lesssim 10^{-20}$. Thus the theory and experiment disagree by about ten orders of magnitude. Of course, there are a number of escapes, and these have been discussed in detail by Kibble, but the situation is still somewhat disquieting.

6. Related Fields.

There were actually few talks that were not directly related to monopoles. However, relationships between monopoles and instantons were discussed by Chakrabarti and between monopoles and CP^n by Dobrov. The topological nature of central charges in supersymmetry was considered by Zizzi and supergravity was discussed at some length by Julia. Gravitation was also discussed in the invited talk by Gibbon, who pointed out the existence of a Bogomolny bound in gravitation. Finally, the physical interpretation of the spinor structure and its connection with affine structure of space-time was discussed by Bujajska.

7. Outlook.

It is always difficult to make predictions (especially about the future!) but it does seem that three broad lines of development seem to emerge from the conference. First, there is the rapid mathematical development in the direction of algebraic geometry, non-linear partial-differential equations etc. Second, there is the still somewhat obscure but physically very important question of the role of monopoles in QCD, confinement etc. Finally there is the all-important phenomenology of monopoles.

To end on a light note, one might say that if the monopole energies prove too large for our machines, and we are reduced to Professor Giacomell's final suggestion, then Physics will end for us where it began — dropping things off the leaning tower of Pisa! Perhaps the angle of the tower is the new angle mentioned by Professor Dirac.

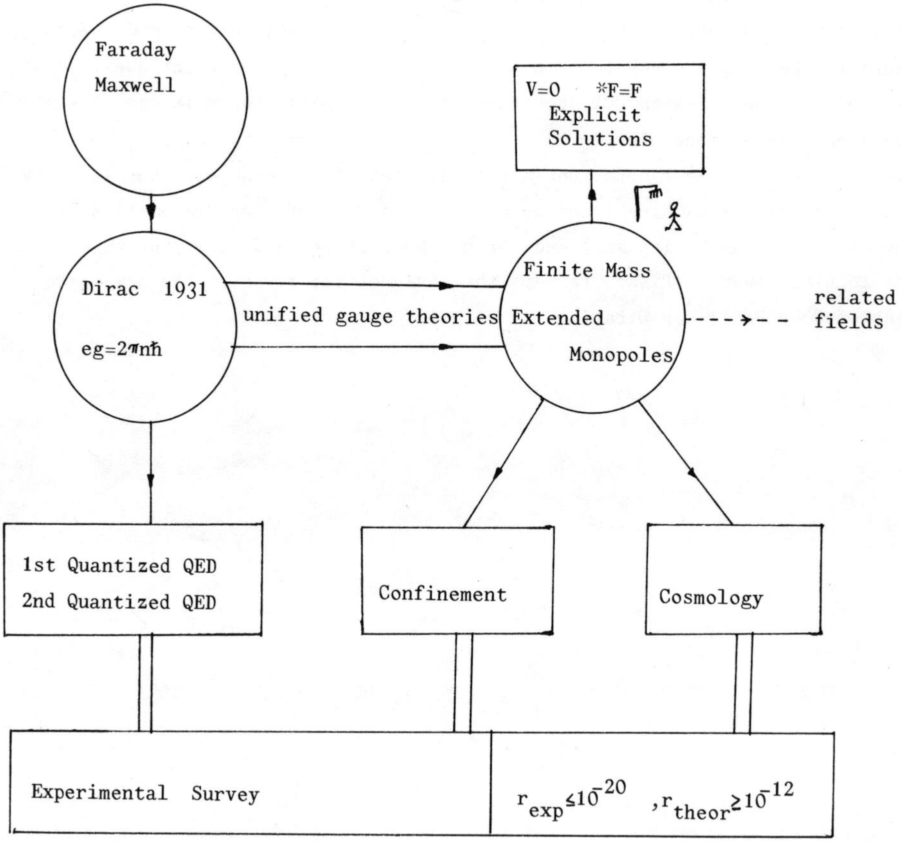
Faraday
Maxwell
V=0 *F=F
Explicit
Solutions
Dirac 1931
eg=2πnħ
unified gauge theories
Finite Mass
Extended
Monopoles
related
fields
1st Quantized QED
2nd Quantized QED
Confinement
Cosmology
Experimental Survey
$r_{exp} \leq 10^{-20}$, $r_{theor} \geq 10^{-12}$

Fig.1.

LIST OF PARTICIPANTS

	Name	Member State
1.	E. AMALDI Istituto di Fisica Piazzale A. Moro 5 00185 Rome Italy	Italy
2.	I. ANDRIC High Energy Physics Group Rudjer Boskovic Institute Bijenicka 54, 41001 Zagreb Yugoslavia	Yugoslavia
3.	S. AOYAMA Division de Physique Théorique Institut de Physique Nucleaire B.P. 1 91406 Orsay, Cedex France	Japan
4.	C. ATHORNE Department of Mathematics University of Durham South Road Durham DH1 3LE UK	UK
5.	M. ATIYAH Mathematical Institute University of Oxford 24-29 St. Giles Oxford OX1 3LB UK	UK
6.	F. BAIS Institute for Theoretical Physics Rijksuniversiteit Utrecht Princetonplein 5 Utrecht 2506 The Netherlands	Fed. Rep. Germany
7.	J. BALOGH Institute of Theoretical Physics Eötvös University VIII Puskin u. 5-7 1088 Budapest Hungary	Hungary
8.	E. BASSILA Physics Department Lebanese University Beirut Lebanon	Lebanon

	Name	Member State
9.	G. BHATTACHARYA Theoretical Physics Group Tata Institute for Fundamental Research Homi Bhabha Road Bombay 400 005 India	India
10.	K. BIALAS-BORGIEL Department of Physics Silesian University Katowice Poland	Poland
11.	N. BILIC High Energy Physics Group Rudjer Boskovic Institute Bijenicka 54 41001 Zagreb Yugoslavia	Yugoslavia
12.	M. BLAGOJEVIC Department of Theoretical Physics Boris Kidrich Institute P.O. Box 522 11001 Belgrade Yugoslavia	Yugoslavia
13.	E. BOGOMOLNY Landau Institute for Theoretical Physics Vorobyevskoye Shossee 2 117 334 Moscow USSR	USSR
14.	H. BOUTALEB Departement de Physique Laboratoire de Physique Théorique Universite Mohammed V B.P. 1014 Rabat Morocco	Morocco
15.	K. BUGAJSKA Department of Physics Silesian University Katowice Poland	Poland
16.	Z. BUJA Institute for Nuclear Physics ul. Radzikowskiego 152 31 342 Krakow Poland	Poland

	Name	Member State
17.	F. CANNATA INFN Via Irnerio 46 40126 Bologna Italy	Italy
18.	C.-Q. CAO c/o International Centre for Theoretical Physics 34100 Trieste Italy	People's Rep. of China
19.	G. CALUCCI Istituto di Fisica c/o ICTP 34100 Trieste Italy	Italy
20.	R. CATENACCI Istituto di Fisica Teorica Via Bassi 4 27100 Pavia Italy	Italy
21.	A. CHAKRABARTI Centre de Physique Théorique Ecole Polytechnique 91128 Palaiseau Cedex France	India
22.	H.-M. CHAN Rutherford & Appleton Laboratories Chilton Didcot OX11 0QX UK	UK
23.	Y. CHO Laboratoire de Physique Théorique et Hautes Energies 4 place Jussieu 75230 Paris, Cedex 05 France	Korea
24.	G. CHRISTOS c/o International Centre for Theoretical Physics 34100 Trieste Italy	Australia

	Name	Member State
25.	A. COMTET Division de Physique Théorique IPN Bâtiment 100 91405 Orsay Cedex France	France
26.	E. CORRIGAN Department of Mathematics University of Durham South Road Durham DH1 3LE UK	UK
27.	N. CRAIGIE INFN and International Centre for Theoretical Physics 34100 Trieste Italy	UK
28.	L. DABROWSKI SISSA c/o ICTP 34100 Trieste Italy	Poland
29.	W. DEANS Université Libre de Bruxelles Campus de la Plaine (CP225) Boulevard du Triomphe 1050 Brussels Belgium	UK
30.	V. DOBREV Institute of Nuclear Resarch and Nuclear Energy Boul. Lenin 72 1184 Sofia Bulgaria	Bulgaria
31.	J. DONEUX Institut de Physique Théorique Université Catholique de Louvain Chemin du Cyclotron 2 1348 Louvain-la-Neuve Belgium	Belgium
32.	B. DRAGOVIC Institute of Physics Studentski Trg 12 V P.O. Box 57 11001 Belgrade Yugoslavia	Yugoslavia

Name	Member State
32. A. DRUKIER Zentralinstitut für Tieftemperatur- forschung Walther-Meissner-Strasse 8046 Garching Fed. Rep. Germany	Fed. Rep. Germany
33. Th.M. EL SHERBINI Department of Physics Cairo University Cairo Egypt	Egypt
34. K. FARAKOS Nuclear Research Center Demokritos Aghia Paraskevi, Attiki Athens Greece	Greece
35. V. FATEEV Landau Institute for Theoretical Physics Vorobyevskoye Shossee 2 117 334 Moscow USSR	USSR
36. D. FAUTH Institut für Theoretische Physik Universität Stuttgart Pfaffenwaldring 57/111 7000 Stuttgart 80 Fed. Rep. Germany	Fed. Rep. Germany
37. L. FERREIRA Blackett Laboratory Imperial College Prince Consort Road London SW7 2BZ UK	Brazil
38. P. FORGACS Central Research Institute for Physics P.O. Box 49 1525 Budapest Hungary	Hungary
39. Y. FUJIMOTO c/o International Centre for Theoretical Physics 34100 Trieste Italy	Japan

Name	Member State

40. N. GANOULIS — Greece
Fachbereich 8 - Physik
Universität Gesamthochschule
Postfach 100127
5600 Wupperthal 1
Fed. Rep. Germany

41. J. GARCIA ESTEVE — Spain
Departamento de Fisica Nucleare
Universidad de Zaragoza
Zaragoza
Spain

42. G. GIACOMELLI — Italy
Istituto di Fisica
Via Irnerio 46
40126 Bologna
Italy

43. G. GIBBONS — UK
Department of Applied Mathematics
and Theoretical Physics
University of Cambridge
Silver Street
Cambridge CB3 9EW
UK

44. P. GNAEDIG — Hungary
Institute of Theoretical Physics
Eötvös University
VIII Puskin u. 5-7
1088 Budapest
Hungary

45. P. GODDARD — UK
Department of Applied Mathematics
and Theoretical Physics
University of Cambridge
Silver Street
Cambridge CB3 9EW
UK

46. J. GOMATAN — India
Department of Mathematics
Glasgow College of Technology
Cowcaddens Road
Glasgow G4 OBA
UK

47. L. GONZALEZ-MESTRES — Spain
Laboratoire de Physique des Particules
IN 2P3, B.P. 909
74019 Annecy-le-Vieux Cedex
France

	Name	Member State
48.	N. HARI-DASS Niels Bohr Institute Blegdamsvej 17 2100 Copenhagen Ø Denmark	India
49.	Z. HLOUSEK High Energy Physics Group Rudjer Boskovic Institute Bijenicka 54 41001 Zagreb Yugoslavia	Yugoslavia
50.	P. HORETZKY Institut für Theoretische Physik Universität Wien Boltzmanngasse 5 1090 Vienna Austria	Austria
51.	M. HORTACSU Physics Department Bogazici University Bebek - Istanbul Turkey	Turkey
52.	Z. HORVATH Department of Physics Eötvös University VIII Puskin u. 5-7 1088 Budapest Hungary	Hungary
53.	P. HORVATHY Fakultat für Physik Universität Bielefeld 4800 Bielefeld 1 Fed. Rep. Germany	Fed. Rep. German
54.	R. IENGO Istituto di Fisica c/o ICTP 34100 Trieste Italy	Italy
55.	G. IMMIRZI Istituto di Fisica Faculta di Ingegneria Piazzale Tecchio 80125 Napoli Italy	Italy
	I. JACK Department of Applied Mathematics and Theoretical Physics University of Cambridge Cambridge CB3 9EW UK	UK
56.	H. JEHLE Sektion Physik Universität München Theresienstrasse 37 8000 Munich 2 Fed. rep. Germany	USA

Name	Member State
57. C.-S. JU Institut für Theoretische Physik Universität Heidelberg Philosophenweg 16 6900 Heidelberg 1 Fed. Rep. Germany	People's Rep. China
58. B. JULIA Laboratoire de Physique Theorique Ecole Normale Superieure 24 rue Lhomond 75231 Paris Cedex 05 France	France
59. T.W. B. KIBBLE Blackett Laboratory Imperial College Prince Consort Road London SW7 2BZ UK	UK
60. G. KOUTSOUBAS Nuclear Research Center Demokritos Aghia Paraskevi Attiki, Athens Greece	Greece
61. B. KOWALCSZYK Department of Physics Silesian University Katowice Poland	Poland
62. G. LEONTARIS Department of Physics University of Ioannina Ioannina Greece	Greece
63. W. LERCHE Sektion Physik Universität München Theresienstrasse 37 8000 Munich 2 Fed. Rep. Germany	Fed. Rep. Germany
64. L. LUSANNA INFN Largo E. Fermi 2 (Arcetri) 50125 Firenze Italy	Italy

Name	Member State
65. J. MAGPANTAY International Centre for Theoretical Physics 34100 Trieste Italy	Philippines
66. E. MAHDAVI SISSA c/o ICTP 34100 Trieste Italy	Iran
67. D. MAISON Max-Planck-Institut für Physik und Astrophysik Föhringer Ring 6 8000 Munich 40 Fed. Rep. Germany	Fed. Rep. Germany
68. S. MANDELSTAM Lawrence Berkeley Laboratory University of California Berkeley, CA 94720 USA	UK
69. P. MANSFIELD Department of Applied Mathematics and Theoretical Physics University of Cambridge Silver Street Cambridge CB3 9EW UK	UK
70. N. MANTON Institute for Theoretical Physics University of California Santa Barbara, CA 93106 USA	UK
71. W. MECKLENBURG SISSA c/o ICTP 34100 Trieste Italy	Fed. Rep. Germany
72. S. MELJANAC High Energy Physics Group Rudjer Boskovic Institute Bijenicka 54 41001 Zagreb Yugoslavia	Yugoslavia
73. F. MIGLIETTA Istituto di Fisica Teorica Via Bassi 4 27100 Pavia Italy	Italy

	Name	Member State
74.	S. MIKOCKI Institute of Nuclear Physics ul. Radzikowskiego 152 31-342 Krakow Poland	Poland
75.	L. MIZRACHI International Centre for Theoretical Physics 34100 Trieste Italy	Israel
76.	S. MUKHI International Centre for Theoretical Physics 34100 Trieste Italy	India
77.	C. MUKKU International Centre for Theoretical Physics 34100 Trieste Italy	India
78.	W. NAHM TH Division CERN 1211 Geneva 23 Switzerland	Fed. Rep. Germany
79.	M.A. NAMAZIE International Centre for Theoretical Physics 34100 Trieste Italy	Singapore
80.	F. NERI Rutgers University Department of Physics New Brunswick, N.J. 08903 USA	Italy
81.	R. NOBLE International Centre for Theoretical Physics 34100 Trieste Italy	USA

Name	Member State

82. J. NUYTS — Belgium
Faculte des Sciences
Université de l'Etat a Mons
Avenue Maistriau 19
7000 Mons
Belgium

83. S.K. OH — Korea
International Centre for
Theoretical Physics
34100 Trieste
Italy

84. P. OLESEN — Denmark
Niels Bohr Institute
Blegdamsvej 17
2100 Copenhagen Ø
Denmark

85. D. OLIVE — UK
Blackett Laboratory
Imperial College
Prince Consort Road
London SW7 2BZ
UK

86. L. O'RAIFEARTAIGH — Ireland
School of Theoretical Physics
Dublin Institute for Advanced Studies
10 Burlington Road
Dublin 4
Ireland

87. H. OSBORN — UK
Department of Applied Mathematics
and Theoretical Physics
University of Cambridge
Cambridge CB3 9EW
UK

88. M. OSTROWSKI — Poland
Obserwatorium Astronomiczne
Uniw. Jagiellonski
ul. Orla 171
30 244 Krakow
Poland

89. M. OZER — Turkey
International Centre for
Theoretical Physics
34100 Trieste
Italy

90. L. PALLA — Hungary
Department of Physics
Eötvös University
VIII Puskin 5-7
1088 Budapest
Hungary

Name	Member State
91. S. PALLUA Theoretical Physics Department Marulicev trg 19/I 41000 Zagreb Yugoslavia	Yugoslavia
92. C. PANAGIOTAKOPOULOS International Centre for Theoretical Physics 34100 Trieste Italy	Greece
93. A. POGREBKOV Steklov Mathematical Institute Vavilov Str. 42 Moscow GSP-1, 117966 USSR	USSR
94. M. POLLOCK Istituto di Fisica Via F. Marzolo 8 35100 Padova Italy	UK
95. M. QUIROS Instituto de Estructura de la Materia Serrano 119 Madrid 6 Spain	Spain
96. S. RANDJBAR-DAEMI International Centre for Theoretical Physics 34100 Trieste Italy	Iran
97. G. RINGWOOD Department of Mathematical Physics University of Birmingham P.O. Box 363 Birmingham B15 2TT UK	UK
98. A. RODRIGUEZ-VARGAS International Centre for Theoretical Physics 34100 Trieste Italy	Colombia
99. J. RAMAO Centro de Fisica de Materia Condensada Av. do Prof. Gama Pinto 2 Lisbon 4 Portugal	Portugal

Name	Member State
100. P. ROSSI Scuola Normale Superiore Piazza dei Cavalieri 56100 Pisa Italy	Italy
101. S. ROUHANI School of Theoretical Physics Dublin Institute for Advanced Studies 10 Burlington Road Dublin 4 Ireland	Iran
102. W. SAYED International Centre for Theoretical Physics 34100 Trieste Italy	UK
103 . E. SEZGIN International Centre for Theoretical Physics 34100 Trieste Italy	Turkey
104. A. SHAYEB Department of Physics Sana'a University Sana'a Yemen	Jordan
105. P. SODANO Istituto di Fisica Via Vernieri 42 84100 Salerno Italy	Italy
106. A. SOPER Department of Applied Mathematics and Theoretical Physics University of Cambrige Silver Street Cambridge CB3 9EW UK	UK
107. J. STRATHDEE International Centre for Theoretical Physics 34100 Trieste Italy	USA

Name	Member State
108. N. STRAUMANN Institut für Theoretische Physik Universität Zurich Schonbergasse 9 8001 Zurich Switzerland	Switzerland
109. D.H. TCHRAKIAN St. Patrick's College Maynooth Co. Kildare Ireland	Ireland
110. P. TOWNSEND Laboratoire de Physique Theorique Ecole Normale Superieure 24 rue Lhomond 75231 Paris Cedex 05 France	UK
111. S. TSOU Mathematical Institute University of Oxford 24-29 St. Giles Oxford OX1 3LB UK	UK
112. H. URBANTKE Institut für Theoretische Physik Universität Wien Boltzmanngasse 5 1090 Vienna Austria	Austria
113. T. VALLON SISSA c/o ICTP 34100 Trieste Italy	Italy
114. G. VENTURI Istituto di Fisica Via Irnerio 46 40126 Bologna Italy	Italy
115. H. VERSCHELDE Seminarie voor Theoretische Vaste Stoff en Lage Energie Kernfysica Rijksuniversiteit Gent Krijgslaan 281/89 9000 Gent Belgium	
116. G. VITIELLO Istituto di Fisica Via Vernieri 42 84100 Salerno Italy	Italy

	Name	Member State
117.	R.S. WARD Department of Mathematics Trinity College Dublin Ireland	Ireland
118.	E. WEINBERG Department of Physics Columbia University New York, N.Y. 10027 USA	USA
119.	A. WIPF Institut für Theoretische Physik der Universität Zurich Schonbergasse 9 8001 Zurich Switzerland	Switzerland
120.	C.N. YANG Institute for Theoretical Physics State University of New York Stony Brook, N.Y. 11794 USA	USA
121.	J.H. YEE International Centre for Theoretical Physics 34100 Trieste Italy	Korea
122.	K. YOSHIDA Istituto di Fisica Via Vernieri 42 84100 Salerno Italy	Japan
123.	A. ZAMOLODCHIKOV Landau Institute for Theoretical Physics Vorobyevskoye Shossee 2 117 334 Moscow USSR	USSR
124.	P. ZIZZI Department of Mathematics King's College Strand London WC2 UK	Italy
125.	D. ZWANZIGER Department of Physics New York University 4 Washington Place New York, N.Y. 10003 USA	USA